INSTRUMENTATION ENGINEER'S HANDBOOK
VOLUME - 1

Process Instrumentation

Ashish Garg

Tata Steel Limited

White Falcon Publishing

www.whitefalconpublishing.com

Instrumentation Engineer's Handbook:
Process Instrumentation
Volume - 1
Ashish Garg

www.whitefalconpublishing.com

ISBN - 978-1-63640-247-5

Preface

The Monitoring and Measuring of industrial processing has become almost an art form. The dipstick and the simple gauge has been given way to high technological innovation of such complexity that the average engineer or technician can easily find himself or herself, at logger heads with management or colleagues as to the true pressure, temperature, flow, level and analysis of liquid and gas in a closed circuit. This is hardly surprising, when there are so many methods and measurement systems on the market. Some are cost effective in the short term, other requires long term investment. Naturally it depends on the degree of accuracy required, the elimination of all misleading and the need for instant data on true flow, pressure temperature, level and gas analysis. For this reason, book on "PROCESS INSTRUMENTATION" has been written as a first source of reference for users which works in process or manufacturing industry. It provides a comprehensive understanding of the fundamentals, standards, principles and detail information on pressure, temperature, flow, level and process gas analyzers measuring instruments, its application and effectiveness.

The book contains vital and useful information, data, tables and charts, to assist engineers, technicians, designers and plant operators in solving their particular problems associated with pressure, temperature, flow, level and process gas analyzer.

This handbook is dedicated to the next generation of instrumentation engineers working in the fields of analysis, measurement, control, and safety. I hope that learning from this book will increase their professional standing around the world.

Acknowledgments

Writing a book is harder than I thought and more rewarding than I could have ever imagined. None of this would have been possible without my wife Neha and my parents, who stood by me during my struggles and successes. That is true friendship, love and care.

At last, it is not without some relief, for it often goes unsaid but relief is concomitant with every finished piece of work. A relief which accompanies joy and precedes praise, if deserved.

Table of Contents

Pressure Measurement

Temperature Measurement

Flow Measurement

Level Measurement

Process Gas Analyzer

1

Pressure Measurement

After completing this chapter, you should be able to:

Know about Pressure Gauges, Selection Guideline of Pressure Gauges

Know about Pressure Switch and Pressure Transmitter

Calibration Procedure of Pressure Sensors (Gauges, Transmitter and Switch)

1.1 What is Pressure

- Pressure is the force exerted per unit area

- Pressure is the action of one force against another force. Pressure is force applied to, or distributed over, a surface. The pressure P of a force F distributed over an area A is defined as P = F/A

1.2 Types of Pressure

- Absolute Pressure

- Atmospheric Pressure

- Barometric Pressure

- Gauge Pressure

- Differential pressure

Now let us understand each term briefly

Absolute Pressure

Measured above total vacuum or zero absolute. Zero absolute represents total lack of pressure.

Atmospheric Pressure

The pressure exerted by the earth's atmosphere. Atmospheric pressure at sea level is 14.696 psia. The value of atmospheric pressure decreases with increasing altitude.

Barometric Pressure

Same as atmospheric pressure.

Gauge Pressure

The pressure above atmospheric pressure. Represents positive difference between measured pressure and existing atmospheric pressure. Can be converted to absolute by adding actual atmospheric pressure value.

Differential Pressure

The difference in magnitude between some pressure value and some reference pressure. In a sense, absolute pressure could be considered as a differential pressure with total vacuum or zero absolute as the reference. Likewise, gauge pressure (defined above) could be considered as Differential Pressure with atmospheric pressure as the reference.

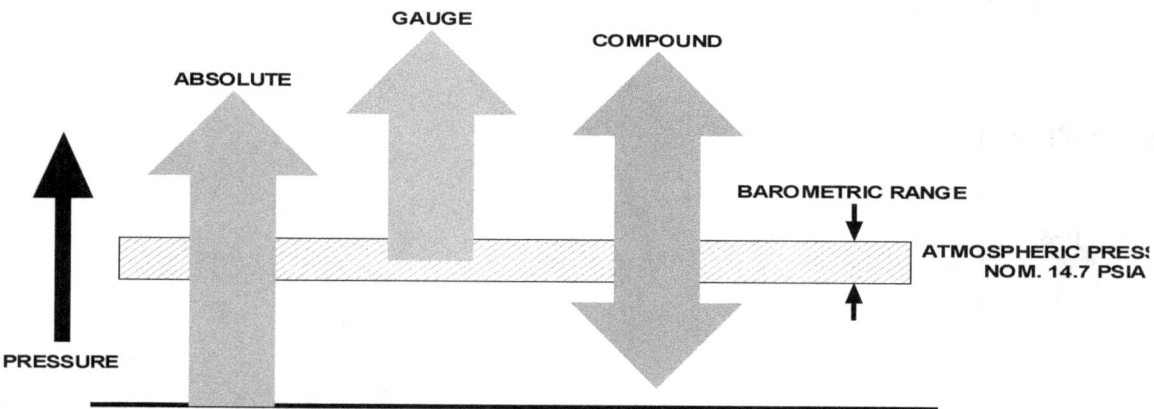

Fig.1 Types of Pressure

1.3 Pressure Measuring Instruments

Types of Pressure Measuring Instruments:

•   Pressure Gauges (Vacuum, Compound, Absolute, Gauge)
•   Differential Pressure Gauge
•   Pressure Switch (Vacuum, Absolute, Gauge)
•   Differential Pressure Switch
•   Pressure Transmitter (Vacuum, Absolute, Gauge)
•   Differential Pressure Transmitter

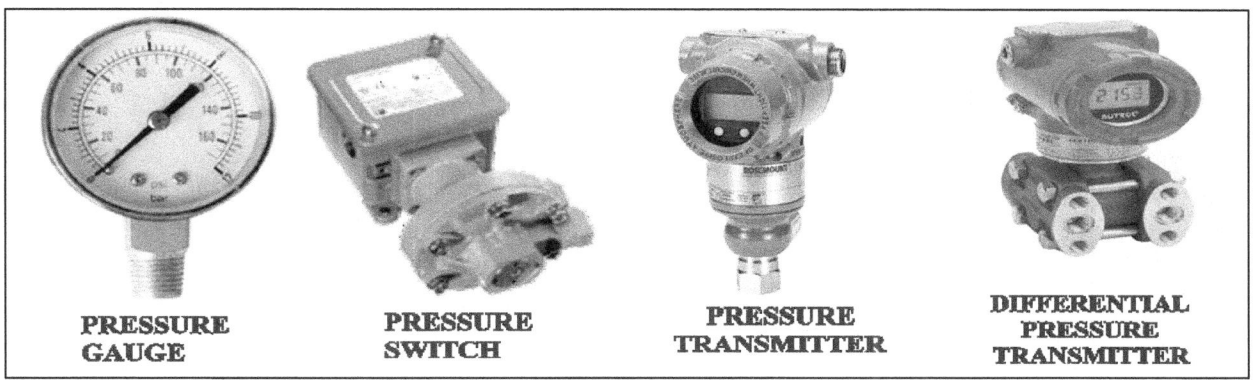

Fig.2 Types of Pressure Measuring Instruments

1.4 Pressure Gauge

•   A Pressure Gauge is used for measuring the pressure of a gas or liquid.

•   A Vacuum Gauge is used to measure the pressure in a vacuum.

•   A Compound Gauge is used for measuring both Vacuum and Pressure.

•   Pressure Gauges are used for Indication only.

1.5 Measuring Principle of Pressure Gauge

Bourdon tube measuring element is made of a thin-walled C-shape tube or spirally wound helical or coiled tube. When pressure is applied to the measuring system through the pressure port (socket), the pressure causes the Bourdon tube to straighten itself, thus causing the tip to move. The motion of the tip is transmitted via the link to the movement which converts the linear motion of the bourdon tube to a rotational motion that in turn causes the pointer to indicate the measured pressure.

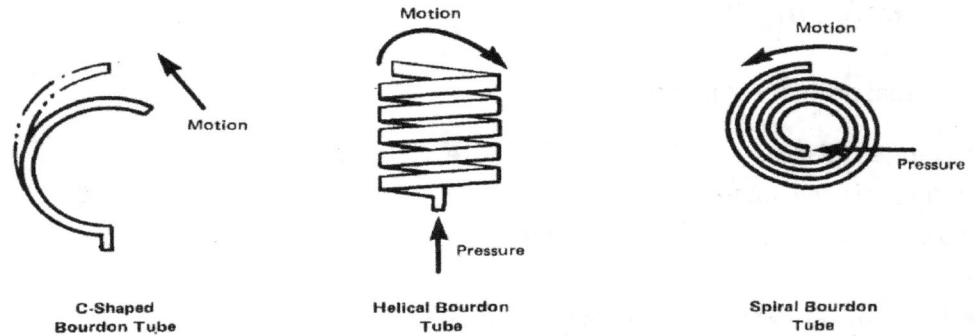

Fig.3 (Left) C Type Bourdon Tube, (Middle) Helical Bourdon, (Right) Coiled Bourdon

1.6 C -Type Bourdon Tube Pressure Gauge

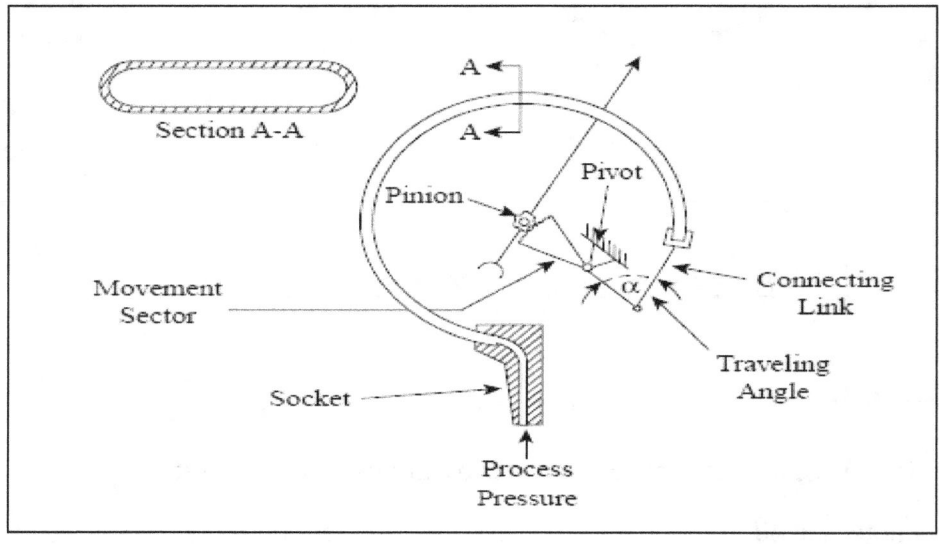

Fig.4 Bourdon Tube Pressure Gauge

Fig.4 illustrates, when pressure is applied, the tube will straighten out until the difference in force is balanced by the elastic resistance of the material composing the tube. The working part of a Bourdon gauge consist of a tube that is bent into a circular arc and is oval in cross-section so that it will tend to straighten more easily when under pressure. The open end of the tube passes through a socket which is threaded (Process Connection/pressure Connection) so that the gauge can be mounted. The closed end of the tube is linked to a pivoted segment gear in mesh with a small rotating gear to which a pointer is attached.

1.7 Construction of C Type Bourdon Tube Pressure Gauge

a) Cases:

Cases may be fabricated from various materials using various manufacturing processes. They may have solid fronts or open fronts, and may or may not employ various case pressure relief means.

a.1) **Cases with Pressure Relief Means**: For gauges used to measure gas pressure of 400 psi and higher and liquid pressure of 1000 psi and higher, cases with pressure relief means are recommended.

a.2) **Solid Front with Pressure Relief Back**: In the event of elastic element failure within its rated pressure range, the solid front (partition between pressure element and window) and the pressure relief back shall be designed to reduce the possibility of window failure and projection of parts outward through the front of the gauge.

a.3) **Open front with Pressure Relief**: In the event of a slow leak of media through the elastic element, the case pressure relief shall be sufficient to vent the case pressure increase before window failure occurs.

b) Dials:

b.1) Common Units: Three basic Classification of Units:

- SI: These unit are recognized by the CIPM (Comite International des Poids et Measures).
- MKSA (Meter, Kilogram- force, second, ampere): The former metric units, which are being replaced by the SI units.
- Customary (inch, pound-force, second, ampere): Customary units are primarily in English Speaking countries and are being replaced in most countries by SI units.

SI Abbreviation	Unit	Definition
Bar	Bar	1 bar = 100 kPa
kPa	Kilopascal	1 kPa = 1000 Pa
Mbar	Millibar	1 mbar = 100 Pa
MPa	Mega pascal	1 MPa= 1000000 Pa
N/m2	Newton per square meter	1 N/m2= 1 Pa

MKSA Abbreviation	Unit	Definition
Kg/cm2	Kilogram per square centimeter	1 kg/cm2= 1 kilogram force per square centimeter
mmHg	Millimeters of mercury	1 mmHg = 1 millimeter of mercury

Customary Abbreviation	Unit	Definition
In H2O (20 deg. C.)	Inches of water	1 in H2O (20 deg. C) = 1 inch of water at 20 degrees Celsius.
Psi	Pounds per square inch	1 psi = 1 pound force per square inch gauge pressure
Psia	Pounds per square inch absolute	1 psia = 1 pound force per square inch absolute pressure

Table 1. Different Units of Pressure

b.2) Dial information: Dial shall indicate the units in which the scale is graduated. Dual scale dials are useful where gauge is employed on equipment that may be used internationally or where users plan to convert from one unit of measure to another.

c) Pointer:

The pointer shall rotate in clockwise direction for increasing positive pressure and counter clockwise for increasing negative pressure.

d) Pressure Connection:

Location of pressure connection

- Stem Mounted: Bottom or Back
- Surface Mounted: Bottom or Back
- Flush Mounted: Back

e) Windows:

e.1) **Laminated Glass**: Laminated Glass shall comply with ANSI Z26.1. It reduces the possibility of glass particles scattering if the pressure element ruptures and windows failure results.

e.2) **Tempered Glass**: Tempered Glass shall comply with ASTM C 1048 (heat treated) or ASTM C 1422 (chemically treated). Tempered Glass is generally 2 to 5 times stronger than plain glass.

e.3) **Plain Glass**: This window material is commonly used due to its abrasion, chemical and wear resistance properties. Careful while selecting in hazardous application.

f) Scales:

The graduations and related numerals on the dial. These may be concentric, eccentric or edgewise. Concentric scales form a major part of a circle with a pointer spindle at the center and the pointer radial to the scale. Eccentric scale forms a small part of a circle and in which the pointer is radial to the scale, but the center of rotation of the pointer is not concentric with the scale. Edgewise scale is a graduation on a rectangular dial bent in the form of an arc with the pointer spindle at the center. Fig. 5 shows different configuration of scales.

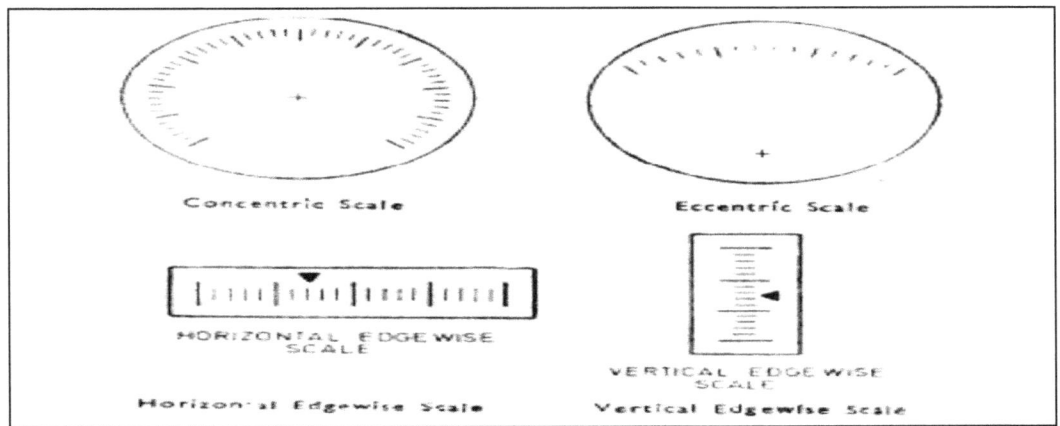

Fig.5 Configuration of Scales

1.8 Differential Pressure Gauge

Differential pressure gauges have two inlet ports, each connected to one of the volumes whose pressure is to be monitored. In effect, such a gauge performs the mathematical operation of subtraction through mechanical means, eliminating the need for an operator or control system to watch two separate gauges and determine the difference in readings.

The gauge shown here utilizes two bourdon tubes one for each pressure measurement and a linkage to determine the differential pressure. A diaphragm or bellows can also be used as the pressure sensing element.

In cases where either input can be higher or lower than the other, a bi-directional differential range should be used. Bidirectional Differential Pressure gauges are constructed such that the zero point is in the middle. The opposing Bourdon tubes are linked to a single pinion gear which rotates a pointer for direct pressure readings. By using two independent Bourdon tubes, the gauge can handle liquids or gases on either or both ports.

Fig. 6 Unidirectional Differential Pressure Gauge Fig.7 Bidirectional Differential Pressure Gauge

1.9 Diaphragm Seals

Diaphragm seals, also known as chemical seals, isolate pressure measuring instruments from the process media. The process pressure is applied to the lower side of diaphragm, while the upper side is at atmospheric pressure. The differential pressure arising across the diaphragm, lifts up the diaphragm and puts the pointer in motion.

Two Types of Diaphragm Available; Flat and Corrugated.

The diaphragm must be made in such a way that the deflection is linear, that is a similar increase in the pressure should always corresponds to a similar deflection of the diaphragm.

A flat diaphragm made of metal will be only linear when deflection is very small. At larger deflections, a flat diaphragm loses its linearity since more and more stress will occur in the diaphragm.

A flexible material, such as a thin sheet of nylon, can however serves as a flat diaphragm. The diaphragm will then be opposed by a calibrated spring which ensures the linearity and pushes the diaphragm back to its starting position.

For industrial application, corrugated diaphragm generally used. The corrugated ensures that the diaphragm will be more elastic and they are arranged such that the deflection of the diaphragm is linear.

The pressure ranges of diaphragm gauges fall between 10 mbar to 40 bar. For smaller measuring ranges (mbar), diaphragm is used with a larger diameter. This increase the sensitivity of the diaphragm for small pressure ranges and also increases the stroke length. Due to increased sensitivity, accuracy will be higher. Below 10 mbar capsule pressure gauges to be used.

Diaphragm pressure gauges are available in accuracy class 0.6 to 1 -1.6-2.5 and 4.

By default, these pressure gauges can resist pressure of about 5 times the full-scale value as compared with 1.3 times full scale value overpressure protection of a Bourdon Tube. This gives an extra advantage over Bourdon Tube Pressure gauge.

Diaphragm sealed gauges should be considered for:

- Process fluids that would clog the pressure elements.
- Process fluids that are toxic, corrosive, slurry and viscous.
- Process fluids that could crystallize or polymerize.
- Materials capable of withstanding the process fluids that are not available as a pressure element, such as high temperature.
- Process fluid that might freeze due to change in ambient temperature and damage the element.
- Hydrocarbon services having a Reid vapor pressure (RVP) of 18 psig and over. (RVP is the absolute vapor pressure exerted by a liquid at 100°F. The higher this value, the more volatile the sample and the more readily it will evaporate).
- Auto-ignitable hydrocarbon services.

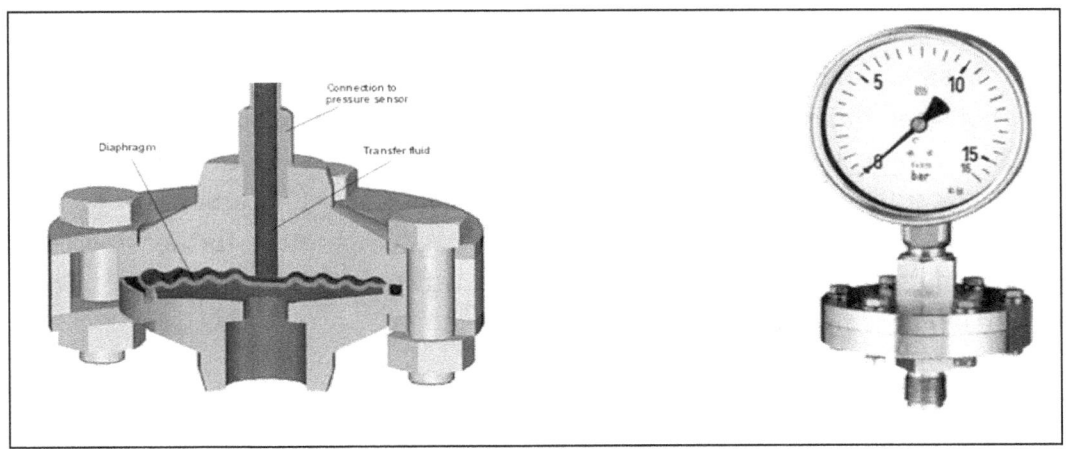

Fig.8 Diaphragm Seal Pressure Gauge

1.10 Pressure Gauge Selection Guideline

When selecting a pressure gauge, it is important to consider the following factors:

1. Pressure Range

2. Temperature Range

3. Conditions affecting wear of the system

4. Pressure Fluid Composition

5. Nominal Size

6. Pressure Connection

7. Accuracy

8. Mounting

9. Overpressure

10. Other Criteria

1. Pressure Range: A gauge range of twice the working pressure is generally recommended for maximum accuracy, safety and for extended gauge life. The range should be such that the maximum working pressure does not exceed 75 % of the maximum scale value for steady pressure or 65 % of the maximum scale value for cyclic pressures.

2. Temperature Range: Ambient temperature for general pressure gauges that are soft-soldered is −40 F to 120 F. Ambient temperature for Process pressure gauges that are silver soldered or welded is −40 F to 190F and for Liquid Filled pressure gauges it is 0 F to 140 F.

3. Conditions Affecting Wear of the System: In applications involving severe pressure fluctuations, vibrations and/or pulsation the use of restrictors or pressure snubbers is recommended. Also, liquid filled gauges should be considered. The fill fluid will lubricate the movement and reduce friction and wear. Liquid filling can also prevent moisture or corrosive atmospheres from affecting the gauge internals. The most common fill fluid is glycerin. Glycerin is non-toxic, biodegradable and not hazardous to water. When lowered to +17 Degrees Celsius, glycerin starts to become viscous. As temperature further decreases, it becomes even more viscous and causes the pointer to move more slowly to the correct position. Silicone can be used in applications with larger temperature extremes.

4. Pressure Fluid Composition: Since the sensing element of a pressure gauge may be exposed directly to the measured medium, consider the characteristics of this medium. It may be corrosive, it may be gas or liquid, it may solidify at various temperatures, or it may contain solids that will leave deposits inside the sensing element.

5. Nominal Size: The size of gauge required shall be selected from EN 837-1 or EN 837-3.

6. Pressure Connection: The pressure connection shall be selected from EN 837-1 or EN 837-3. Other connections specific to certain industries and applications shall be specified.

7. Accuracy: The accuracy class required shall be selected from EN 837-1 or EN 837-3 and ASME B40.1.

8. Mounting: Type of mounting required shall be selected from EN 837-1 or EN 837-3.

9. Overpressure: Measuring range and overpressure range as per EN 837-1.

Measuring Range	Overpressure Range
-1 to 0 Bar	3 Bar
0 to 0.6 Bar	2.5 Bar
0 to 1 Bar	4 Bar
0 to 1.6 Bar	6 Bar
0 to 2.5 Bar	10 Bar
0 to 4 Bar	16 Bar
0 to 6 Bar	25 Bar
0 to 10 Bar	40 Bar
0 to 16 Bar	60 Bar
0 to 25 Bar	80 Bar
0 to 40 Bar	100 Bar

Table 2. Measuring Pressure range and Overpressure Pressure Range

Blow Out Device: This is part of pressure gauge, usually a plug located at the back or top of the cases. When the pressure inside the casing, as a result of a failure of the bourdon tube, the blow out device will be blown away. The pressure inside the case must not become higher than window burst pressure.

When pressure gauge is liquid filled, a blow-out device is mandatory. Fig.9 (a) shows blow-out device.

Energy Release Test: Safety pattern gauge are subjected to an energy release test. This test simulates a fatal rupture of Bourdon tube and the release of high pressure gas into the casing of pressure gauge. Fig 9 (b) safety pattern pressure gauges.

S1 design: Pressure gauge built according to S1 design just need to have a blow-out device.

S2 design: Safety pattern gauges having a diameter between 40 mm to 80 mm, without a baffle wall, are built according to S2 design. These should successfully have passed the energy release test, have a window with laminated safety glass or non-splintering plastic and a have blow-out device.

S3 design: Safety pattern gauges having a diameter between 40 to 250 mm, with a baffle wall are built according to S3 design, in addition to the requirement of S2 design they must be equipped with a blow-out back.

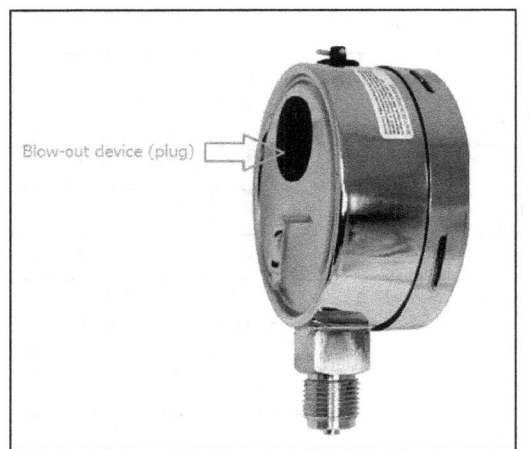 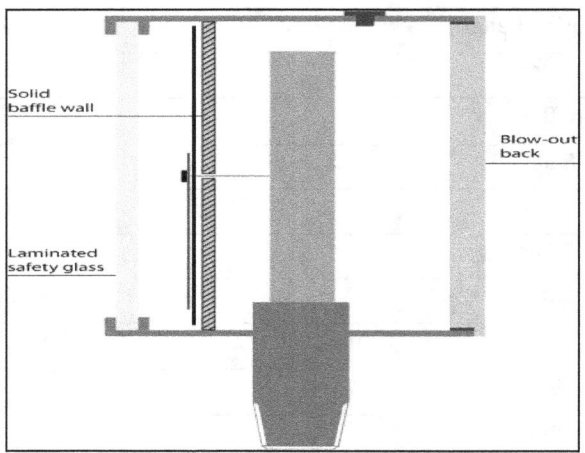

Fig.9 a) Blow-out Device b) Safety Pattern Gauge

Table 3 contains Safety Pattern Pressure Gauges Selection Criteria.

Pressure Fluid	Liquid								Gas or Steam							
Case Filling	Dry				Liquid				Dry				Liquid			
Nominal Size	<100		>=100		<100		>=100		<100		>=100		<100		>=100	
Pressure Range (in bar)	<=25	>25	<=25	>25	<=25	>25	<=25	>25	<=25	>25	<=25	>25	<=25	>25	<=25	>25
Minimum Safety Design	0	0	0	0	S1	S1	S1	S1	0	S2	S1	S3	S1	S2	S1	S3

Safety Design Codes :
0 – Gauge without blow out device
S1- Blow out device gauge
S2 – Safety pattern gauge without baffle wall
S3 – Safety pattern gauge with baffle wall
Note : All oxygen and acetylene gauges shall be safety pattern gauges
User must have cognizance of their special requirements and may use safety pattern gauges at pressure lower than 25 bar

Table 3. Sélection Criteria for Pressure Gauge on Safety Pattern

10. Other Criteria : If the application involves pressure pulsations, vibrations, extremes of temperature, shock loading, solids in suspension, viscous or chemically aggressive pressure fluid, hostile environment, or requires correction for a static head, the manufacturer shall be consulted.

1.11 Pressure Gauge Installation

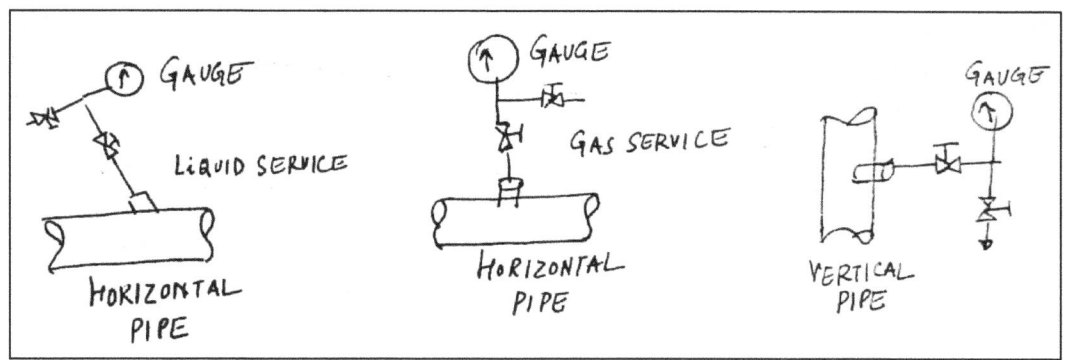

Fig.10 Diaphragm Seal Pressure Gauge

Top connection preferred for gas installations & side connection preferred for liquid installations.

The pressure gauge can be connected to the pipe by individual block and bleed valves or a two-way manifold.

1.12 Special Conditions shall be referring before Installation

• **Mechanical shocks**: Pressure gauges shall not be subject to mechanical shocks. If installations are subject to mechanical shocks, gauges shall be mounted remotely and connected by flexible pipe.

• **Vibrations**: Use of liquid filled (silicon or glycerin) pressure gauges.

• **Overpressure**: Any overpressure creates stress in the pressure responsive element and consequently reduces its life and accuracy. It is therefore always preferable to use an instrument whose maximum scale value is greater than the maximum working pressure and which will consequently absorb overpressure and surges more easily.

• **Temperature**: To protect a pressure gauge from a fluid which is too hot, a siphon or a similar device may be inserted so as to provide condensed fluid in the pressure responsive element. A siphon or a similar device shall always be placed close to the pressure gauge and be filled with condensate before the installation is pressurized to avoid the hot fluid reaching the gauge on the initial pressurisation.

• **Gauges for use with Oxygen**: Burdon Tubes and other parts in contact with oxygen shall be free of oil and grease. Only lubricants suitable for use in oxygen at maximum working pressure shall be used. The Dial shall be marked with the word "oxygen" written in English. Cleaning for oxygen service gauges shall be per MIL-STD-1330.

• **Gauge for use with Acetylene**: A gauge designed to indicate acetylene pressure. The gauge may bear the inscription ACETYLENE on the dial.

• **Gauge for use with Ammonia**: A gauge designed to indicate ammonia pressure and to withstand the corrosive effect of ammonia. The gauge may bear the inscription AMMONIA or NH3 on the dial. Material such as copper, brass and silver brazing alloys should not be used.

1.13 Accuracy w.r.t EN 873-1 & ASME B40.1

Accuracy Classes

The Accuracy Class stating the limits of permissible error is expressed as a percentage of the span.

The following accuracy classes are defined: 0.1, 0.25,0.6, 1, 1.6, 2.5 and 4.

For gauges with pointer stop, the accuracy class will be cover 10% to 100% of the range. For gauges with free zero, the accuracy class will cover 0 % to 100% of the range and zero shall be used as an accuracy check point.

Nominal Size (in mm)	Accuracy Class						
	0.1	0.25	0.6	1	1.6	2.5	4
40 & 50					X	X	X
63			X	X	X	X	
80			X	X	X	X	
100			X	X	X	X	
150 & 160		X	X	X	X		
250	X	X	X	X	X		

Accuracy Class	Limits of Permissible Error (% of Span)
0.1	+/- 0.1%
0.25	+/- 0.25%
0.6	+/- 0.6
1	+/- 1%
1.6	+/- 1.6%
2.5	+/- 2.5%
4	+/- 4%

Table 4. Accuracy Class Based on EN 873-1 (Left) Nominal Size compared to the accuracy class, (Right) Maximum Permissible Errors

Accuracy Grade	Permissible Error (± Percent of Span) (Excluding Friction)			Maximum Friction (Percent of Span)	Minimum Recommended Gauge Size (270 deg Dial Arc)
	Lower ¼ of Scale	Middle ½ of Scale	Upper ¼ of Scale		
4A	<————————	0.1	————————>	[Note (1)]	8½
3A	<————————	0.25	————————>	0.25	4½
2A	<————————	0.5	————————>	0.5	2½
1A	<————————	1.0	————————>	1.0	1½
A	2.0	1.0	2.0	1.0	1½
B	3.0	2.0	3.0	2.0	1½
C	4.0	3.0	4.0	3.0	1½
D	5.0	5.0	5.0	3.0	1½

Fig.11 Accuracy Grade Based on ASME B40.1

Grade 4A: Gauges offer the highest accuracy and are calibrated to ±0.1% of span over the entire range of the gauge. The gauges are called laboratory precision test gauges and are generally 8 1/2″, 12″ or 16″ dials. These high-accuracy gauges may be temperature compensated. They must be handled carefully to retain accuracy.

Grade 3A: Gauges are calibrated to an accuracy of ±0.25% of span over the entire range of the gauge. The gauges are called test gauges and are generally 4 1/2″, 6″ or 8 1/2″ dials.

Grade 2A: Gauges are calibrated to an accuracy of ±0.5% of span over the entire range of the gauge. These gauges are generally used by the petrochemical industry for process pressure measurement. They are often referred to as process gauges and are usually supplied as 4 1/2″ and 6″ cases.

Grade 1A: Gauges are calibrated to an accuracy of ±1% over the entire range of the gauge. These gauges are high-quality industrial gauges and are supplied in 2 1/2″, 3 1/2″ and 4 1/2″ sizes.

Grade A: Gauges are calibrated to an accuracy of ±1% of span over the middle half of the scale and ±2% of span over the first and last quarters of the scale. These gauges are often referred to as industrial gauges and are usually supplied in 2 1/2″, 3 1/2″ and 4 1/2″ case sizes.

Grade B: Gauges are calibrated to an accuracy of ±2% of span over the middle half of the scale and ±3% of span over the first and last quarters of the scale. This accuracy of gauge represents the majority of those manufactured and used for pressure measurement on water pumps, swimming pool filters, air compressors, filter regulations, etc. These gauges are often referred to as commercial or utility gauges and are supplied in 1 1/2″, 2″, 2 1/2″, 3 1/2″ and 4 1/2″ case sizes.

Grade C: Gauges are calibrated to an accuracy of ±3% of span over the middle half of the scale and ±4% of span over the first and last quarters of the scale. These are used in similar applications as Grade B gauges except that they are less accurate.

Grade D: Gauges are calibrated to an accuracy of ±5% of span over the entire scale. These 5% gauges are used as indicators when minimal accuracy is required for application on water pumps and pool filters.

1.14 Pressure Gauge Limits

Over Pressure:

The gauge shall withstand the overpressure for a period of 15 mins as shown in table 5.

Maximum Scale Value of Pressure Gauge (bar)	Overpressure to be applied
<= 100	1.25 * Maximum Scale Value
>100 to <= 600	1.15 * Maximum Scale Value
>600 to <=1600	1.10 * Maximum Scale Value

Table 5. Over-Pressure based on EN 873-1

Cyclic Pressure:

The gauge shall withstand a pressure fluctuating from 30 % to 60 % of the maximum scale value for the number of pressure cycle shown in table 6.

Maximum Scale Value of Pressure Gauge (bar)	Number of Pressure Cycles
<= 25	100000
>25 to <= 600	50000
>600 to <=1600	15000

Table 6. Cyclic Pressure based on EN 873-1

Degree of Protection:

Recommended minimum protection rating in accordance with EN 60529.

For Indoor Use: IP 31

For Outdoor Use: IP 44

Pressure/ Process Connection:

For thread forms and sizes , see the attached table 7.

Parallel Pipe Threads	Taper Pipe Threads
G 1/8 B	1/8 – 27 NPT EXT
G 1/4 B	1/4 – 18 NPT EXT
G 3/8 B	
G 1/2 B	1/2 – 14 NPT EXT
NOTE: G 3/8 B is not preferred	

Table 7. Thread Forms and Sizes based on EN 873-1

1.15 Modes of Pressure Gauge Failure

•	Fatigue Failure: Fatigue Failure caused by pressure induced stress generally occurs from the inside to the outside along a highly stressed edge radius of a Bourdon Tube, appearing a small crack that propagates along the edge radius. Such failures are usually more critical with compressed gas media than with liquid media.
A snubber placed in the gauge pressure inlet will reduce pressure surges and fluid flow from partially open elastic element.

•	Overpressure: Overpressure failure is caused by the application of internal pressure greater than the rated limits of the elastic element and can occur when a low range pressure gauge is installed in high pressure system.
Placing a snubber in the pressure gauge the immediate effect of failure, but will help control flow of escaping fluid.

•	Corrosion Failure: Corrosion failure occurs when the elastic element has been weakened through attacks by corrosive chemicals present in either media inside or the environment outside it. A Diaphragm seal should be considered for use with pressure media that may have a corrosive effect on the elastic element.

•	Vibration Failure: The most common mode of vibration failure is wear of mechanical components because of high cyclic loading caused by vibration. This is characterized by gradual loss of accuracy and ultimately failure of the pointer to indicate any pressure change.

1.16 Mounting of Pressure Gauge

Different mounting types of gauges as shown in fig. 12. When mounting, ensure that there is enough free space for the blow out devices if any.

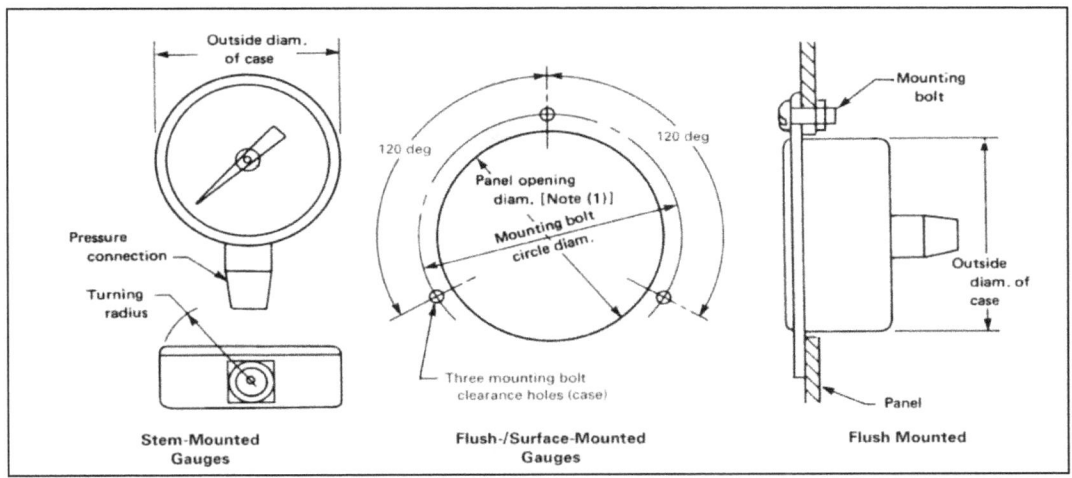

Fig.12 Different Mounting for Pressure Gauges

1.17 Pressure Gauge – Accessories

• **Snubber**

Placing a snubber or a restrictor between a process connection and elastic element will not reduce the immediate effect of failure but will help reduce flow of escaping fluid following rupture and reduce the potential of secondary effects.

Fig.13 Snubber

- **Pulsation Damper (Adjustable Snubber)**

Pulsation dampers are frequently utilized to reduce the magnitude of line pressure pulsations. Rapidly pulsating pressure can quickly destroy gage accuracy by producing abnormal wear on moving parts. The elastic element is a metal spring member and rapid pulsations can cause excessive reversing stresses and eventual metal-fatigue failure.

Fig.14 Pulsation Damper

- **Pressure Limit Valve**

Protects pressure instruments against surges and pulsations. Provides automatic positive protection and accurate, repeatable performance. Automatic pressure shut-off, built in snubber enhances instrument protecting performance.

Fig.15 Pressure Limit Valve

- **Siphon Tubes**

Used to dissipate heat by trapping condensed liquid to keep high temperature steam or condensing vapor from damaging the pressure gauge.

PIG TAIL

COIL PIPE

Fig.16 Siphon Tubes Types

1.18 Calibrating Pressure Gauge

Pressure gauges are often used as local indicators of process pressure. An analog pressure gauge, because of its links, levers, and elastic pressure sensing element, requires periodic calibration checks. Pressure changes applied to the gauge cause the elastic element to expand and contract. The movement of the element is translated into movement of the pointer through links, levers, and gears. The measurement values of the gauge are read directly on the gauge scale from the position of the pointer.

Calibrating a pressure gauge includes 'adjustment of these components until the gauge accurately represents the input.

Select the proper input standard for the pressure range tested. The pressure standard may be a precision gauge or digital test standard with the correct pressure module, manometer, or dead weight tester.

Use a tee to connect the input *test* standard to the pressure source and the gauge under test. Be sure the gauge under test is mounted in the same orientation as in the process.

Determine the five test points used for the upscale and downscale checks of the gauge under test. With any mechanical instrument, it is important to accurately determine whether hysteresis is present in the instrument. This means that you will begin your upscale check from 0%and approach the first test point, 10% from below. Similarly, start the downscale check by increasing the input to 100%, then approach 90% from above.

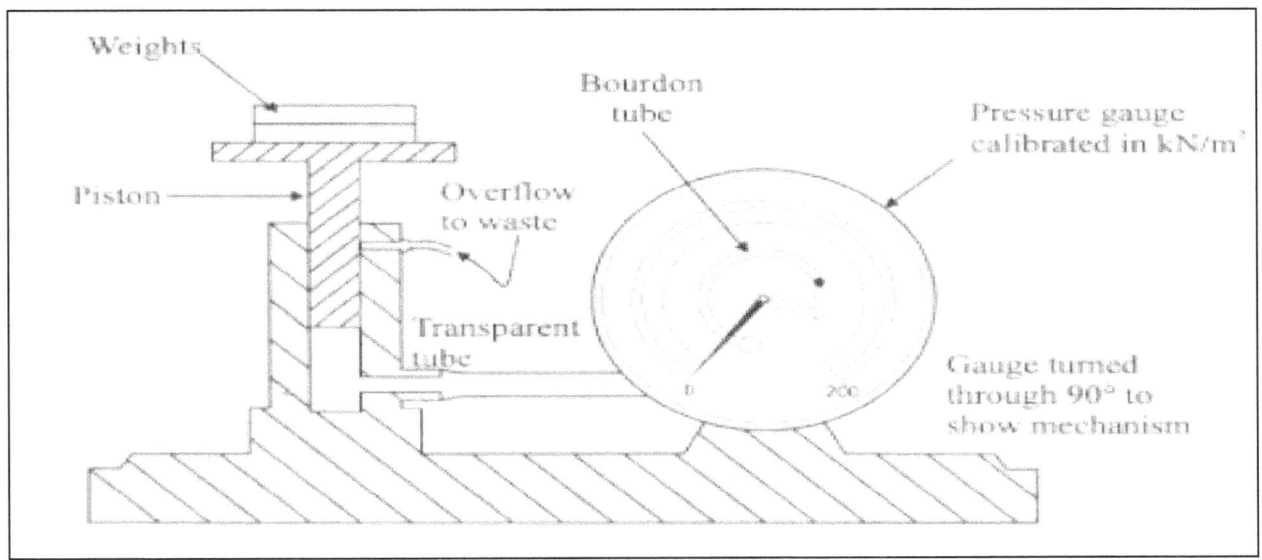

Fig.17 Pressure Gauge Arrangement

On most motion balance instruments, adjust linearity first. The movement of the elastic element in a pressure gauge causes a proportional movement in the linkage. On a properly calibrated gauge, the linkage angle will be exactly 90° when the input to the gauge is at 50% of total range. Therefore, apply 50% input and use a template to check the 90-degree angle. With linearity adjusted, position the pointer so the gauge reads midscale.

You may need to remove the pointer and reposition it on the shaft. If removing the pointer is necessary, be sure to use the proper tool. Now lower the input to 10% and adjust the zero so the gauge reading equals the applied input. Now correct the span error. Increase the-input pressure to 90% and adjust the gauge to read the input value. For span adjustments, loosen the screws and rotate the entire adjustment mechanism.

Repeat the zero and span adjustment until the readings at 10% and 90% are accurate. Zero and span interact in the mechanical device, so rechecking them is necessary. When zero and span require no further adjustments, recheck the linearity to make sure it is still properly adjusted. There are no adjustments for hysteresis. After adjusting linearity, zero and span, perform another full-scale check and record as left calibration data.

1.19 Data Sheet of Pressure Gauge

	RESPONSIBLE ORGANIZATION	PRESSURE GAUGE		SPECIFICATION IDENTIFICATIONS
1			6	
2	(ISA)	Device Specification	7	Document no
3			8	Latest revision / Date
4			9	Issue status
5			10	

OPERATING PARAMETERS

				DIAL AND POINTER
11			60	
12	Project number / Sub project no		61	Dial scale type
13	Project		62	Pointer adjustment
14	Enterprise		63	Graduations and color
15	Site		64	Scale range type
16	Area / Cell / Unit		65	Dial material
17	Related equipment		66	
18	Service		67	
19			68	
20	P&ID/Reference dwg number		69	**SIPHON**
21	Material name		70	Siphon type
22	Minimum pressure		71	End conn nominal size / Style
23	Normal pressure		72	Overall length
24	Maximum pressure		73	Siphon material
25	Normal temperature		74	
26	Maximum temperature		75	
27	Material phase		76	
28			77	**PERFORMANCE CHARACTERISTICS**
29			78	Max press at design temp / At
30			79	Min working temperature / Max
31	**PROCESS CONNECTION AND CASE**		80	Min ambient working temp / Max
32	Case type		81	
33	Case style		82	
34	Gauge size		83	
35	Process conn nominal size / Style		84	
36	Process conn location		85	**ACCESSORIES**
37	Case pressure relief type		86	Pressure limit valve matl
38	Ring style		87	Restrictor style
39	Mounting type		88	Pressure snubber matl
40	Case material		89	Pulsation dampener matl
41	Ring material		90	
42	Exterior treatment-color		91	
43	Window material		92	**SPECIAL REQUIREMENTS**
44	Stem material		93	Custom tag
45	Gasket/O ring material		94	Reference specification
46	Liquid fill material		95	Special preparation
47			96	Compliance standard
48			97	Service design
49			98	
50	**PRESSURE ELEMENT AND MOVEMENT**		99	
51	Elastic element type		100	
52	Movement style		101	**PHYSICAL DATA**
53	Nominal accuracy grade		102	Estimated weight
54	Joint type		103	Maximum thickness
55	Element material		104	Max case outside dia
56	Movement material		105	Mfr reference dwg
57			106	
58			107	
59			108	

CALIBRATIONS AND TEST

	TAG NO/FUNCTIONAL IDENT	PRESSURE OR SCALE		SCALE	
110				LRV	URV
111					
112		Pressure/Scale 1			
113		Pressure/Scale 2			
114					
115					
116					
117					

COMPONENT IDENTIFICATIONS

	COMPONENT TYPE	MANUFACTURER	MODEL NUMBER
118			
119			
120			
121			
122			
123			
124			
125			

Rev	Date	Revision Description	By	Appv1	Appv2	Appv3	REMARKS

Form: 20P2001 Rev 0

© 2001 ISA

1. Pressure Gauge Data Sheet by ISA

1.20 Pressure Switch

Measuring Principle:

•   The device contains a micro switch, connected to a mechanical lever and set pressure spring. The contacts get actuated when process pressure reaches the set pressure of the spring.

•   It can be used for alarming or interlocking purposes, on actuation.

•   It can be used for high / high-high or low / low-low actuation of pressure in the process. The set range can be adjusted within the switch range.

•   The sensing element may be a Diaphragm or a piston.

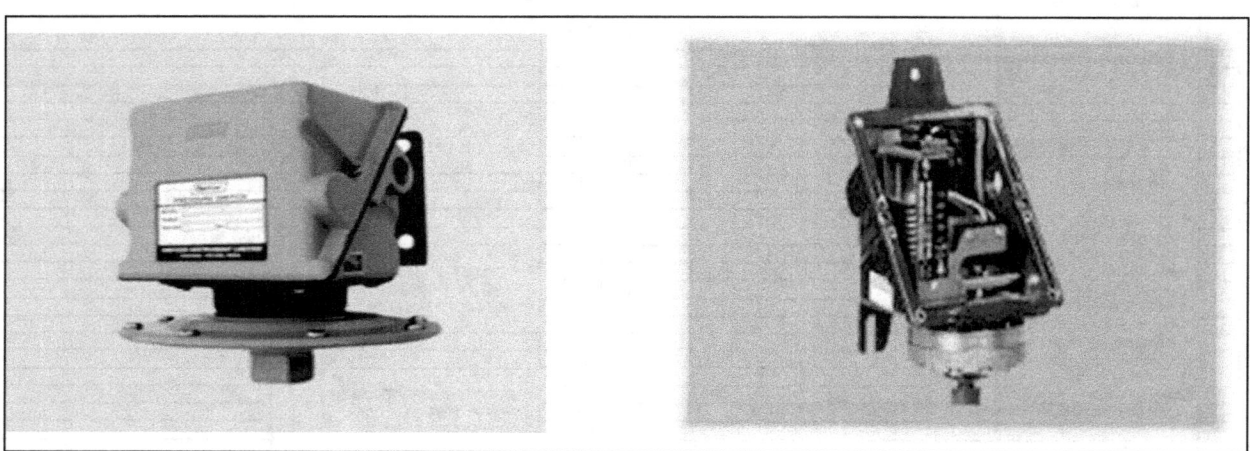

Fig.18 Pressure Switch

1.21 Types of Pressure Switch

Pressure switches can be used to sense either positive pressure or vacuum. They can either   energize or de-energize a circuit depending on how they are configured. They can also be provided with a single or dual switches. Some dual switch pressure switches can be set up so that each switch can operate at a different pressure set point. The electric contacts can be configured as single pole double throw (SPDT), in which case the switch is provided with one normally closed (NC) and one normally open (NO) contact. Alternately, the switch can be configured as double pole double throw (DPDT), in which case two SPDT switches are furnished, each of which can operate a separate electric circuit.

Gauge Pressure Switch:

A Gauge Pressure Switch with a single pressure connection and responsive to the difference   between the pressure applied to that connection and the surrounding atmospheric pressure. The switch is responsive only to pressure greater than the atmospheric pressure.

Vacuum Pressure Switch:

A Vacuum Pressure Switch is like gauge pressure switch but responsive to pressure less than the atmospheric pressure.

Absolute Pressure Switch:

An Absolute Pressure Switch with a single pressure connection and responsive to the difference between the pressure applied to that connection and a reference high vacuum.

Differential Pressure Switch:

A Differential Pressure Switch with two pressure connection and responsive to the difference between the two-applied pressure.

1.22 Types of Switch Contacts

Pressure Switch are classified according to their types of switch contacts as follows:

- Air Break Switch
- Mercury Tilt Switch
- Magnetically Operated Switch
- Contactless Switch

Any of these types of switch contacts are further classified in terms of number of poles and configuration.

1.23 Types of Pressure Elements

Pressure Switches are classified according to their type of pressure elements as follows:

- Bourdon Type
- Diaphragm
- Capsule
- Capsule Stack
- Bellows
- Piston

If a pressure switch operating between 30-70% of working range and the electric load is not severe, then some generalization can be made about the service life. If service life of one million pressure cycle a diaphragm, bourdon or bellows element is satisfactory. If greater service life is expected, a piston switch should be used. However, where pressure variations are small (20 % or less of the adjustable range of the switch), a Bourdon Tube, bellows or diaphragm switch can be expected to provide a useful life of 2-5 million cycles before fatigue failure of the sensing element. Fig 19. Show the operating range of pressure switch.

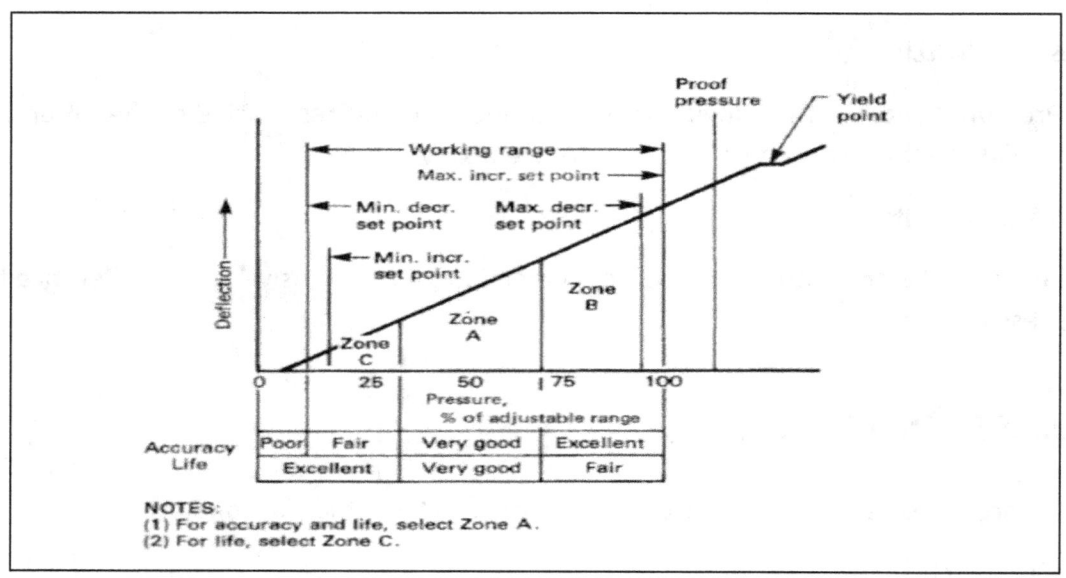

Fig.19 Operating Range of Pressure Switch

1.24 Accuracy Class

Pressure Switches are classified according to their accuracy class. Accuracy Classes are 1 %, 2 % and 5 %.

1.25 Calibrating Pressure Switch

The calibration setup of a pressure switch is like a pressure gauge and pressure transmitter, except a multimeter set to measure voltage or resistance, as applicable, is connected to the applicable set of contacts (such as the common and NO terminals). The applied input pressure is increased to the setpoint. The multimeter will read the change of voltage or resistance when the switch changes state. In other words, the multimeter will read 0 volts with the switch closed and supply voltage with the switch open. If the contact is a "dry" contact (no voltage), the multimeter will read close to 0 ohms with the contact closed and infinite ohms with the switch open.

1.26 Data Sheet of Pressure Switch:

	RESPONSIBLE ORGANIZATION	PRESSURE SWITCH w/wo TRANSMITTER Device Specification		SPECIFICATION IDENTIFICATIONS
1			6	
2	(ISA)		7	Document no
3			8	Latest revision — Date
4			9	Issue status
5			10	

	SWITCH BODY			PERFORMANCE CHARACTERISTICS
11			59	
12	Body type		60	Max press at design temp — At
13	Process conn nominal size — Rating		61	Min working temperature — Max
14	Process conn termn type — Style		62	Output accuracy rating
15	Body material		63	Repeatability
16	Seal/O ring material		64	Press Lower Range-limit — URL
17			65	Max overrange limit
18			66	Dead band rating
19			67	Min ambient working temp — Max
20	SENSING ELEMENT		68	Contacts ac rating — At max
21	Sensor element type		69	Contacts dc rating — At max
22	Adjustable LRL — URL		70	
23	Diaphragm/Wetted material		71	
24			72	
25			73	
26			74	
27	SWITCH MECHANISM w/wo TRANSMITTER		75	
28	Housing type		76	
29	Element Style		77	
30	Output signal type		78	
31	Enclosure type no/class		79	
32	Reset style		80	
33	Set point Adjustment type		81	
34	Signal power source		82	
35	Measurement type		83	ACCESSORIES
36	Contacts arrangement — Quantity		84	Sealed leads adapter
37	Failsafe style		85	Breather/Drain style
38	Integral indicator style		86	
39	Signal termination type		87	
40	Cert/Approval type		88	
41	Mounting type		89	
42	Dead band type		90	SPECIAL REQUIREMENTS
43	Enclosure material		91	Custom tag
44	Exterior treatment matl		92	Reference specification
45			93	Special preparation
46			94	Compliance standard
47			95	
48			96	
49			97	
50			98	PHYSICAL DATA
51			99	Estimated weight
52			100	Overall height
53			101	Removal clearance
54			102	Signal conn nominal size — Style
55			103	Mfr reference dwg
56			104	
57			105	
58			106	

	CALIBRATIONS AND TEST		INPUT/SETPOINT/TEST			OUTPUT OR SCALE	
110	TAG NO/FUNCTIONAL IDENT	MEAS/SIGNAL/TEST	LRV	URV	ACTION	LRV	URV
111							
112		Press setpoint 1-Output					
113		Press setpoint 2-Output					
114		Pressure-Analog output					
115		Pressure-Scale					
116							
117							
118							

	COMPONENT IDENTIFICATIONS		
119			
120	COMPONENT TYPE	MANUFACTURER	MODEL NUMBER
121			
122			
123			
124			
125			
126			

Rev	Date	Revision Description	By	Appv1	Appv2	Appv3	REMARKS

Form: 20P2401 Rev 0

© 2001 ISA

2. Pressure Switch Data Sheet by ISA

1.27 Pressure Transmitter

What is Pressure Transmitter?

The equipment which are used for detection of pressure (gauge or absolute) or differential pressure generated in the field then convert them into 4 to 20 ma of unified electrical signal or Field Bus Protocol and send the signals to remote indicator, recorder or controller etc.

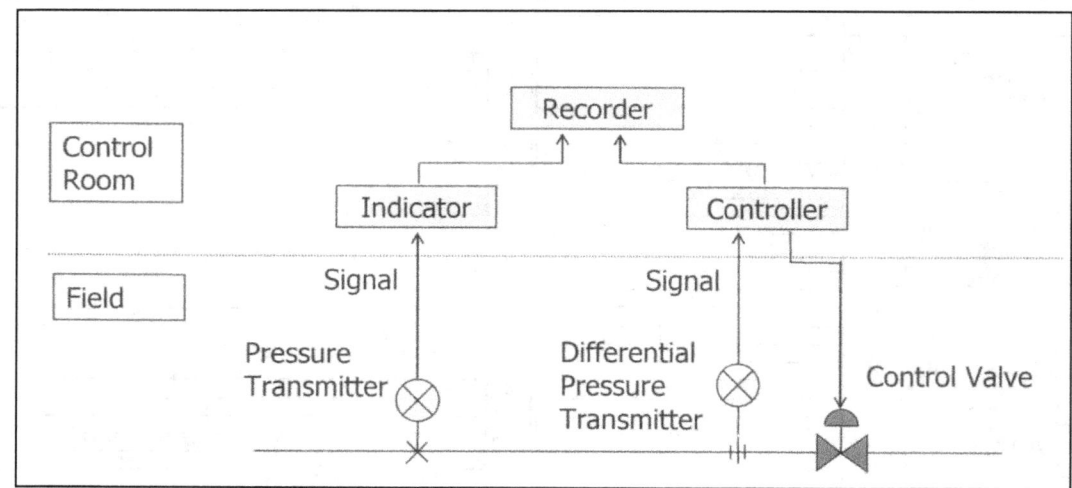

Fig.20 Inferface of Pressure Transmitter with PLC/DCS representation

1.28 Sensing Technology of Pressure Transmitter

Different Sensing Technologies of Pressure Transmitter comes in market over the period.

* ❖ Capacitance Type
* ❖ Piezo Electric Type
* ❖ Silicon Resonant Type

How's Capacitance Type Sensor works?

Sensor which converts diaphragm displacement to measurable signal is a capacitance sensor. Process pressure is transmitted through an isolating diaphragm and silicone oil fill fluid to a sensing diaphragm in the center of the cell. The position of the sensing diaphragm is detected by the capacitance plates on both sides of the sensing diaphragm. Differential Capacitance between the sensing diaphragm and the capacitor plates is converted electronically to 2 wires 4-20 ma signal.

The approach is based on the following concepts:

$P1-P2 = K (C1- C2)$

Where P1= Process Pressure at Higher Side

P2= Process Pressure at Lower Side

K = Constant

C1= Capacitance between the high-pressure side and sensing diaphragm

C2= Capacitance between the low-pressure side and sensing diaphragm

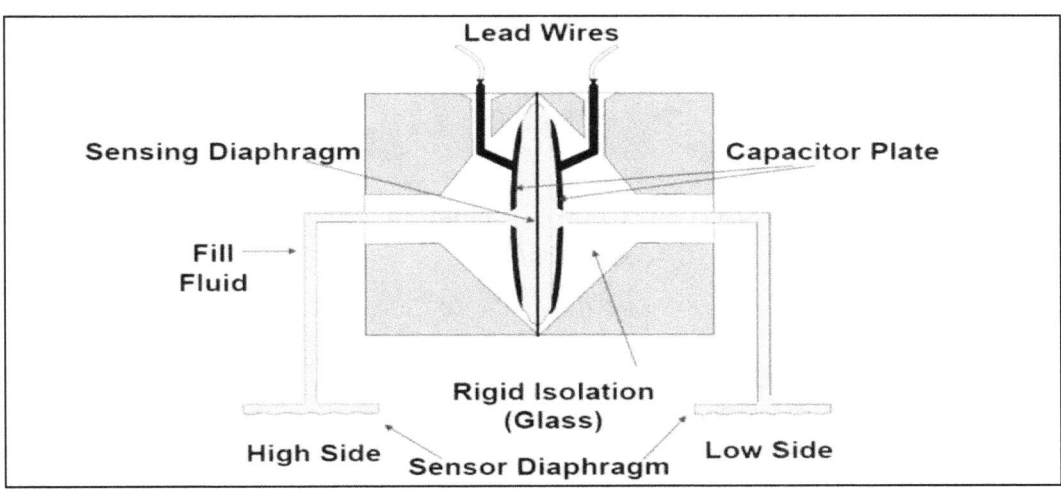

Fig.21 Capacitance Type Sensing Element

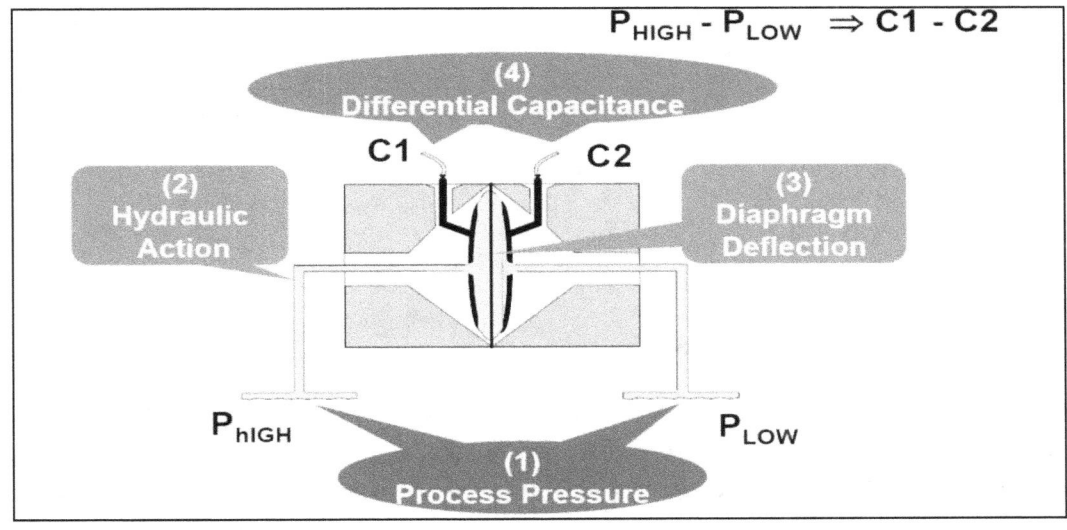

Fig.22 Capacitance Type Principle

How's Piezo Electrical Type Sensor works?

Piezoelectrical Crystals form effective secondary elements for dynamic pressure measurement. Under the action of compression, tension or shear, a piezoelectrical crystal deforms and develops a surface charge q, which is proportional to the force acting to bring about the deformation.

In a piezoelectrical pressure transducer, a preloaded crystal is mounted to the diaphragm sensor as indicated in fig. 23. Pressure acts normal to the crystal axis and changes the crystal thickness t by small amount Δ t. This sets up a charge, q=KpA, where p is the pressure acting over the electrode area A and K is the crystal charge sensitivity, a material property. A charge amplifier is used to convert charge to voltage so that the voltage developed across the electrode is

$$E= q/C$$

Where C is the capacitance of the crystal electrode combination. The crystal sensitivity for the quartz, the most common material used is K= $2.2 * 10^{-9}$ coulombs/N.

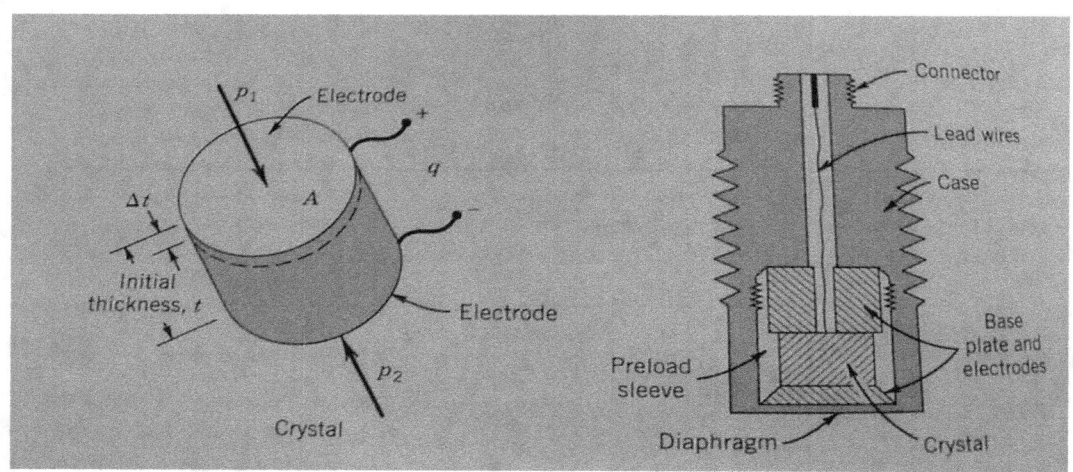

Fig.23 Piezoeletrical Pressure Transducer

How's Silicon Resonant Type Sensor works?

Silicon resonant sensors are fabricated from a single crystal silicon using 3D semi-conductor micromachining techniques. Two "H" shaped resonators are patterned on the sensor, each operating at a high frequency output. As pressure is applied, the bridges are simultaneously stressed, one in compression and one in tension. The resulting change in resonant frequency produces a high differential output (kHz) directly proportional to the applied pressure. This simple time-based function is managed by a microprocessor. The microprocessor can receive the digital signal directly from the sensor without having to go through an A/D converter. This improves the overall accuracy of the transmitter since, though small, there is a certain probability for error in each stage of conversion. In a DP application, the microprocessor can also use the two frequencies to determine the Static Pressure. Therefore, this sensor can measure two different process attributes with a single sensor.

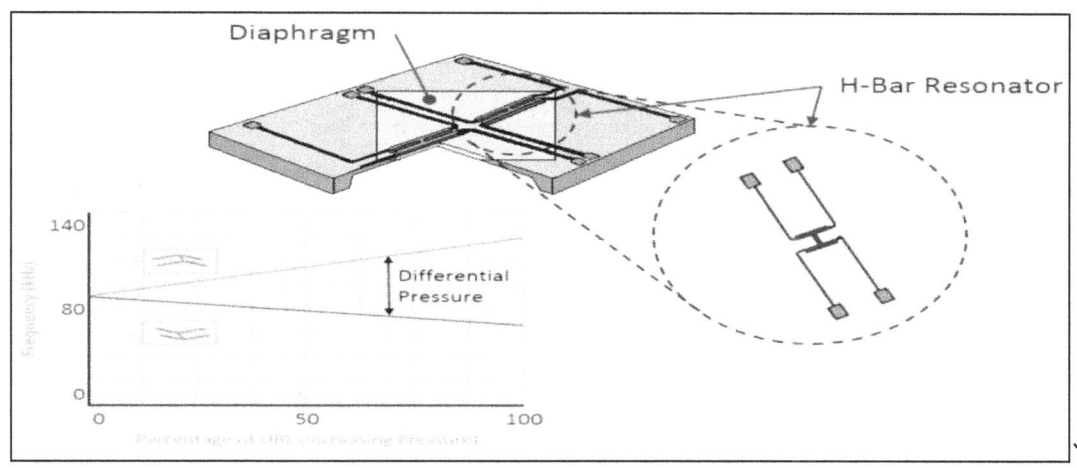

Fig.24 Silicon Resonant Pressure Transducer

1.29 Comparison of Sensing Technology:

Principle	Capacitance Type	Piezo Electrical Type	Silicon Resonant Type
Construction			
Merits	Simple Construction, Long experience	Small Hysteresis, Multi Sensing	Small Hysteresis, Small Temperature Effect Multi Sensing Differential Pressure, Static pressure, Temperature
Demerits	Big Hysteresis, Low Signal to Noise Ratio, No temperature Compensation	Big Temperature Effect (Before Compensation)	Cost High.

Table 8. Comparsion Between Sensing Technology

1.30 Essential Terms to Know

Lower Range Limit (LRL): Minimum value that a sensor can measure.

Upper Range Limit (URL): Maximum value that a sensor can measure.

Range (URL + |LRL|): Total pressure that the sensor can measure.

Lower Range Value (LRV): The lowest value that the transmitter has been adjusted to measure. This value corresponds to the 4mA analog signal.

Upper Range Value (URV): The highest value that the transmitter has been adjusted to measure. This value corresponds to the 20mA analog signal.

Span (URV − LRV): The difference between the URV and LRV. Sensors have a minimum span requirement that must be considered while assigning the 4 to 20 mA analog output.

Transfer function: Pressure Transmitter provides a selection of output function, as follows

a) Linear for Pressure, Differential Pressure, Absolute Pressure, Vacuum Pressure and Level Measurement

b) Square root for Flow Measurement: Using this function, the output is proportional to the square root of the input signal. Fig. 25 shows, for an input signal variation from 0 to 4% the output varies linearly. If input values greater than 4% the output follows the square root function. For eg. If transmitter calibration range 0-400 mbar, with 196 mbar input pressure applied, percentage of flow is determined as follows:

(196/400) *100= 49% of calibrated pressure

Sq. rt (49) *10= 70% of calibrated flow

To convert from a percentage of the calibrated flow to the equivalent output current, first divide the percentage of flow by 100, then multiply by the 16 mA, result will add with 4 mA.
(70% calibrated flow/100) * 16 mA +4 mA= 15.2 mA

Input (% of Span)	Output (% of Span)	mA
0.0%	0.0%	4 mA
0.5%	7.1%	5.13 mA
1%	10%	5.6 mA
25%	50%	12 mA
64%	80%	16.8 mA
100%	100%	20 mA

Table 9. Typical DP Transmitter with square root extraction

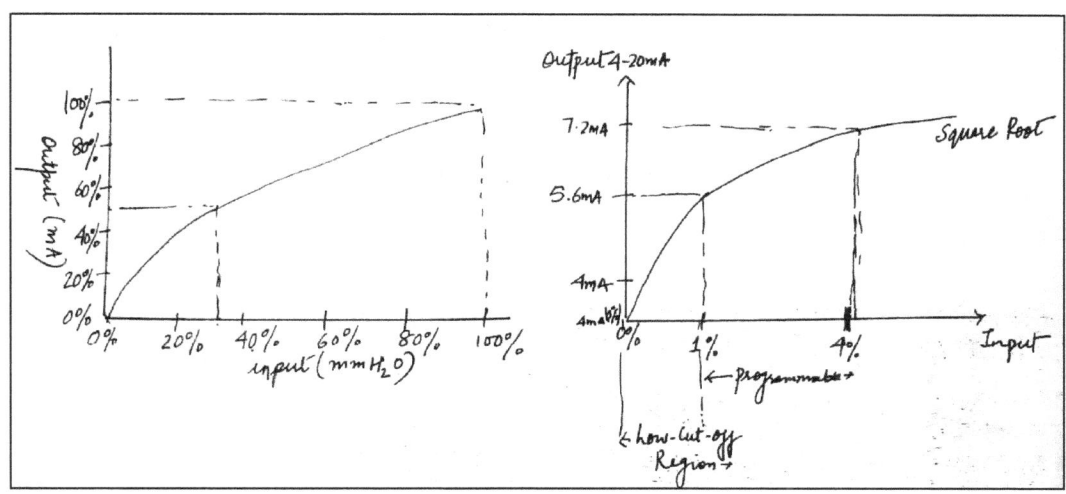

Fig.25 Output mA and % Input Pressure Variation Curve when Transfer Function is Square Root

Generally from 0% input to 1% input the area defined under " Low Flow Cut- Off" region.

Reference Accuracy: In process instrumentation, a number or quantity that defines a limit that errors will not exceed when a device is used under specified operating conditions.

Reference Accuracy can be expressed in the following ways:

a) Accuracy rating expressed in terms of the measured variable.

b) Accuracy rating expressed in percent of span.

c) Accuracy rating expressed in percent of the upper range value.

Wetted Part Material: It includes diaphragm, process connector, vent plug and cover flange. Selection of material for wetted part of transmitter depends on the application. Following materials used for wetted part of transmitter:

a) SS 316L

b) Hastealloy- C

c) Monel

d) Tantalum

e) PFA

Turn Down Ratio: Is the ratio between the maximum transmitter range and minimum transmitter range which stated accuracy is maintained.

Rangeability: Is the ratio between the maximum URL and minimum span of the transmitter.

Zero Error: Zero Error is a shift of a constant magnitude between the measured variable and the ideal variable. It is normally measured at the zero-reference point.

Span Error: Span Error is the difference between the actual span and ideal span. With this type of error, the actual span may be slightly larger or smaller than the ideal span.

Precision or Repeatability: Agreement of the reading among themselves. If the same value of the measured variable is measured many times and all the results agree very closely then the instrument is said to have a high degree of precision. A high degree of precision means an instrument has no drift i.e calibration of an instrument does not gradually shift over a period. Difference between repeatability and reproducibility is that repeatability does not include hysteresis.

Overpressure Limits: The maximum pressure that a transmitter can withstand without damages is called Overpressure. For Differential Pressure Transmitter, the overpressure is also called Static Pressure and is usually applied to both the sides of the transmitter (High and Low).

Upper Range Limit	Over Range Pressure
1 MPa (10 Bar) Upto 14 MPa (140 Bar)	Upper Range Limit +25% or 15kPa (1.5 Bar) whichever is greater
14 MPa to 70 MPa (140 Bar to 700 Bar)	Upper Range Limit +15%
Over 70 MPa (Over 700 Bar)	Upper Range Limit +10%

Table 10. Over-Range Pressure as per BS 6447

Response Time: Response time is the amount of time it takes for the output signal to reach 63.2% of the actual pressure change from the time the input change occurs. Response time consists of two important components Dead Time and Time Constant. (See. Fig.26)

Dead Time: Dead time is the amount of time the transmitter takes to initially respond to the change in pressure.

Time Constant: Time constant consist of the mechanical and electronic response time. Mechanical time is the time it takes for the process pressure change to be transmitted by the hydraulic force to the sensor. Electronic response is the time from the sensor detecting the pressure change to the transmitter electronics producing an output signal.

▪ Test Procedure

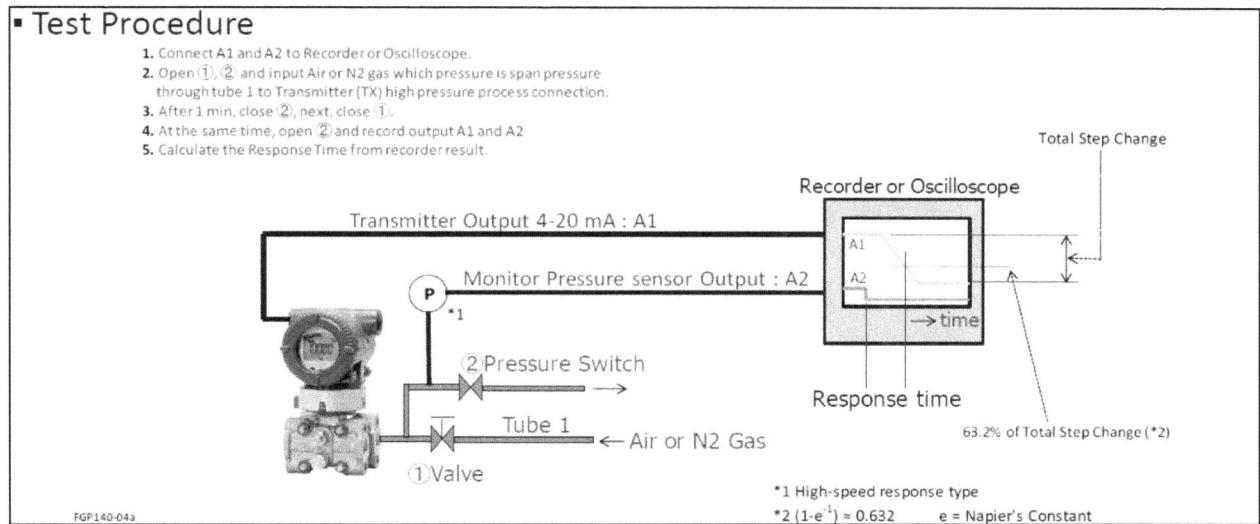

1. Connect A1 and A2 to Recorder or Oscilloscope.
2. Open ①, ② and input Air or N2 gas which pressure is span pressure through tube 1 to Transmitter (TX) high pressure process connection.
3. After 1 min, close ②, next, close ①.
4. At the same time, open ② and record output A1 and A2
5. Calculate the Response Time from recorder result.

*1 High-speed response type
*2 $(1-e^{-1}) \approx 0.632$ e = Napier's Constant

Fig.26 Response Time of Pressure Transmitter

Damping: Damping is the progressive reduction or suppression of oscillation in the output of the transmitter. Damping allows the user to filter out unwanted noise by increasing the damping.

Drift: An undesired change in output over a period of time, the change being unrelated to the input, environment or load.

Accuracy Class	Limits of Average Long Term Drift
0.2	% Output Span per 30 days +/- 0.075%
0.5, 1 & 2	+/- 0.15%
5	+/- 0.3%

Table 11. Limits of average long term drift as per BS 6447

Error: The algebraic difference between the indication and the ideal value of the measured signal.

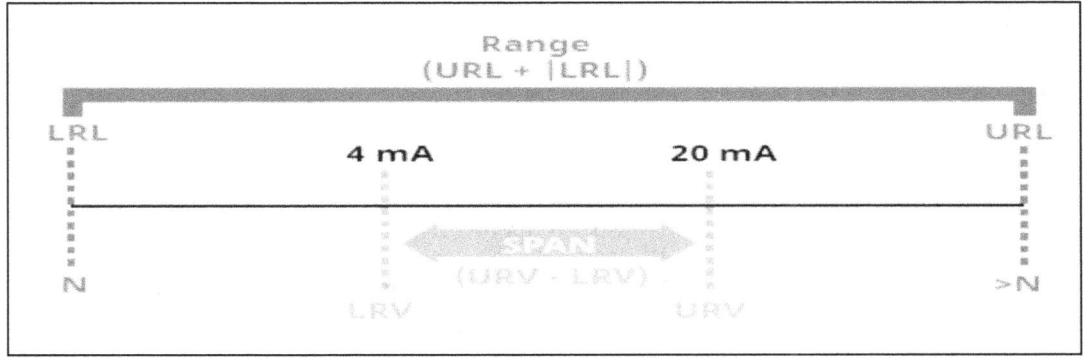

Fig.27 Pressure Transmitter Terms

1.31 Protocols Types

Analog / Digital

BRAIN® Protocol: A 4 to 20mA analog signal that corresponds to the primary variable. Analog signal has a digital signal superimposed using BRAIN® protocol. BRAIN® protocol stores the driver in the transmitter. This gives the device a plug-n-play type capability. BRAIN® is proprietary to Yokogawa.

HART® Protocol: A 4 to 20mA analog signal that corresponds to the primary variable. Analog signal has a digital signal superimposed (via frequency shift keying) using HART® protocol. HART® protocol requires that the proper driver be loaded into the controller / monitor / DCS / communicator to communicate properly with the transmitter. HART® is the most widely used protocol for pressure transmitters.

HART® Protocol: A 1 to 5 VDC analog signal that corresponds to the primary variable. Analog signal has a digital signal superimposed using HART® protocol. HART® protocol requires that the driver be loaded into the controller / monitor / DCS to work properly. This option has very low power consumption and is primarily with alternate power sources.

FOUNDATION™ Fieldbus: All-digital communication protocol that does not have an analog component. FOUNDATION™ Fieldbus greatly expands the capability of the transmitters but is a complex protocol that some users may find difficult to use. This protocol eliminates the D/A converter needed for the analog signal.

PROFIBUS® PA: All-digital communication protocol that does not have an analog component. PROFIBUS® PA has a very small install base in North America. This protocol eliminates the D/A converter needed for the analog signal.

1.32 Communication Conditions

Power Supply Voltage: 10.5 Vdc to 42 Vdc.

Load Resistance: Details mentioned in Loop Load Capacity

Communication Distance: 2 Kms when polyethylene insulated PVC sheathed control cables are used.

Spacing from power lines: 15 cm or more.

1.33 Standard Analog Output Signal vs NAUMER NE 43 Standard

Standard Analog Output transmitters sets to an Analog Output Lower Limit (AO-LL) and Analog Output Upper Limit (AO-UL) of 3.6 ma and 21.6 ma. During operation, if the AO-LL or AO-UL limits are reached, the analog signal locks to the respective limits. This locked value indicates an "out of range" event.

The AO-LL and AO-UL can be set to any value between 3.6 ma to 21.6 ma.

NAUMER NE 43 is a standard used to define the operating AO-LL and AO-UL values. NAUMER NE 43 complaint Transmitters have AO-LL and AO-UL of 3.8 ma and 20.5 ma respectively. These values are set and cannot be changed.

1.34 Loop Load Capacity

Two wire must have a certain minimum voltage at the terminal in order to function. Typically, this value is 12 Vdc. With a 24 Vdc power supply, and the transmitter requiring 12 Vdc at its terminals, it leaves only the difference or 12 Vdc for voltage drop around the loop.

By applying Ohms's law and the maximum analog signal is 20 mA, the maximum resistance around in the loop.

R= E/I

R= 12 Vdc/20 mA

R=600 Ohms

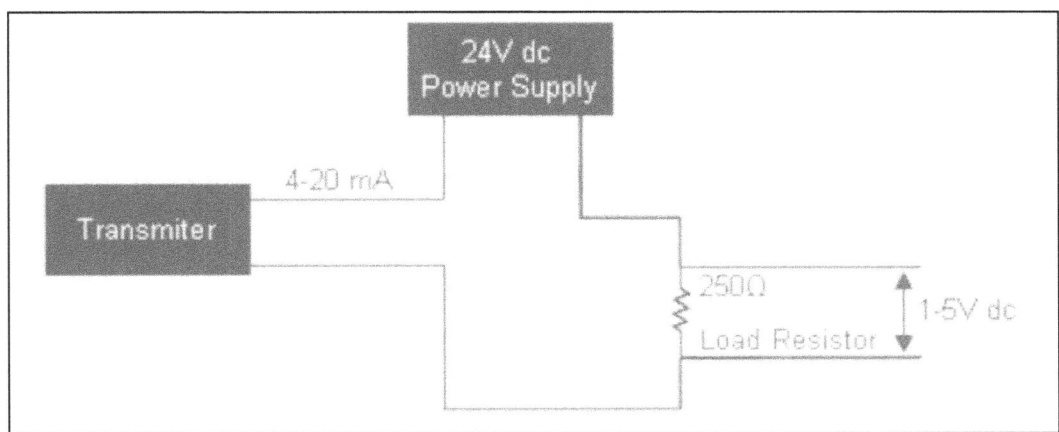

Fig.28 Load Loop Capacity

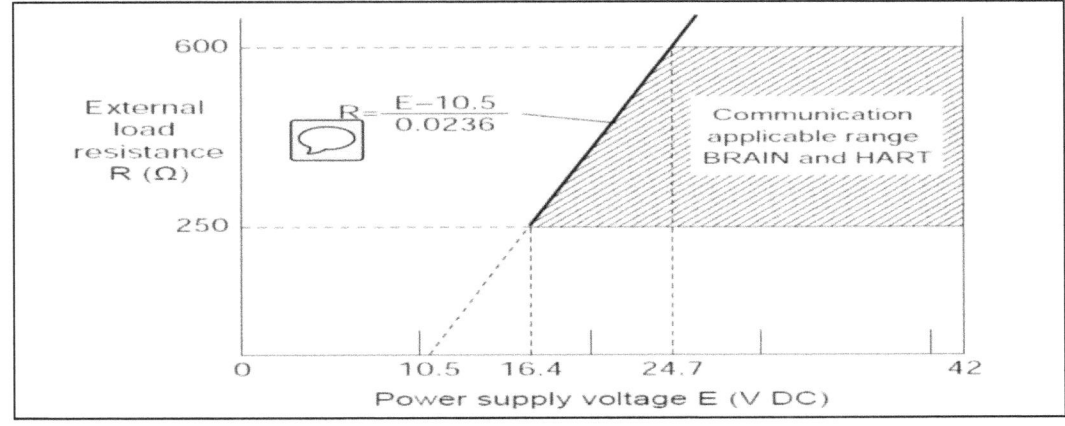

Fig.29 Relationship between power supply voltage and load resistance

1.35 Typical Applications of Pressure/ Differential Pressure Transmitters

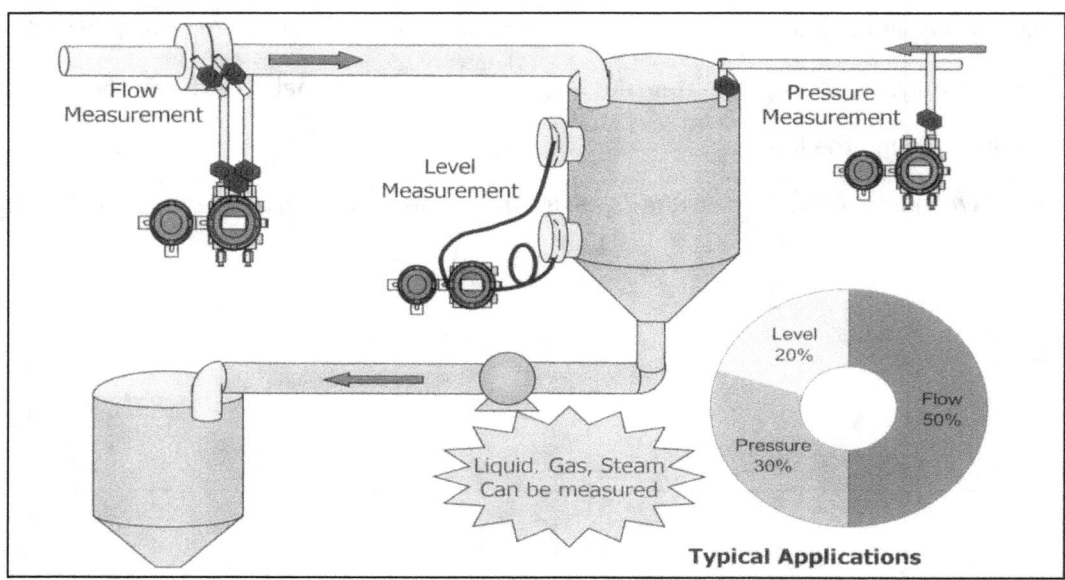

Fig.30 Applications where Pressure and Differential Pressure Transmitter Use

Flow Measurement through Differential Pressure Transmitter:

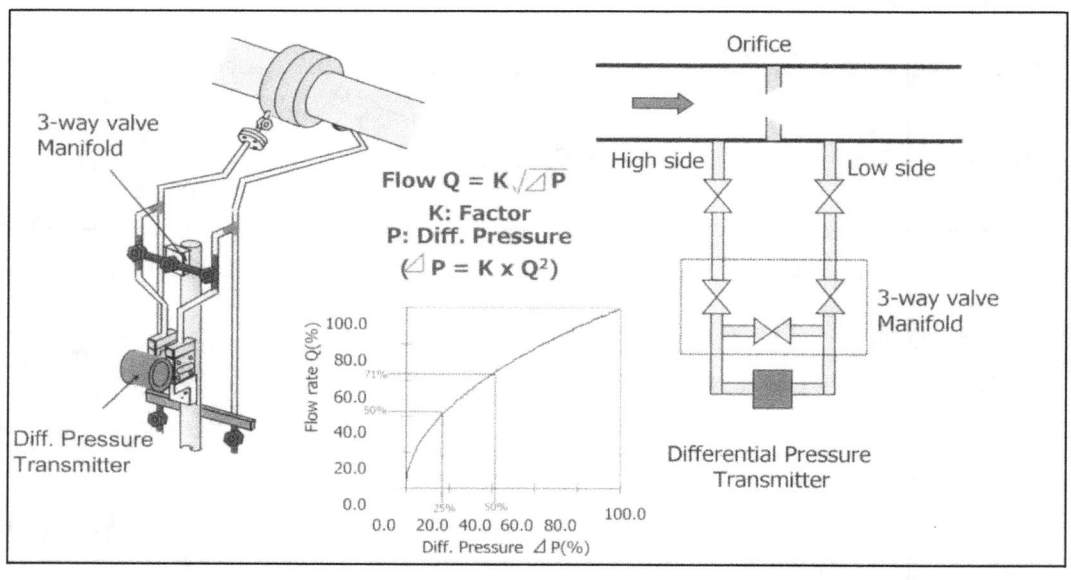

Fig.31 Flow Measurement

Pressure Measurement through Pressure Transmitter:

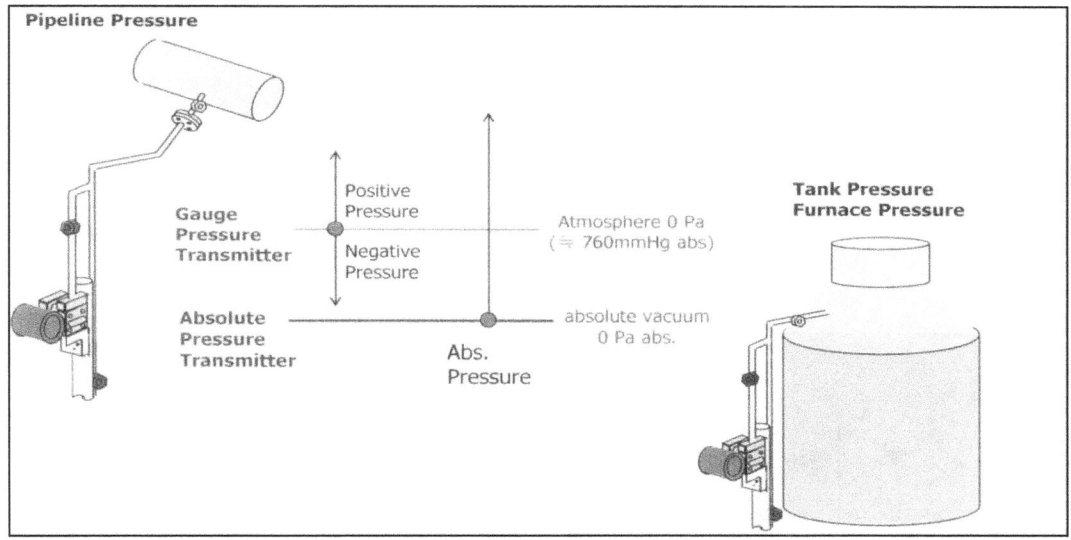

Fig.32 Pressure Measurement

Level Measurement through Pressure/ Differential Pressure Transmitter:

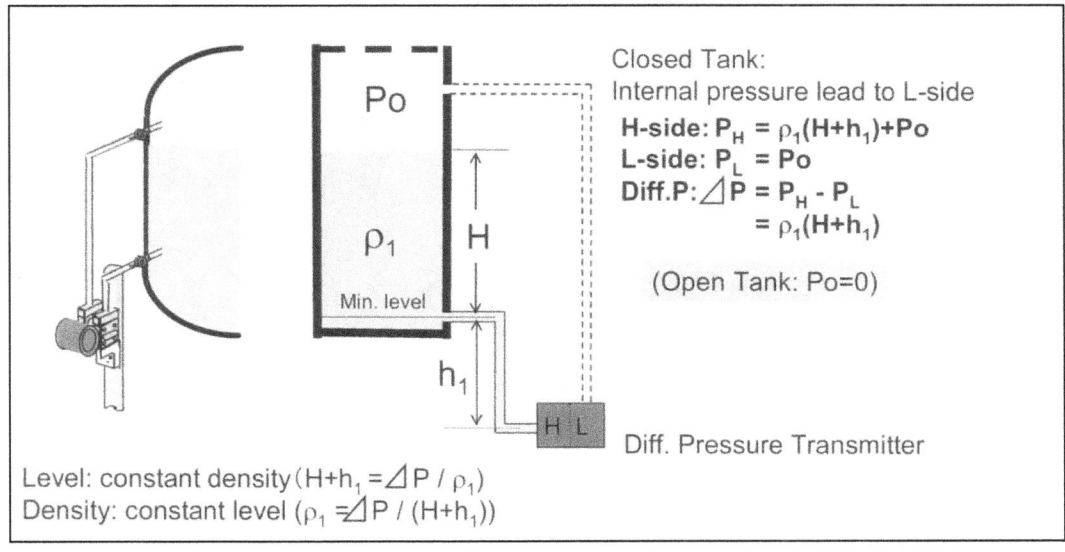

Fig.33 Level Measurement

1.36 Mounting Configuration for Pressure and Differential Pressure Transmitter

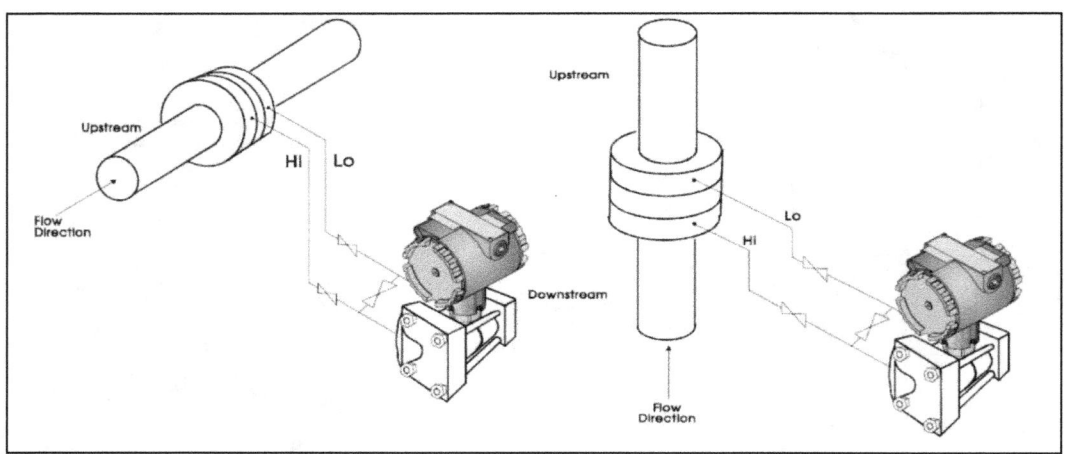

Fig.34 Differential Pressure Transmitter Installation Liquid Application

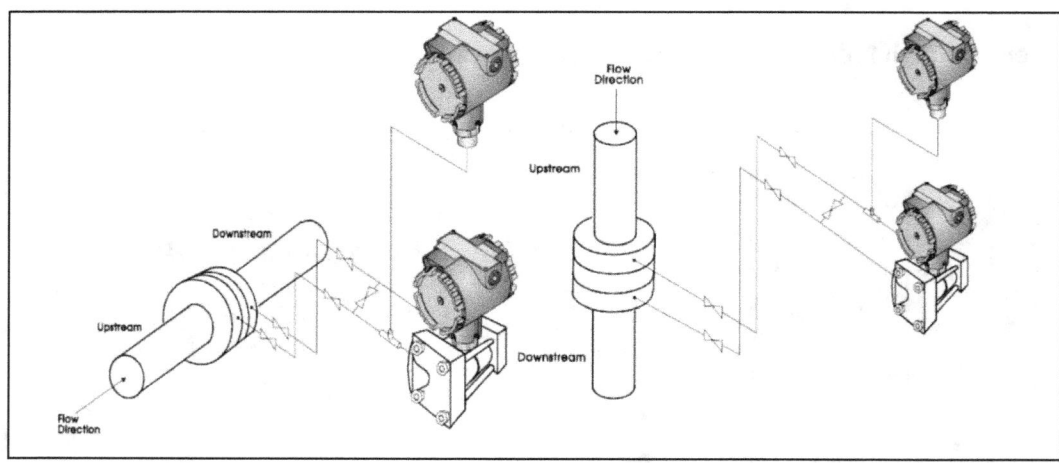

Fig.35 Differential Pressure Transmitter Installation Gas Application

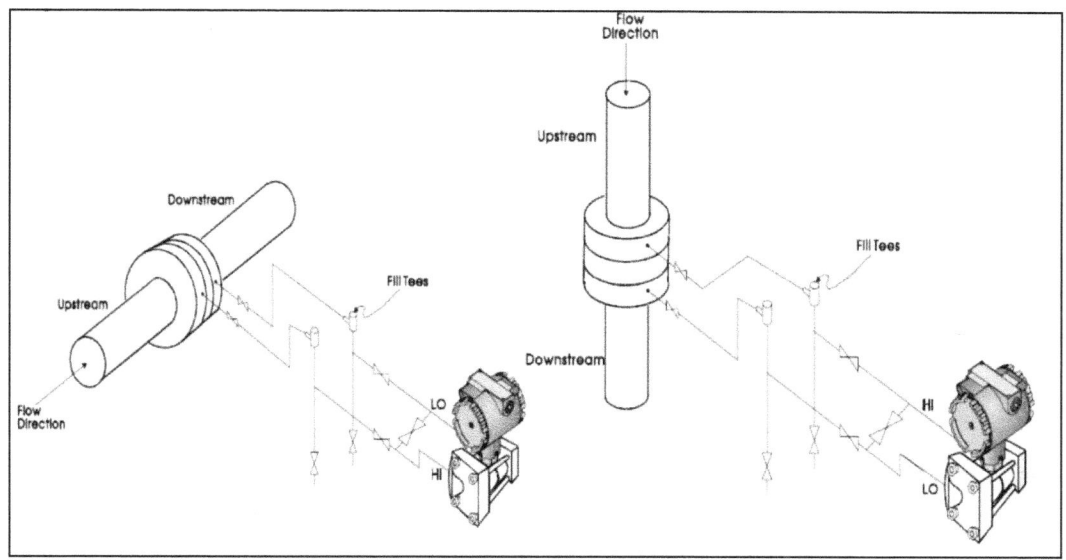

Fig.36 Differential Pressure Transmitter Installation Steam Application

1.37 Impulse Piping

Using impulse lines cools the process before it comes in contact with the pressure transmitter. Impulse piping may also allow the user to relocate the transmitter to a more convenient location for maintenance.

For higher ambient temperatures, longer impulse lines are required. For lower ambient temperatures, shorter impulse lines are required.

If the impulse lines are too long, other problems may present themselves:

• Damping of the pressure signal

• Blockage of the pressure signal

• Leakage at couplings

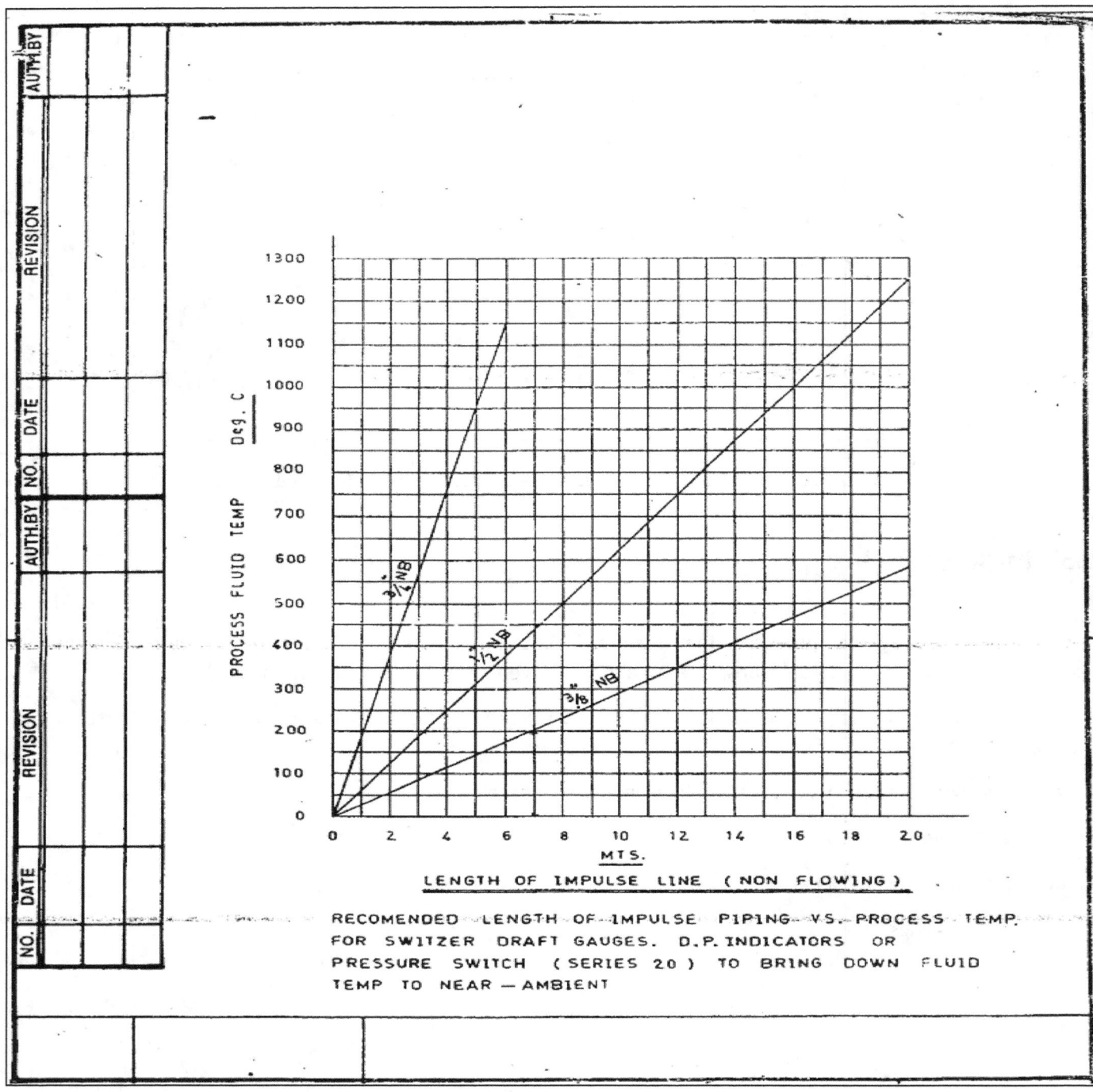

Fig.37 Piping Nomogram Curve

1.38 Remote Seal Pressure Transmitter

A remote seal consists of remote element made up of transmitter, a process connection and a capillary connecting both.

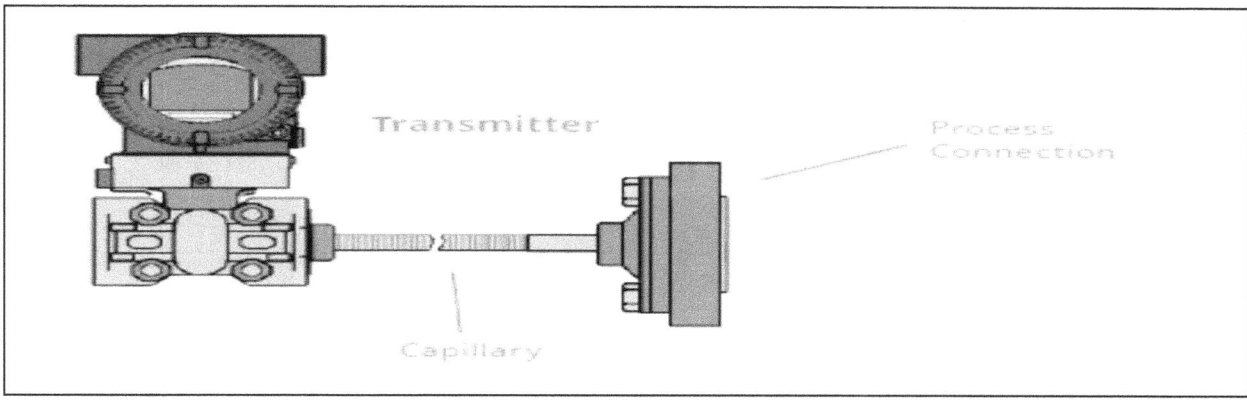

Fig.38 Remote Seal Pressure Transmitter

Transmitter: The part of the system that measures and converts the measured variable to a signal to be transmitted.

Process Connection: The part of the system that include the diaphragm seal, the process connection is what is in contact with the process. Table 12 is a guideline of recommended sizes for different medium span.

Transmitter Span	Suggested Diaphragm Size
10 to 100 in H_2O	≥ 3 inch
100 to 1,000 in H_2O	≥ 2.4 inch
30 to 200 psi	≥ 2 inch
200 to 1,000 psi	≥ 1 inch

Table 12. Diaphragm Size for Different Transmitter Span

How Remote Seal Transmitter works?

The remote seal can be integral with the transmitter or remote with a capillary length up to some meters. The capillary is protected by some armours and may or may not be coated in PVC. Once connected the individual components, the system is evacuated from the air and filled with incompressible fluid.

When process pressure is applied to the sealed diaphragm, this one deflects and exerts force against the fill fluid. Since the liquid is incompressible, this force is transmitted hydraulically to the sensing diaphragm in the transmitter body, causing in it turn to deflect. The deflection of the sensing diaphragm is the basis for the pressure measurement.

Diaphragm Seal systems shall be used for:

• For process fluid that would clog the pressure elements.

• For process fluids that are toxic, corrosive, slurry and viscous.

• For process fluids that could freeze or solidify.

- For process temperatures outside the normal operating range and cannot be brought to those limits by impulse piping.

- For process that needs frequent cleaning.

- For processes that need replacement of wet legs, to reduce maintenance.

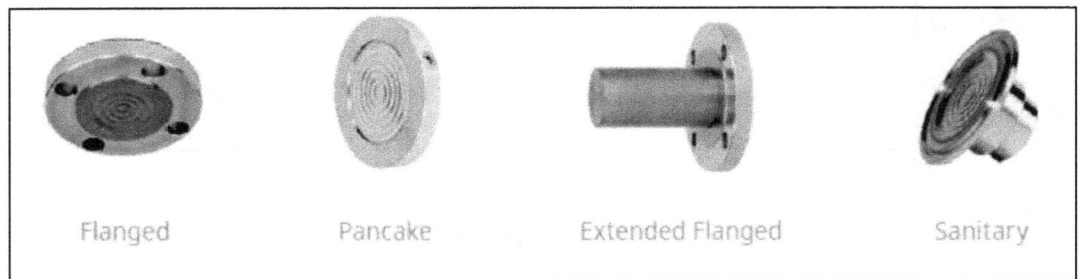

Fig.39 Different Types of Remote Seals

1.39 Selection Guidelines for Remote Seal Pressure Transmitter

Following Parameters kept in mind while selecting Remote Seal Type Pressure Transmitter.

a) **Size**
b) **Filling Fluid**
c) **Capillary**
d) **Response Time**
e) **Accuracy**

Size: A seal membrane must be capable of transferring the required volume of liquid to actuate the measuring element through its full range. The seal membrane must be able to accommodate changes in volume that results from the thermal expansion and contraction of the filling liquid.

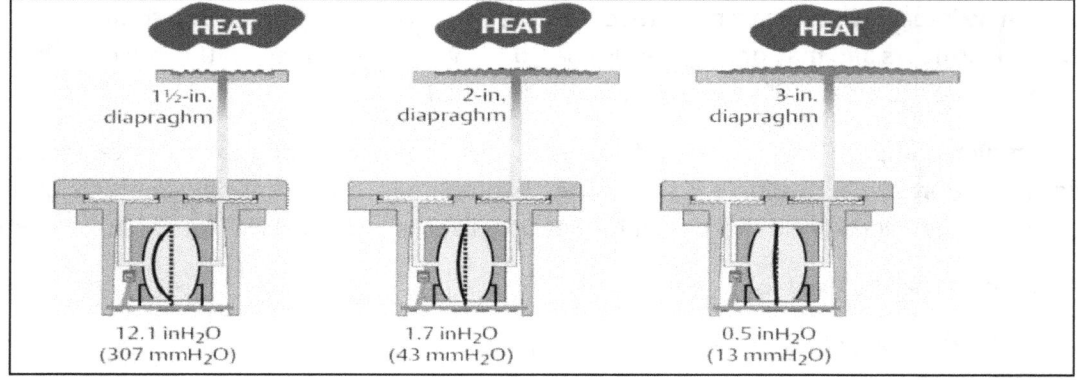

Fig.40 Back Pressure on Diaphragm Causing Error

Fig. 40 shows how diaphragm size can affect the measurement reading at the transmitter. For smaller seal sizes, such as the 11/2-in. size, the amount of back pressure on the transmitter causes an additional 12.1 inH2O error. Moving to the 2-in. size gives 1.7 inH2O and the largest 3-in. size shown only has 0.5 inH2O error. Using a larger diaphragm can drastically improve performance and provides a more stable reading.

Filling Fluid: Selection of Filling liquid should take into account its coefficient of thermal expansion, compressibility, viscosity, its freezing and boiling point, tendencies to decompose at the maximum operating temperature, compatibility with the material of the measuring element and seal system.

Capillary: Capillary length is often dictated by the application, but when there is a choice, length should be kept to minimum. With Length established, volume transfer requirement, viscosity of filling liquid and capillary bore size determines the response time.

Response Time: The response time of a measurement system having a remote seal with capillary and a transmitter is defined as the time the transmitter pressure takes to read 63% of the pressure variation value applied on a 10% to 90% range of the measured pressure. Fig. 41 shows the response curve.

The response time is the result of the resistance of the oil displacement in the capillary, so that, the bigger the capillary and the oil viscosity are, the longer will be the response time. The transmitter range influences the response time due to the rigidity of the sensor diaphragm, so that the wider the range, the quickest the response time.

The response time is also influenced by the viscosity of the filling fluid, which varies with the temperature. The higher the temperature, the lesser the viscosity of the filling fluid, which, consequently, reduces the response time.

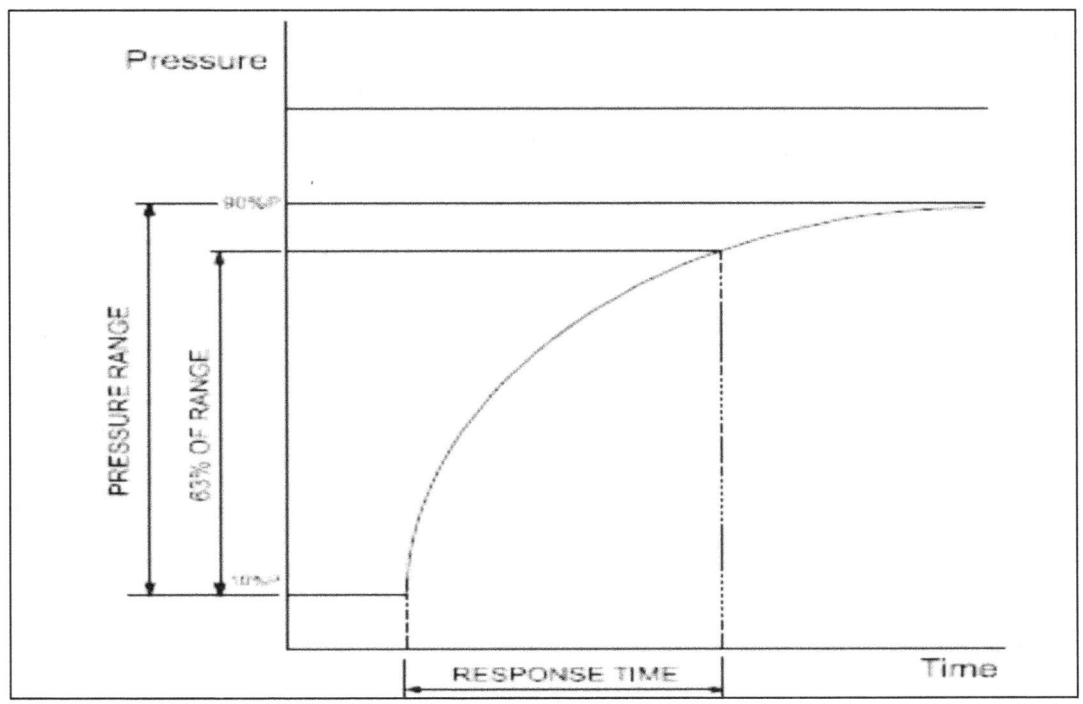

Fig.41 Response Time Curve with Pressure

Accuracy: the accuracy and response time for the entire system are the same as the transmitter used in the system. The accuracy of the entire system is equal to the transmitter accuracy plus the accuracy of the capillary and diaphragm seal. Therefore, the accuracy of the entire system is less than the individual transmitter. The same is for the response time. The system response time is equal to the transmitter response time plus the response time of the capillary/diaphragm seal.

System Accuracy = Transmitter Accuracy + Capillary/Diaphragm Seal Accuracy
System Response Time = Transmitter Response + Capillary/Diaphragm Response

TRANSMITTER RANGE	CAPILLARY TEMPERATURE E(°C)	Response Time in seconds/meter of capillary (s/m)[6]								
		DC 200	DC 704	FLUOROLUBE	SYLTHERM 800	NEOBEE M20	Glycerin 50% + Water 50%	FOMBLIM	KYTROX	HALOCARBOM
2	100[4]	2.69E-01	3.99E-01	1.72E-01	1.09E-01	5.59E-02	2.16E-02	2.66E-01	3.08E-01	8.37E-02
	75	3.38E-01	5.75E-01	2.84E-01	1.52E-01	8.89E-02	4.32E-02	4.46E-01	4.83E-01	1.19E-01
	50	4.55E-01	9.29E-01	7.86E-01	2.23E-01	1.57E-01	1.19E-01	1.00E+00	9.85E-01	1.88E-01
	25	6.98E-01	1.72E+00	3.78E+00	3.47E-01	3.15E-01	5.32E-01	3.41E+00	2.90E+00	3.80E-01
	10	9.87E-01	2.69E+00	1.28E+01	4.67E-01	5.09E-01	2.01E+00	9.03E+00	6.92E+00	6.90E-01
	0	1.30E+00	3.74E+00	3.26E+01	5.77E-01	7.21E-01	6.20E+00	1.94E+01	1.37E+01	1.13E+00
	-10	1.78E+00	N.A.	9.18E+01	7.21E-01	1.04E+00	N.A.	4.62E+01	2.99E+01	1.99E+00
	-20	2.56E+00	N.A.	2.86E+02	9.12E-01	N.A.	N.A.	1.22E+02	7.19E+01	3.86E+00
	-40	6.10E+00	N.A.	N.A.	1.51E+00	N.A.	N.A.	N.A.	5.76E+02	1.98E+01
3	100[4]	5.39E-02	7.97E-02	3.45E-02	2.18E-02	1.12E-02	4.32E-03	5.32E-02	6.17E-02	1.67E-02
	75	6.75E-02	1.15E-01	5.68E-02	3.03E-02	1.78E-02	8.65E-03	8.91E-02	9.67E-02	2.38E-02
	50	9.09E-02	1.86E-01	1.57E-01	4.45E-02	3.15E-02	2.38E-02	2.01E-01	1.97E-01	3.76E-02
	25	1.40E-01	3.45E-01	7.56E-01	6.94E-02	6.30E-02	1.06E-01	6.81E-01	5.80E-01	7.60E-02
	10	1.97E-01	5.38E-01	2.56E+00	9.34E-02	1.02E-01	4.02E-01	1.81E+00	1.38E+00	1.38E-01
	0	2.60E-01	7.48E-01	6.53E+00	1.15E-01	1.44E-01	1.24E+00	3.88E+00	2.75E+00	2.25E-01
	-10	3.57E-01	N.A.	1.84E+01	1.44E-01	2.09E-01	N.A.	9.24E+00	5.98E+00	3.98E-01
	-20	5.12E-01	N.A.	5.72E+01	1.82E-01	N.A.	N.A.	2.45E+01	1.44E+01	7.73E-01
	-40	1.22E+00	N.A.	N.A.	3.03E-01	N.A.	N.A.	N.A.	1.15E+02	3.96E+00
4	100[4]	4.86E-03	7.19E-03	3.11E-03	1.97E-03	1.01E-03	3.90E-04	4.80E-03	5.56E-03	1.51E-03
	75	6.09E-03	1.04E-02	5.13E-03	2.74E-03	1.60E-03	7.80E-04	8.04E-03	8.72E-03	2.14E-03
	50	8.20E-03	1.68E-02	1.42E-02	4.02E-03	2.84E-03	2.15E-03	1.81E-02	1.78E-02	3.39E-03
	25	1.26E-02	3.11E-02	6.82E-02	6.26E-03	5.68E-03	9.60E-03	6.15E-02	5.24E-02	6.86E-03
	10	1.78E-02	4.85E-02	2.31E-01	8.42E-03	9.18E-03	3.62E-02	1.63E-01	1.25E-01	1.25E-02
	0	2.35E-02	6.75E-02	5.89E-01	1.04E-02	1.30E-02	1.12E-01	3.50E-01	2.48E-01	2.03E-02
	-10	3.22E-02	N.A.	1.66E+00	1.30E-02	1.89E-02	N.A.	8.33E-01	5.39E-01	3.59E-02
	-20	4.61E-02	N.A.	5.16E+00	1.65E-02	N.A.	N.A.	2.21E+00	1.30E+00	6.97E-02
	-40	1.10E-01	N.A.	N.A.	2.73E-02	N.A.	N.A.	N.A.	1.04E+01	3.57E-01
5	100[4]	2.11E-04	3.13E-04	1.35E-04	8.54E-05	4.38E-05	1.69E-05	2.09E-04	2.42E-04	6.56E-05
	75	2.65E-04	4.50E-04	2.23E-04	1.19E-04	6.96E-05	3.39E-05	3.49E-04	3.79E-04	9.31E-05
	50	3.56E-04	7.28E-04	6.16E-04	1.75E-04	1.23E-04	9.32E-05	7.87E-04	7.72E-04	1.47E-04
	25	5.47E-04	1.35E-03	2.96E-03	2.72E-04	2.47E-04	4.17E-04	2.67E-03	2.27E-03	2.98E-04
	10	7.74E-04	2.11E-03	1.00E-02	3.66E-04	3.99E-04	1.57E-03	7.08E-03	5.42E-03	5.41E-04
	0	1.02E-03	2.93E-03	2.56E-02	4.52E-04	5.65E-04	4.86E-03	1.52E-02	1.08E-02	8.82E-04
	-10	1.40E-03	N.A.	7.20E-02	5.65E-04	8.19E-04	N.A.	3.62E-02	2.34E-02	1.56E-03
	-20	2.00E-03	N.A.	2.24E-01	7.15E-04	N.A.	N.A.	9.59E-02	5.64E-02	3.03E-03
	-40	4.78E-03	N.A.	N.A.	1.19E-03	N.A.	N.A.	N.A.	4.52E-01	1.55E-02
6	100[4]	1.66E-04	2.46E-04	1.06E-04	6.71E-05	3.44E-05	1.33E-05	1.64E-04	1.90E-04	5.16E-05
	75	2.08E-04	3.54E-04	1.75E-04	9.34E-05	5.48E-05	2.66E-05	2.75E-04	2.98E-04	7.32E-05
	50	2.80E-04	5.73E-04	4.84E-04	1.37E-04	9.70E-05	7.33E-05	6.19E-04	6.07E-04	1.16E-04
	25	4.30E-04	1.06E-03	2.33E-03	2.14E-04	1.94E-04	3.28E-04	2.10E-03	1.79E-03	2.34E-04
	10	6.08E-04	1.66E-03	7.89E-03	2.88E-04	3.13E-04	1.24E-03	5.56E-03	4.26E-03	4.25E-04
	0	8.01E-04	2.31E-03	2.01E-02	3.56E-04	4.44E-04	3.82E-03	1.20E-02	8.46E-03	6.93E-04
	-10	1.10E-03	N.A.	5.66E-02	4.44E-04	6.44E-04	N.A.	2.85E-02	1.84E-02	1.23E-03
	-20	1.58E-03	N.A.	1.76E-01	5.62E-04	N.A.	N.A.	7.54E-02	4.43E-02	2.38E-03
	-40	3.76E-03	N.A.	N.A.	9.33E-04	N.A.	N.A.	N.A.	3.55E-01	1.22E-02

Fig.42 Remote Seal Response Time

1.40 Selection Criteria of Fluid for Remote Seal Pressure Transmitter

Temperature – the operating temperature ranges for a fill fluid depend on the measured pressure. A given grade of silicone oil, for example, can operate at –90 to 80 °C (–130 to +176 °F) under vacuum conditions, but instead at –90 to 180 °C (–130 to +356 °F) when the pressure is above 1.03 bar (15 psia).

Expansion coefficient – the various fill fluids in use have different coefficients of thermal expansion. Wide fluctuations in ambient temperature can cause the fill fluid to expand or contract significantly. This effect becomes particularly important in unbalanced systems where capillaries of the High and Low connections have lengths that differ or they are exposed to different ambient temperatures. The smaller the expansion coefficient, the less the measurement will be affected.

Compatibility with measured fluid – because the seal diaphragm is usually thin, there is always the chance that it could be damaged and the fill fluid could leak into the measured medium. This possibility should particularly be considered if contamination of the measured fluid is covered by government regulations (as with foods processing) or where inert fluid is required in a combustive process like wet chlorine or oxygen service.

Viscosity – the fill fluid's viscosity can significantly influence the response time of the measurement. The less viscous the fluid, the faster the response. Remember also that if the temperature around the capillary decreases, the viscosity increases.

1.41 Level Measurement through Remote Seal Pressure Transmitter

The basic measurement of level follows the same principle as pressure transmitters without seals: pressure is proportional to level. The head pressure of the liquid corresponds to its height multiplied by the specific gravity.

Open Tank – Single Seal System with Transmitter below tap

An open tank single seal system with transmitter below the tap is very similar to a transmitter system that uses impulse piping going to the transmitter filled with process fluid.

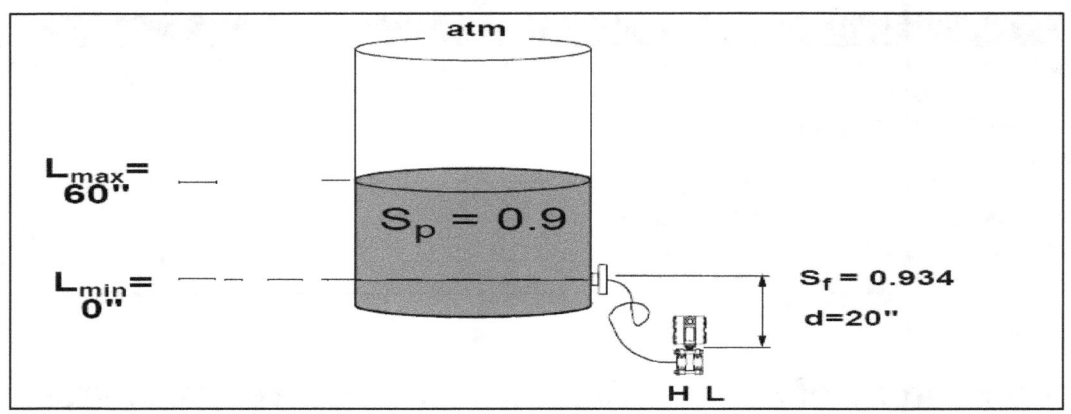

Fig.43 Open Tank with Single Seal System, transmitter below the tap

The Calibrated Span set points are:

$4 \text{ mA} = L_{min} S + d S_f$

$4 \text{ mA} = (0*0.9) + (20*0.934)$

$4 \text{ mA} = 18.7 \text{ inH}_2\text{O}$

$20 \text{ mA} = L_{max} S + d S_f$

$20 \text{ mA} = (60*0.9) + (20*0.934)$

$20 \text{ mA} = 72.7 \text{ in H}_2\text{O}$

Calibration Range for Transmitter = 18.7 in H_2O to 72.7 in H_2O

❖   Silicone 200 has a specific gravity of 0.934

Open Tank – Single Seal System with Transmitter above tap:

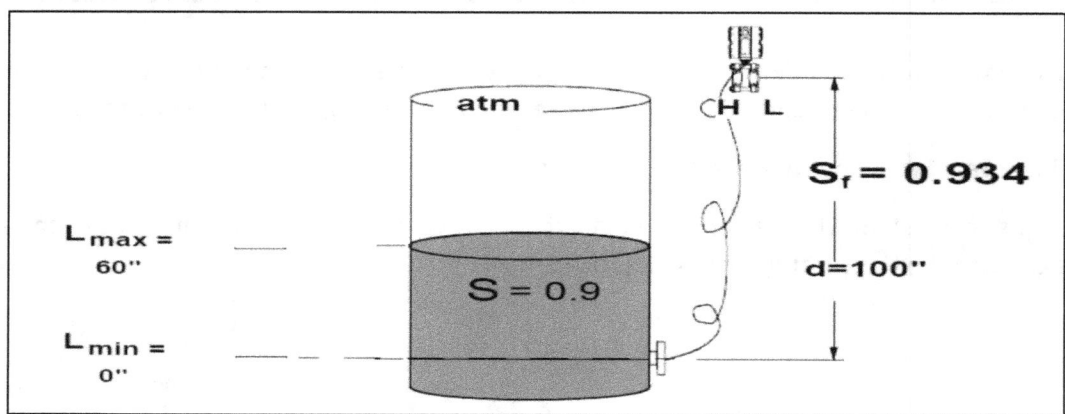

Fig.44 Open Tank with Single Seal System, transmitter above the tap

The Calibrated Span set points are:

4 mA= L_{min} S - d S_f

4 mA= (0*0.9) - (100*0.934)

4 mA= - 93.4 inH_2O

20 mA= L_{max} S - d S_f

20 mA= (60*0.9) - (100*0.934)

20 mA= - 39.4 in H_2O

Calibration Range for Transmitter = - 93.4 in H_2O to - 39.4 in H_2O

Closed Tank – Two Seal System:

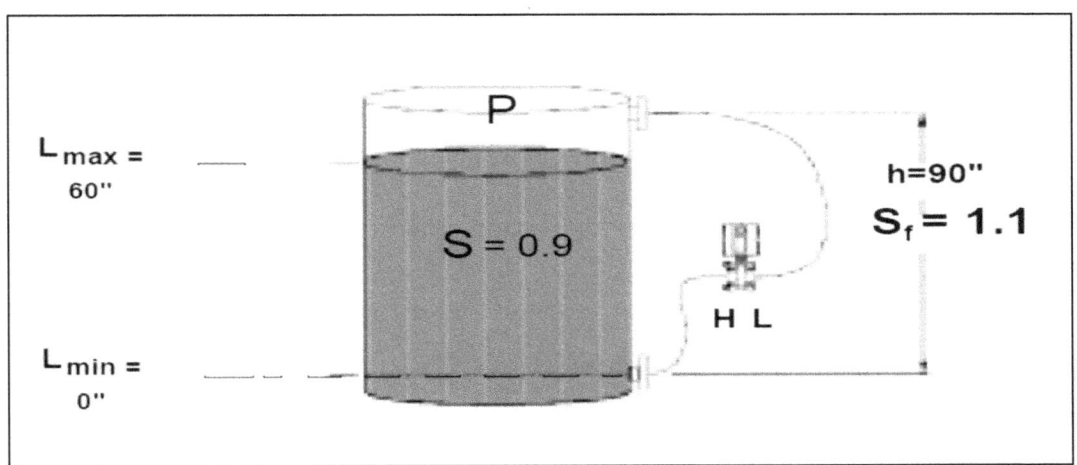

Fig.45 Closed Tank with Two Seal System

The Calibrated Span set points are:

4 mA= L_{min} S - h S_f

4 mA= (0*0.9) - (90*1.1)

4 mA= - 99 in H_2O

20 mA= L_{max} S - h S_f

20 mA= (60*0.9) - (90*1.1)

20 mA= - 45 in H_2O

Calibration Range for Transmitter = - 99 in H_2O to - 45 in H_2O

1.42 Determining Calibration Interval for Pressure Transmitter

Real World Performance vs Laboratory Performance:

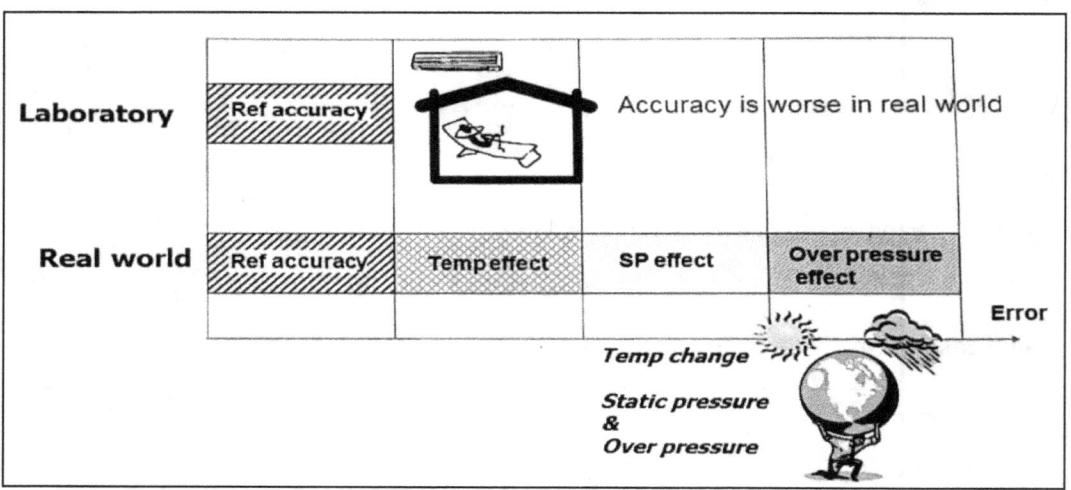

Fig.46 Performance Characteristics

> ➤ Traditional Definition:
> $$TPE = Sq.rt\ (\ E_1{}^2 + E_2{}^2 + E_3{}^2)$$
>
> Where TPE is **Total Probable Error**
> E_1 is Reference Accuracy of Calibrated Span
> E_2 is Ambient Temperature Effects per 50 degree Fahrenheit change (Zero and Span)
> E_3 is Static Span Effects per 1000 psi change

> ➤ Better Definition:
> $$TA = Sq.rt\ \{\ E_1{}^2 + E_2{}^2 + (E_3{}^2 + E_4{}^2) + E_5{}^2\}$$
>
> Where E_1 is Reference Accuracy of Calibrated Span
> E_2 is Ambient Temperature Effects per 50 degree Fahrenheit change (Zero and Span)
> E_3 is Static Span Effects per 1000 psi change
> E_4 is Static Pressure Zero effects per 100 psi change
> E_5 is Overpressure Effects up to MWP (maximum working pressure)
>
> This is the **complete Total Accuracy Definition**.

Consider an Example of X Manufacture:

> E_1: Reference Accuracy of Calibrated Span
> - X "Medium (M)" Capsule Span 0-100 in H_2O

E_1 is +/- 0.04 % of Span
Span = 100 inH_2O
E_1 = +/- 0.04% * 100 inH_2O
E_1 = +/- (0.004) * 100 inH_2O

E_1 = +/- 0.04 inH_2O

> E_2: Ambient Temperature Effects per 50 degree Fahrenheit Change (Zero and Span)
> - Ambient Temperature 40-90 Degree Fahrenheit (50 Degree Fahrenheit Change)

E_2 is +/- (0.04 % of Span + 0.009% of URL)
Span = 100 in H_2O
URL = 400 inH_2O
E_2 = +/- {(0.04% * 100 inH_2O) + (0.009% * 400 inH_2O)}
E_2 = +/- {(0.004) * 100 inH_2O) + (0.00009 * 400 inH_2O)}

E_2 = +/- (0.04 inH_2O + 0.036 inH_2O)
E_2 = +/- 0.076 inH_2O

> E_3: Static Pressure Span Effects per 1000 psi change
> - Line pressure 0-1000 psig

E_3 is +/- 0.075% of span
Span = 100 in H_2O
E_3 = +/- 0.075% * 100 in H_2O
E_3 = +/- (0.00075*100 inH_2O)
E_3 = +/- 0.075 inH_2O

> E_4: Static Pressure Zero Effects per 1000 psi change
> - Line pressure 0-1000 psig

E_4 is +/- 0.02% of URL
URL = 400 in H_2O
E_4 = +/- 0.02% * 400 in H_2O
E_4 = +/- (0.0002*400 inH_2O)
E_4 = +/- 0.08 inH_2O

> E₅: Overpressure Effects up to MWP

E_5 is +/- 0.03% of URL
URL = 400 in H_2O
E_5 = +/- 0.03% * 400 in H_2O
E_5 = +/- (0.0003*400 inH_2O)
E_5 = +/- 0.12 inH_2O

TA = Sq.rt { $E_1^2 + E_2^2 + (E_3^2 + E_4^2) + E_5^2$ }

By putting the results in the equation:
TA = +/- 0.21 in H_2O

Reflect in % of Span

TA = +/- 0.21 in H_2O/100 in H_2O = +/- 0.21% of Span

Acceptable Performance = +/- 0.5 % of Span

Total Accuracy +/- 0.21 % of Span

Stability of Transmitter per Month as per X = 0.1 % of URL per 120 Months (10 years)/120

URL is 400 inH_2O

Stability of Transmitter per Month as per X = (0.001 *400 in H_2O)/120 = +/- 0.003 inH_2O

Reflect in % of Span, Span is 100 in H_2O

Stability per Month = +/- (0.003 in H_2O/100 in H_2O) = +/- 0.003% of Span

---

Calibration Intervals, Months = (Acceptable Performance – Total Accuracy)/ Stability per Month

Acceptable Performance is +/- 0.5 % of Span

Total Accuracy is +/- 0.21 % of Span

Stability per Month is +/- 0.003% of Span

Calibration Intervals, Months = 96.7 Months = 8 years

1.43 Calibration Management System

Calibration refers to the process of checking and accurate adjustment of an instrument response with respect to a device with an established accuracy standard. So, the output of an instrument accurately corresponds to its input throughout a specified range. It is important to calibrate an instrument for the following reason:

a. Even the best instruments drift and lose their ability to give accurate measurements. The drift makes calibration necessary.

b. Environment conditions, elapsed time, and type of application can all affect the stability of an instrument.

c. Even instruments of the same manufacturer, type and range can show varying performance from one unit to another.

d. To maintain the credibility of measurements.

e. To maintain the quality of process instruments at a good as new level.

f. Safety and environmental regulations.

Calibration is performed by comparing or applying a known signal to the instrument under test and errors are detected by performing a calibration. An Error is the algebraic difference between the indication and the actual value of the measured variable. Types of Error; Zero Error, Span Error, Linearization error and Combined Zero and Span error.

Traceability: Traceability is defined by ANSI/NCSI Z540-1-1994 as "the property of a result of a measurement whereby it can be related to appropriate standards, generally national or international standards, through an unbroke chain of comparisons." Traceability is accomplished by ensuring the test standards we use are routinely calibrated by "higher level" reference standards.

Calibration Standards: A calibration standard refers to a substance or device used as a reference to compare against an instrument's response.

Types of calibration standard:

a) Primary Standard

b) Secondary Standard

Primary Calibration Standards: Primary calibration standards refers to the calibration of instruments against primary standards. The calibrating process is difficult and proper care should be taken. Calibrating a secondary device is time consuming but accuracy is very good. Different type of standards with different types of accuracies and tolerances are used for different purpose. The standards maintained by NIST (USA) are used to calibrate and provide authority to other facilities in calibration certification. These facilities use their standards in turn to verify the standards of industrial plants.

Secondary Calibration Standards: The devices routinely used for instrument calibration are called secondary instruments. Secondary standards are extremely precise for calibration measurement and controlled instrumentation. For eg. In manometers, the physical relationship between hydrostatic head and pressure provides the values for testing and adjusting secondary pressure standards.

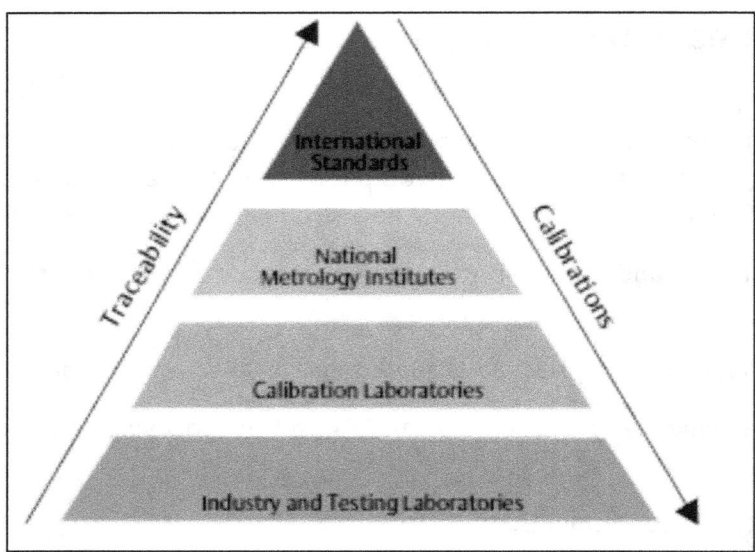

Fig.47 Traceability Flow Diagram

Calibration Concepts: The Instrument Calibration Series defines calibration as "determination of the experimental relationship between the quantity being measured and the output of the device which measures it; where the quantity measured is obtained through a recognised standard of measurement." There are two fundamental operations involved in calibrating an instrument.

a) testing the instrument to determine its performance.

b) adjusting the instrument to perform within specification

1.44 Calibration of Transmitter

Transmitter are of two types:

a) Analog Transmitters

b) Digital or SMART Transmitters

In analog transmitter, the term calibration is nothing but adjusting the zero and span settings. Thus, calibration of analog transmitter includes reranging the instrument. Fig.51 illustrates the analog pressure transmitter, which consists of three blocks. Low pass filter is used to reduce the noise obtained by the sensor. Amplifier is used to amplify the original signal that comes from the low pass filter within the specified range. This is done by adjusting zero and span of the amplifier. The output section takes the signal from the amplifier and gives the output in the form of 4-20ma.

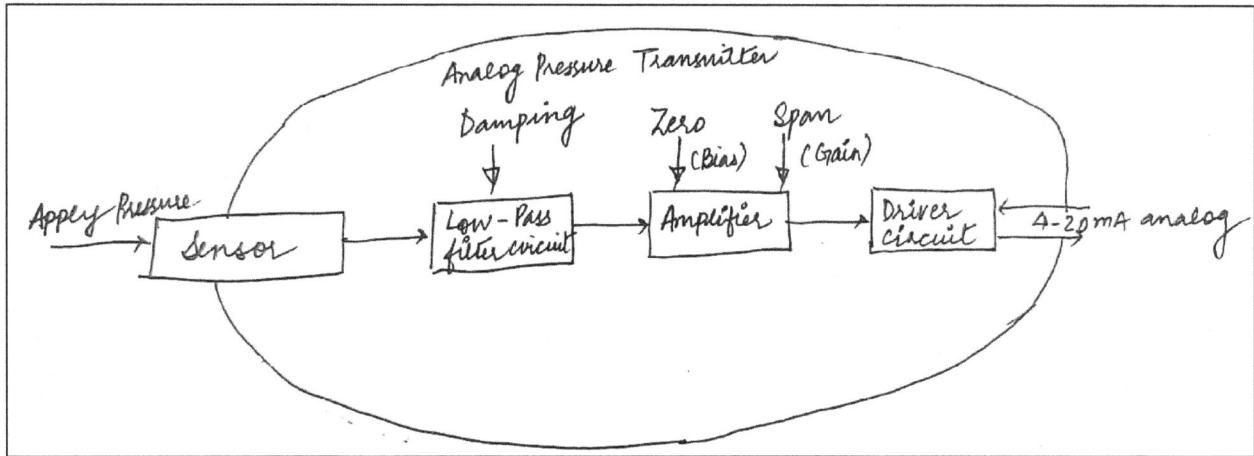

Fig.48 Block Diagram of Analog Pressure Transmitter

The advantage of "SMART" field instruments containing microprocessors. These devices have built in diagnosis ability, greater accuracy and ability to connect digitally. In Smart transmitters, the calibration process is divided into three parts:

a) Sensor Trim

b) Range Setting

c) Current Trim

The reason for separating these functions is that the range can be changed without applying a physical input.

Sensor Trim: Over a lone span of time, all sensors show some drift depending on the type of sensor. Sensor reading may also be offset due to mounting position. Sensor trim is used to correct the digital reading as seen in the device local indicator and received over the digital communication. For eg. If a pressure is 0 bar but transmitter reading shows 0.03 bar, then sensor trim is used to adjust it back to 0 bar. Sensor trims requires a physical input to the transmitter. Typically, three forms of sensor trim are available in transmitter:

a) Zero Sensor Trim

b) Lower Sensor Trim

c) Upper Sensor Trim

Zero trim requires the physical input applied to be zero, most often used with pressure transmitter. For best accuracy achieved, sensor trim is performed at two points, close to the lower range and upper range. A known physical input is applied to the transmitter to perform the sensor trim with the help of HART, this would be allowing the transmitter to correct itself.

Range Setting: It is defined as setting of scale for 4ma to 20ma.

Output Current Trim: It is used when any transmitter shows wrong current reading compared to input and output values using standard instruments. For eg., if the measured current output is 4.13ma as compared to actual 4ma, then current trim is used to adjust it to 4ma if range and applied input all ok.

1.45 Data Sheet of Differential Pressure Transmitter

	RESPONSIBLE ORGANIZATION	DIFFERENTIAL PRESSURE TRANSMITTER Device Specification		SPECIFICATION IDENTIFICATIONS
1	(ISA)		6	Document no
2			7	Latest revision
3			8	Date
4			9	Issue status
5			10	

	TRANSMITTER BODY			PERFORMANCE CHARACTERISTICS
11			60	
12	Body/Flange type		61	Max press at design temp — At
13	Process conn nominal size — Rating		62	Min working temperature — Max
14	Process conn termn type — Style		63	Accuracy rating
15	Vent/Drain location		64	Diff pressure LRL — URL
16	Mounting type		65	Min ambient working temp — Max
17	Body/Flange material		66	
18	Vent/Drain material		67	
19	Bolting material		68	
20	Flange adapter material		69	
21	Gasket/O ring material		70	
22	Mounting kit material		71	
23			72	
24			73	
25			74	
26	SENSING ELEMENT		75	
27	Detector type		76	
28	Min diff press span — Max		77	
29	Diaphragm/Wetted material		78	
30	Fill fluid material		79	
31			80	
32			81	
33			82	
34	TRANSMITTER		83	
35	Output signal type		84	ACCESSORIES
36	Enclosure type no/class		85	Air set filter style
37	Characteristic curve		86	Air set gauges
38	Digital communication std		87	Heating kit style
39	Signal power source		88	Remote indicator style
40	Transient protection		89	Manifold valve style
41	Integral indicator style		90	
42	Signal termination type		91	SPECIAL REQUIREMENTS
43	Cert/Approval type		92	Custom tag
44	Span-Zero adjust lct		93	Reference specification
45	Failure/Diagnostic action		94	Special preparation
46	Enclosure material		95	Compliance standard
47			96	Software configuration
48			97	
49			98	
50			99	
51			100	PHYSICAL DATA
52			101	Estimated weight
53			102	Overall height
54			103	Removal clearance
55			104	Signal con nominal size — Style
56			105	Mfr reference dwg
57			106	
58			107	
59			108	

	CALIBRATIONS AND TEST		INPUT			OUTPUT OR SCALE	
110	TAG NO/FUNCTIONAL IDENT	MEAS/SIGNAL/TEST	LRV	URV	ACTION	LRV	URV
111							
112		Diff press-Analog output					
113		Diff pressure-Scale					
114		Diff press-Digital output					
115		Press-Digital output					
116		Temp-Digital output					
117							

	COMPONENT IDENTIFICATIONS		
118	COMPONENT TYPE	MANUFACTURER	MODEL NUMBER
119			
120			
121			
122			
123			
124			
125			

Rev	Date	Revision Description	By	Appv1	Appv2	Appv3	REMARKS

Form: 20P2301 Rev 0

© 2001 ISA

Data Sheet For Differential Pressure Transmitter

1.46 Data Sheet of Pressure Transmitter

RESPONSIBLE ORGANIZATION	PRESSURE TRANSMITTER Device Specification	6 7 8 9 10	SPECIFICATION IDENTIFICATIONS Document no / Latest revision — Date / Issue status

(ISA)

#	TRANSMITTER BODY		#	PERFORMANCE CHARACTERISTICS	
11			60		
12	Body/Flange type		61	Max press at design temp	At
13	Process conn nominal size	Rating	62	Min working temperature	Max
14	Process conn temn type	Style	63	Accuracy rating	
15	Vent/Drain location		64	Pressure LRL	URL
16	Mounting type		65	Min ambient working temp	Max
17	Body/Flange material		66		
18	Vent/Drain material		67		
19	Bolting material		68		
20	Flange adapter material		69		
21	Gasket/O ring material		70		
22	Mounting kit material		71		
23–25			72–74		
26	SENSING ELEMENT		75		
27	Detector type		76		
28	Min pressure span	Max	77		
29	Diaphragm/Wetted material		78		
30	Fill fluid material		79		
31–33			80–82		
34	TRANSMITTER		83		
35	Output signal type		84	ACCESSORIES	
36	Enclosure type no/class		85	Air set filter style	
37	Characteristic curve		86	Air set gauges	
38	Digital communication std		87	Heating kit style	
39	Signal power source		88	Remote indicator style	
40	Transient protection		89	Manifold valve style	
41	Integral indicator style		90		
42	Signal termination type		91	SPECIAL REQUIREMENTS	
43	Cert/Approval type		92	Custom tag	
44	Span-Zero adjust lct		93	Reference specification	
45	Failure/Diagnostic action		94	Special preparation	
46	Enclosure material		95	Compliance standard	
47–50			96	Software configuration	
51			97–99		
52			100	PHYSICAL DATA	
53			101	Estimated weight	
54			102	Overall height	
55			103	Removal clearance	
56			104	Signal conn nominal	Style
57			105	Mfr reference dwg	
58–59			106–108		

	CALIBRATIONS AND TEST		INPUT			OUTPUT OR SCALE	
110	TAG NO/FUNCTIONAL IDENT	MEAS/SIGNAL/TEST	LRV	URV	ACTION	LRV	URV
111		Pressure–Analog output					
112		Pressure–Scale					
113		Pressure–Digital output					
114		Temp–Digital output					
115–117							

	COMPONENT IDENTIFICATIONS		
118	COMPONENT TYPE	MANUFACTURER	MODEL NUMBER
119–125			

Rev	Date	Revision Description	By	Appv1	Appv2	Appv3	REMARKS

Form: 20P2201 Rev 0

© 2001 ISA

Data Sheet For Pressure Transmitter

1.47 Data Sheet of Remote Seal Pressure Transmitter

	RESPONSIBLE ORGANIZATION	PRESSURE TRANSMITTER WITH DIAPHRAGM SEAL Device Specification			SPECIFICATION IDENTIFICATIONS	
1	(ISA)		6		Document no	
2			7		Latest revision	Date
3			8		Issue status	
4			9			
5			10			

	TRANSMITTER BODY			PERFORMANCE CHARACTERISTICS	
11		60		
12	Body/Flange type	61	Max press at design temp	At
13	Process conn nominal size	Rating	62	Min working temperature	Max
14	Process conn termn type	Style	63	Accuracy rating	
15	Vent/Drain location	64	Pressure Lower Range-Limit	URL
16	Mounting type	65	Ambient temperature error	
17	Body/Flange material	66	Min ambient working temp	Max
18	Vent/Drain material	67		
19	Bolting material	68		
20	Flange adapter material	69		
21	Gasket/O ring material	70		
22	Mounting kit material	71		
23		72		
24	SENSING ELEMENT	73		
25	Detector type	74		
26	Min pressure span	Max	75		
27	Diaphragm/Wetted material	76		
28	Fill fluid material	77		
29		78		
30	TRANSMITTER	79		
31	Output signal type	80		
32	Enclosure type no/class	81		
33	Characteristic curve	82		
34	Digital communication std	83		
35	Signal power source	84	ACCESSORIES	
36	Transient protection	85	Air set filter style	
37	Integral indicator style	86	Air set gauges	
38	Signal termination type	87	Heating kit style	
39	Cert/Approval type	88	Remote indicator style	
40	Span-Zero adjust lct	89		
41	Failure/Diagnostic action	90		
42	Enclosure material	91	SPECIAL REQUIREMENTS	
43		92	Custom tag	
44	DIAPHRAGM SEAL	93	Reference specification	
45	Seal type	94	Special preparation	
46	Process conn nominal size	Rating	95	Compliance standard	
47	Process conn termn type	Style	96	Software configuration	
48	Diaphragm extension lg	97		
49	Flushing conn quantity	98		
50	Instrument conn nom size	99		
51	Capillary/Fitting dia	Length	100	PHYSICAL DATA	
52	Diaphragm material	101	Estimated weight	
53	Capillary-armor matl	102	Overall height	
54	Bolting material	103	Removal clearance	
55	Upper housing material	104	Signal conn nominal size	Style
56	Lower housing/Flange mat	105	Mfr reference dwg	
57	Gasket/O ring material	106		
58	Fill fluid material	107		
59		108		

	CALIBRATIONS AND TEST		INPUT			OUTPUT OR SCALE	
110							
111	TAG NO/FUNCTIONAL IDENT	MEAS/SIGNAL/TEST	LRV	URV	ACTION	LRV	URV
112		Pressure-Analog output					
113		Pressure-Scale					
114		Pressure-Digital output					
115		Temp-Digital output					
116							
117							

	COMPONENT IDENTIFICATIONS		
118			
119	COMPONENT TYPE	MANUFACTURER	MODEL NUMBER
120			
121			
122			
123			
124			
125			

Rev	Date	Revision Description	By	Appv1	Appv2	Appv3	REMARKS

Form: 20P2211 Rev 0

© 2001 ISA

5. Data Sheet For Remote Seal Pressure Transmitter

1.48 Basic Ordering Information Must Know for Pressure Transmitter

a)   Measurement Span
b)   Measurement Range
c)   Output Signal
d)   Power Supply
e)   Accuracy
f)   Process Temperature Limit
g)   Ambient Temperature Limit
h)   Process Connection
i)   Wetted Part Material
j)   Integral Indicator
k)   Electrical Connections
l)   Housing

1.49 References

A)   Pressure Measurement, ISA, by Mark Murphy, PE Technical Director, Fluor Corp.
B)   Selection and installation recommendations for pressure gauges, BS EN 837-2:1998
C)   Pressure Measurement, Part 2, ASNI/ASME PTC 19.2- 1987
D)   BS 6134:1991 - Specification for pressure and vacuum switches
E)   Differential Pressure Transmitter, Omega
F)   Introduction of DPharp Transmitter, Yokogawa
G)   Pressure Handbook, A Basic Guide to Understanding Pressure, Yokogawa
H)   Flow Measurement Installation Details, PIP PCIFL000 Process Industry Practices
I)   2600T Series Pressure Transmitters Basic Transmitter Theory, ABB
J)   Burdon Tube Pressure Gauges- Dimension, Metrology, Requirements and Testing Part1, EN 837-1
K)   Pressure Gauge and Gauge Attachments, ASME B40.100- 2005
L)   Specification forms for process Measurement and Control Instruments: Part 1 General Consideration ISA TR.20.00.01-2001
M) Specification for Absolute and gauge pressure transmitter with electrical outputs, BS 6447: 1984
N) Specification for Pressure and Vacuum Gauges, IS 3624- 1987.

2

Temperature Measurement

After completing this chapter, you should be able to:

Know about Temperature, Temperature Measuring Devices

Know about RTD and Thermocouple Standards

Selection of Temperature Sensor and Radiation Pyrometry

2.1 What is Temperature

The temperature of a substance is the degree of hotness or coldness of the substance. A hot substance is said to have a high temperature whereas a cold substance is said to have a low temperature. Therefore, the temperature of a substance is an indication of the average kinetic energy of the molecules of the substance. Heat always flow from a body at higher temperature to the body at lower temperature. So, we can also say that temperature of a body is the property which governs the flow of heat.

Temperature is one of the most widely measured physical quantities. Its accurate measurement is essential for the safe and efficient operation and control of a vast range of industrial processes.

Temperature can be measured through various types of sensors. According to the type of application, measurement of temperature can be divided into following parts:

a)   Contact Method
b)   Non-Contact Method

Contact Method: This method is used when the body and the sensor can remain in contact with each other.

Non- Contact Method: This method is used when the body and the sensor are not allowed to remain in contact with each other.

Contact Method uses three types of thermometers:
a)   Expansion Thermometers
b)   Filled System Thermometers
c)   Electrical Temperature Instruments

Non -Contact Method uses three types of thermometers:
a)   Infrared Sensors and Pyrometers
b)   Thermal Imagers

The most commonly used sensors in process industry is Electrical Temperature Instruments.

Electrical Temperature Instruments:
These types of instruments sense the temperature in the terms of electrical quantities like voltage, resistance etc. Devices are required which converts temperature in to another form of signal. The most common devices used in these types of temperature instruments are:

a)   Thermocouple
b)   RTD
c)   Thermistors

2.2 RTD (Resistance Temperature Detector)

A resistance temperature detector or platinum resistance thermometer works on the principle that the electrical resistance of a metal changes in a significant and repeatable way when temperature changes. This resistance is inversely proportional to cross sectional area and proportional to length.

Structure and Measuring Methods:
Structure: Metal wire that changes its electric resistance to changes in temperature are utilized is called "Resistance Wire". This resistance wire normally of Platinum, due to chemical stability, availability in pure form and highly reproducible electrical properties.

Measuring Method:
a)   2 Wire RTD
b)   3 Wire RTD
c)   4 Wire RTD

Equation of RTD (Pt 100):

At 0° C, a platinum RTD has a resistance of 100 ohm and a temperature co-efficient of about 0.00385 $\Omega/\Omega/°C$. These non- linearities are described in Callender- Van Duesen equation. This equation consists of both linear portion and nonlinear portion.

Range: -200 to 0° C: $R(t) [\Omega] = R_0 (1 + At + Bt^2 + C (t-100° C) t^3)$

Range 0 to 850° C: $R(t) [\Omega] = R_0 (1 + At + Bt^2)$

With R_0 is resistance at 0 ° C
A= $3.903 * 10^{-3} °C^{-1}$
B= $-5.775 * 10^{-7} °C^{-2}$
C= $-4.183 * 10^{-12} °C^{-4}$

Temperature Coefficient of Resistance (α):
Temperature Coefficient of Resistance is normally defined as the average resistance change per ° C over the range 0 to 100 ° C divided by R_0 ° C.

$\alpha (\Omega/ \Omega /° C) = (R_{100} - R_0) / (100 * R_0)$
where α= Temperature Coefficient of Resistance
R_{100} = RTD resistance at 100 ° C
R_0 = RTD resistance at 0 ° C

Temperature coefficient resistance values for the common element are:
Copper= 0.00427 $\Omega/ \Omega /° C$
Nickel- Iron = 0.00518 $\Omega/ \Omega /° C$
Nickel = 0.00672 $\Omega/ \Omega /° C$
Platinum = 0.00385 $\Omega/ \Omega /° C$

Table 1 indicates some common RTD standards

ORGANIZATION	STANDARD	ALPHA	NOMINAL RESISTANCE AT 0 Deg. C
British Standards Association	B.S. 1904-1964	0.003850	100
FachnormenausschuB Elektrotchnek im Deutschen NormenausschuB	DIN 43760	0.003850	100
International Electrotechnical Commission	IEC 751: 1983	0.003850	100
US Department of Defense	MIL-T-24388	0.00392	100

Table 1. RTD STANDARDS

Constructional features of metallic resistance thermometer sensors

Platinum is predominantly used for the sensing resistors of industrial metallic resistance thermometer sensors because its refinement and properties are well established, its temperature/resistance characteristic is reproducible and it can be used up to about 850 °C. Nickel is sometimes used on the grounds of economy or because of its better sensitivity, but its characteristic is less linear. Copper, which has good linearity but sensitivity poorer than nickel, is also sometimes used, but neither of these base metals is normally suitable for sensing resistors which are to be used outside the range – 100 °C to + 180 °C.

Sensing Resistor	Minimum Operating Temperature	Maximum Operating Temperature	Special Maximum Operating Temperature
Metallic sensing resistor	°C	°C	°C
Copper	-100	+100	+150
Nickel	-60	+180	+350
Platinum	-200	+600	+850
Semiconductor sensing resistor			
Mixed Metal Oxides	-100	+200	+600
Silicon	-160	+160	+200

Table 2. Operating Temperature for RTD sensing resistor as per BS 1041

Types of RTD (Based on Tolerance):

Platinum RTD's typically are provided in two accuracy as per IEC 751; Class A and Class B.

Class A is considered high accuracy and has an ice point tolerance of +/- 0.06 ohms. Class B is standard accuracy and has an ice point tolerance of +/- 0.12 ohms. Class B is widely used by most industries.

The accuracy will decrease with temperature. Class A have an accuracy of +/- 0.43 ohms (+/- 1.35 °C) at 600 °C and class B will be +/- 1.06 ohms (+/- 3.3 °C) at 600 °C. The chart below shows the tolerance vs temperature (IEC 751).

Tolerances for PT 100 RTD

Temperature	Class A		Class B	
(°c)	(±°C)	(±Ω)	(±°C)	(±Ω)
-200	0.55	0.24	1.3	0.56
-100	0.35	0.14	0.8	0.32
0	0.15	0.06	0.3	0.12
100	0.35	0.13	0.8	0.30
200	0.55	0.20	1.3	0.48
300	0.75	0.27	1.8	0.64
400	0.95	0.33	2.3	0.79
500	1.15	0.38	2.8	0.93
600	1.35	0.43	3.3	1.06
650	1.45	0.46	3.6	1.13

Fig.1 Tolerance Table for Type of RTD according to IEC 751

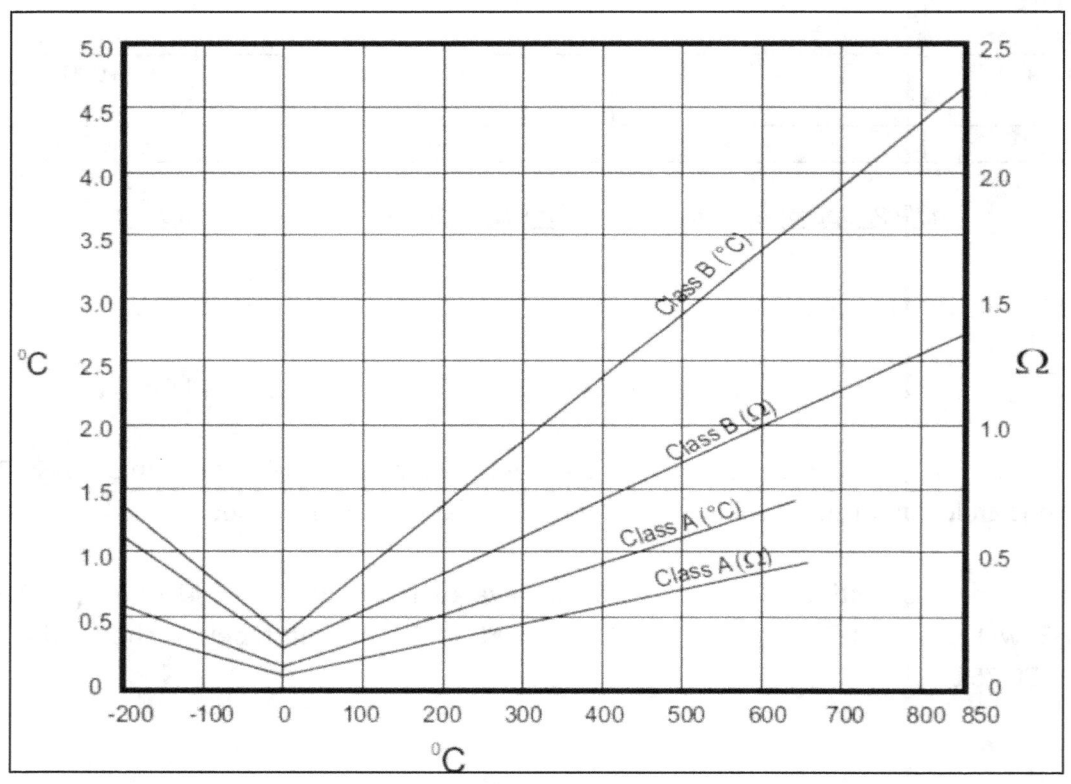

Fig.2 Tolerance values as a function of temperature for 100 Ω Thermometers

Platinum RTD's typically are provided in two accuracy as per ASTM E 1137; Grade A and Grade B.

Temperature	Grade A		Grade B	
	°C	Ω	°C	Ω
-200	0.47	0.20	1.1	0.47
-100	0.30	0.12	0.67	0.27
0	0.13	0.05	0.25	0.10
100	0.30	0.11	0.67	0.25
200	0.47	0.17	1.1	0.40
300	0.64	0.23	1.5	0.53
400	0.81	0.28	1.9	0.66
500	0.98	0.33	2.4	0.78
600	1.15	0.37	2.8	0.88
650	1.24	0.40	3.0	0.94

Table 3. Classification of Tolerance as per ASTM E 1137

2.3 RTD Standard

There are two standards for platinum RTD's: The European Standard (also known as DIN or IEC Standard) and American Standard.

RTD's specified in IEC 60751:

Class AA or 1/3 DIN = 100 +/- 0.04 Ω at 0 °C
Class A = 100 +/- 0.06 Ω at 0 °C
Class B = 100 +/- 0.12 Ω at 0 °C

Also, one Special Class not included in DIN/ IEC 60751:
1/10 DIN = 100 +/- 0.012 Ω at 0 °C

2.4 Lead Wire Configuration

RTD's are available with three different lead wire configurations.

a)   Two Wire RTD; b) Three Wire RTD; c) Four Wire RTD

The two wires RTD is the simplest wire configuration; The three wire RTD is the most popular configuration for use in industrial application; A four wire RTD is the most accurate method to measure the temperature. It is primarily used in laboratories.

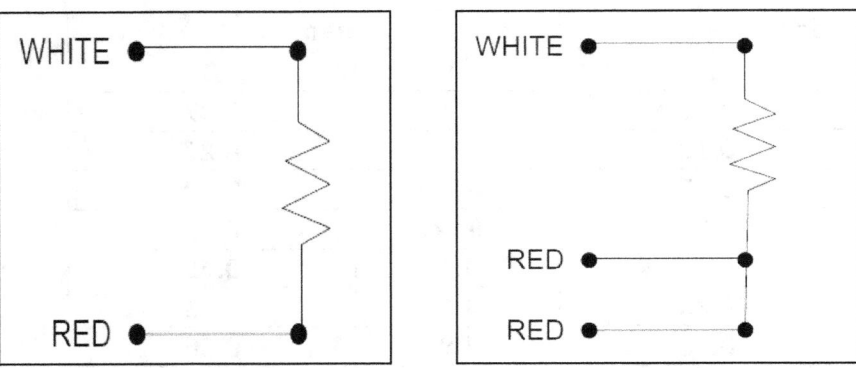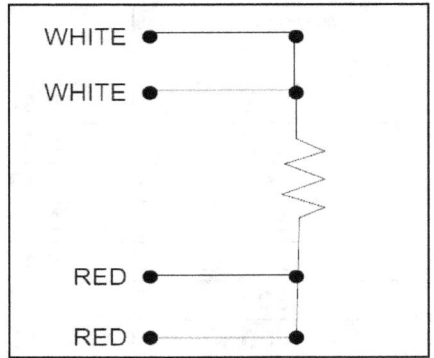

Fig.3 (a)Two Wire Configuration ; (b) Three Wire Configuration ; (c) Four Wire Configuration

2.5 Thermal response time for RTD

Sheath Outside Diameter		63.2% Step Response Time
INCH.	MM	Secs
0.125	3.2	3
0.250	6.4	8

Table 4. Thermal Response Time w.r.t Sheath Diameter as per ASTM E 1137

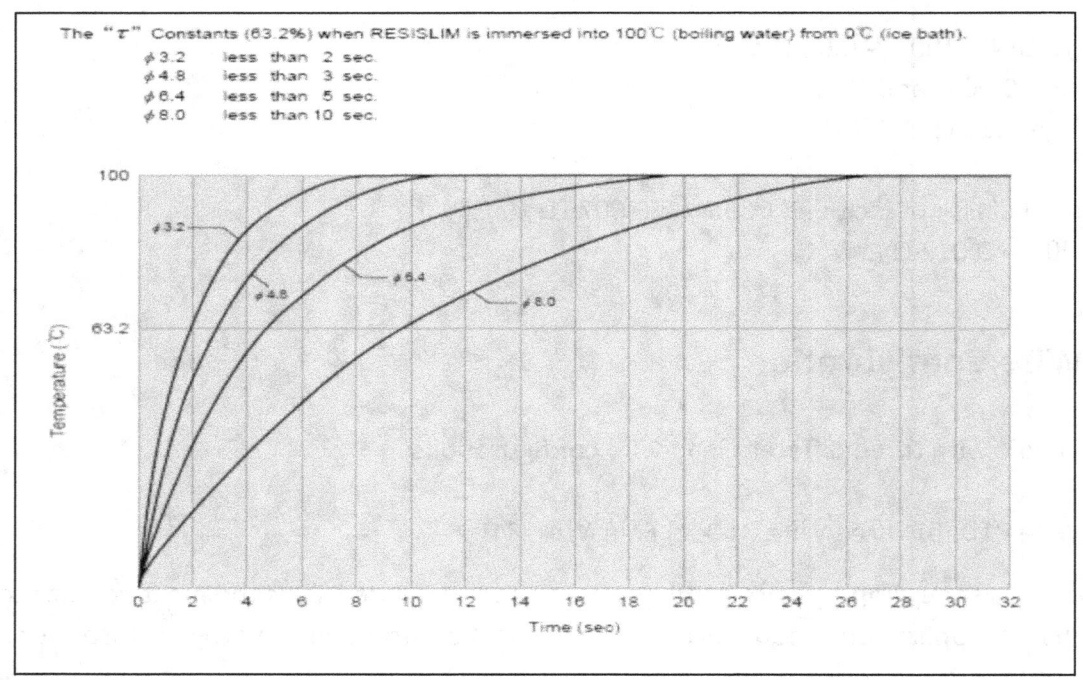

Fig.4 Thermal Response Time w.r.t Sheath Diameter as per JIS C 1604

2.6 Ordering Information for RTD

a)   Sheath Diameter and overall length
b)   Sheath Material
c)   Sensing Element
d)   Insulation
e)   Minimum and Maximum sensed temperature
f)   Connection Configuration: 2 wire, 3 wire or 4 wire
g)   Tolerance: Class A, Class B, 1/3 DIN, 1/10 DIN, Grade A, Grade B
h)   Nominal resistance at 0°C
i)   Process Connection

2.7 Calibration of RTD

Comparison Method:

Most common method Comparison of unknown to known sensors. Multiple sensors can be calibrated at the same time. Comparison calibrations can be performed in a laboratory.

Equipment required for Comparison Method:

Meter, Standard PRT (Platinum Resistance Thermometer), Recorder, etc. (system)

a)   All add to uncertainty level.
b)   The standard PRT should have an accuracy at least four times greater than the unit under test.

2.8 Thermocouple

The thermocouple is an abbreviation of an earlier term "thermo- electric couple" is a pair of two dissimilar metals joined at one place and isolated electrically from each other. It produces a small electrical signal that may lead to good estimate of temperature from one of the points of contact.

The junction that is put into the process in which temperature is being measured is called "HOT JUNCTION or Measuring Point". The other junction at the end of thermocouple material is called "COLD JUNCTION".

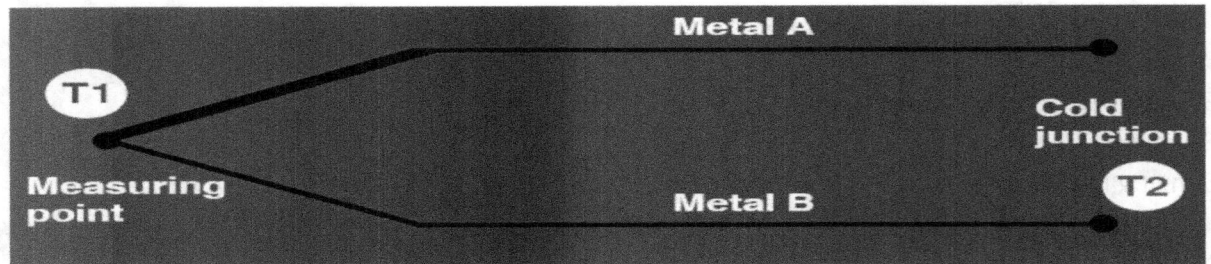

Fig.5 Thermocouple (Seeback Effect)

Since the voltage is measured at ambient temperature, the displayed voltage value would be subtracted by the voltage value at ambient temperature. To obtain the value for the absolute measuring point temperature, the so called "Cold Junction Compensation" is used.

In current instruments with thermocouple input (transmitter) an electronic cold junction compensation is included in the circuity of the instrument.

Basis of Temperature Measurement using Thermocouple

Different conductive metals will produce different level of emf. Thomas Seeback discovered this principle and it is known as "Seeback Effect".

The change in material EMF w.r.t change in temperature is called "Seeback Coefficient or Thermoelectric sensitivity".

Law of Successive Thermocouple

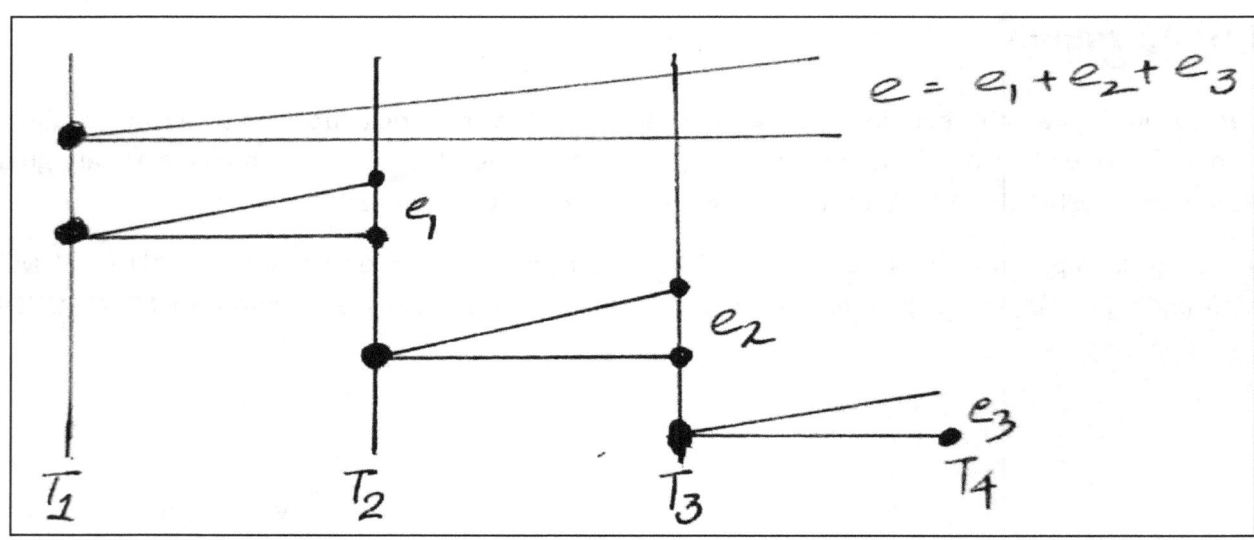

Fig.6 Law of Successive Thermocouple

Fig.6 shows one thermocouple has its measuring point at T1 and open end at T2. The second thermocouple has its measuring point at T2 and its open end at T3. The third Thermocouple has its measuring point at T3 and its open end at T4. The emf level for the thermocouple that is measuring T1 is e1.; that for the other thermocouple is e2; and for the last thermocouple is e3. The sum of the three emf's e1, e2, e3 equals the emf that would be generated by the combined thermocouple operating between T1 & T4.

2.9 Measuring Junctions

Three alternative tip configurations are usually offered:

a)   Exposed Junction
b)   Insulated/ Ungrounded Junction
c)   Grounded Junction

 Exposed Junction: An Exposed Junction is recommended for the measurement of flowing or static non-corrosive gas temperature. It has greatest sensitivity and quickest response.

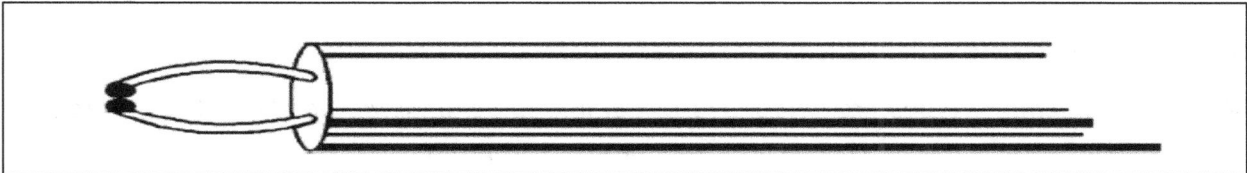

Fig.7 Exposed Junction

Insulated/ Ungrounded Junction: An insulated junction is more suitable for corrosive media although the thermal response is slower.

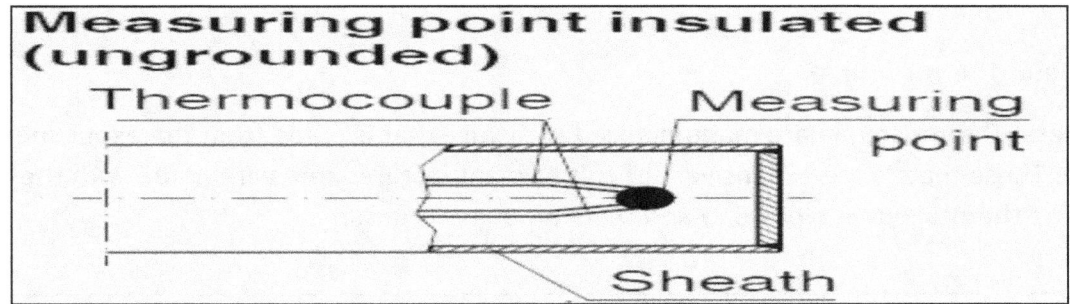

Fig.8 Ungrounded Junction

Grounded Junction: A grounded junction is also suitable for corrosive media and for high pressure applications. It provides faster response than the insulated junction and protection not offered by the exposed junction.

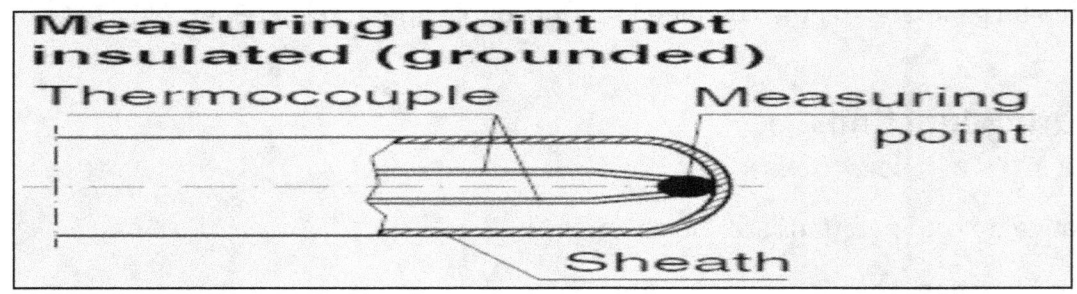

Fig.9 Grounded Junction

2.10 Types of Thermocouple

The standard covers eight specified and most commonly used thermocouples. These thermocouple types can be subdivided in 3 groups, base metal, noble metal and refractory metal thermocouple.

Base Metal Thermocouple:

Base Metal Thermocouple types are composed of common, inexpensive metals such as nickel, iron and copper. The thermocouple types E, J, K, N and T are among this group.

Nobel Metal Thermocouple:

Noble Metal Thermocouple are manufactured with wire that is made with precious or noble metals like Platinum and Rhodium. Nobel Metal Thermocouples are use in oxidizing or inert applications and must be used in ceramic protection tube surrounding the thermocouple element. The thermocouple types S, B and R are among this group.

Refractory Metal Thermocouple:

Refractory Metal Thermocouple are manufactured with wire that is made from the exotic metals tungsten and Rhenium. These metals are expensive, difficult to manufactured and wire made with these metals are very brittle. The thermocouple types C, G and D are among this group.

Thermocouple Type	Material	Temperature Range
B	**PLATINUM-30 PERCENT RHODIUM versus platinum-6 percent rhodium**	0 to 1820 °C

R	PLATINUM-13 PERCENT RHODIUM versus platinum	-50 to 1767 °C
S	PLATINUM-10 PERCENT RHODIUM versus platinum	-50 to 1768 °C
J	IRON versus copper-nickel	-210 to 760 °C
K	NICKEL-10 PERCENT CHROMIUM versus nickel-5 percent (aluminum, silicon)	-270 to 1372 °C
E	NICKEL-10 PERCENT CHROMIUM versus copper-nickel	-270 to 1000 °C
T	COPPER versus copper-nickel	-270 to 400 °C
N	Nicrosil & Nisil	-270 to 1300 °C
C	95% Tungsten/5% Rhenium & 74% Tungsten/26% Rhenium	0 to 2320 °C
G	Tungsten & 74% Tungsten / 26 % Rhenium	0 to 2320 °C
D	97% Tungsten /3 % Rhenium & 75% Tungsten /25% Rhenium	0 to 2320 °C

Table 5. Types of Thermocouple

2.11 Connecting Cables for Thermocouples

a)   **Extension Cable:** The internal leads of the extension cable are made of same material type as the thermocouple.

b)   **Compensating Cable:** The internal leads of the compensating cable are made of materials which correspond to the thermoelectric properties of the original thermocouples.

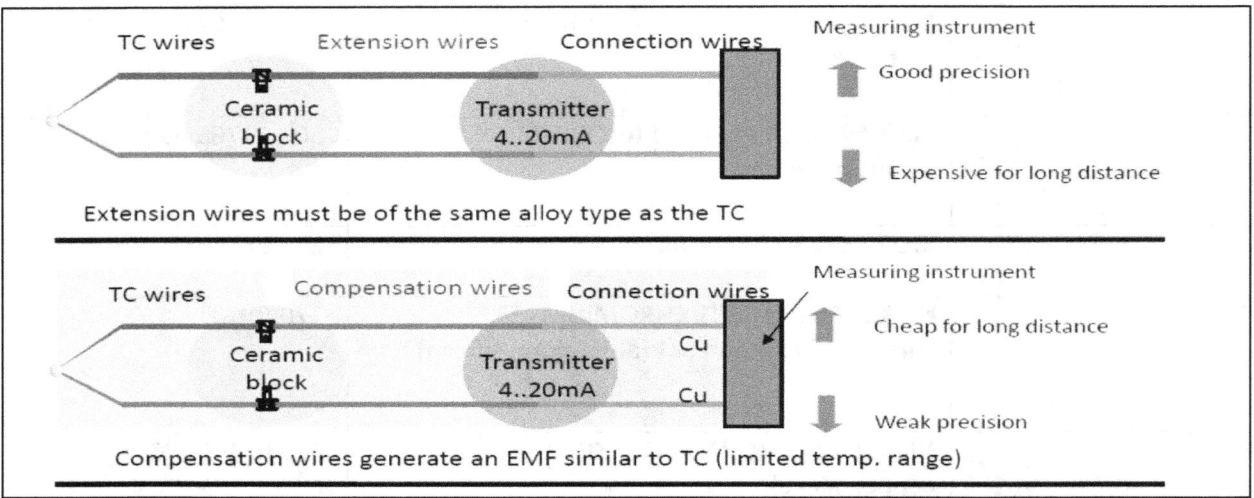

Fig.10 Extension Wire vs Compensation Wire

2.12 Thermocouple Tolerances

Tolerances denotes the maximum allowable value obtained by subtracting the temperature reading or the temperature at the hot junction from the standard temperature converted from the applicable temperature EMF table.

Thermocouple Type	Temperature Range °C	Reference Junction	
		Standard Tolerance (whichever is greater)	Special Tolerance (whichever is greater)
B	870 to 1700	+/- 0.5%	-
E	0 to 900	+/- 1.7 °C or +/-0.5%	+/- 1 °C or +/- 0.4%
J	0 to 750	+/- 2.2 °C or +/- 0.75%	+/- 1.1 °C or +/- 0.4%
K	0 to 1250	+/- 2.2 °C or +/- 0.75%	+/- 1.1 °C or +/- 0.4%
R or S	0 to 1450	+/- 1.5 °C or +/- 0.25%	+/- 0.6 °C or +/- 0.1%
T	0 to 350	+/- 1 °C or +/- 0.75%	+/- 0.5 °C or +/- 0.4%

Table 6. Calibration tolerances for thermocouples as per ANSI MC 96.1 standard

Type	Material	Tolerance Value Per	Class	Range	Tolerance Value
K N	Cr-Al CrSi-Si-Mg	IEC 60584	1	-40 To 1000 ° C	+/- 1.5 ° C
			2	-40 To 1200 ° C	+/- 2.5 ° C
		ASTM E230	Special	0 To 1260 ° C	+/- 1.1° C
			Standard	0 To 1260 ° C	+/- 2.2° C
			1	-40 To 750 ° C	+/- 1.5 ° C

J	Fe- CuNi	IEC 60854	2	-40 To 750 $^\circ$ C	+/- 2.5 $^\circ$ C
		ASTM E230	Special	0 To 760 $^\circ$ C	+/- 1.1° C
			Standard	0 To 760 $^\circ$ C	+/- 2.2° C
R S	Pt 13% Rh-Pt Pt10%Rh-Pt	IEC 60584	1	0 To 1600 $^\circ$ C	+/- 1.0° C
			2	0 To 1600 $^\circ$ C	+/- 1.5° C
		ASTM E230	Special	0 To 1480 $^\circ$ C	+/- 0.6° C
			Standard	0 To 1480 $^\circ$ C	+/- 1.5° C
B	Pt30%Rh-Pt6%Rh	IEC 60584	2	+600 To +1700 $^\circ$ C	+/- 0.0025*t
			3	+600 To +1700 $^\circ$ C	+/- 4° C
		ASTM E230	Special	-	-
			Standard	+870 To +1700 $^\circ$ C	+/- 0.5° C

Table 7. Operating Limit and Accuraies of Thermocouple as per IEC and ASTM Standards

2.13 Maximum Operating Temperature for Thermocouple

Operating temperature limit means the upper temperature where thermocouple can be used continuously in air. Maximum limit means the upper temperature where thermocouple can be used temporarily for a short period of time.

Principal Factors that affect the life of a thermocouple are:
a)   Temperature: Thermocouple life decreases by 50% when an increase of 50 $^\circ$ C occurs.
b)   Diameter: By double the diameter of conductor wire, the life increases by 2-3 times.
c)   Protection: When thermocouple is covered by a protective sheath and placed into ceramic insulators, their life is considerably extended.

Recommended Upper Temperature Limit as per JIS C 1602 and ASTM E 988 Standard

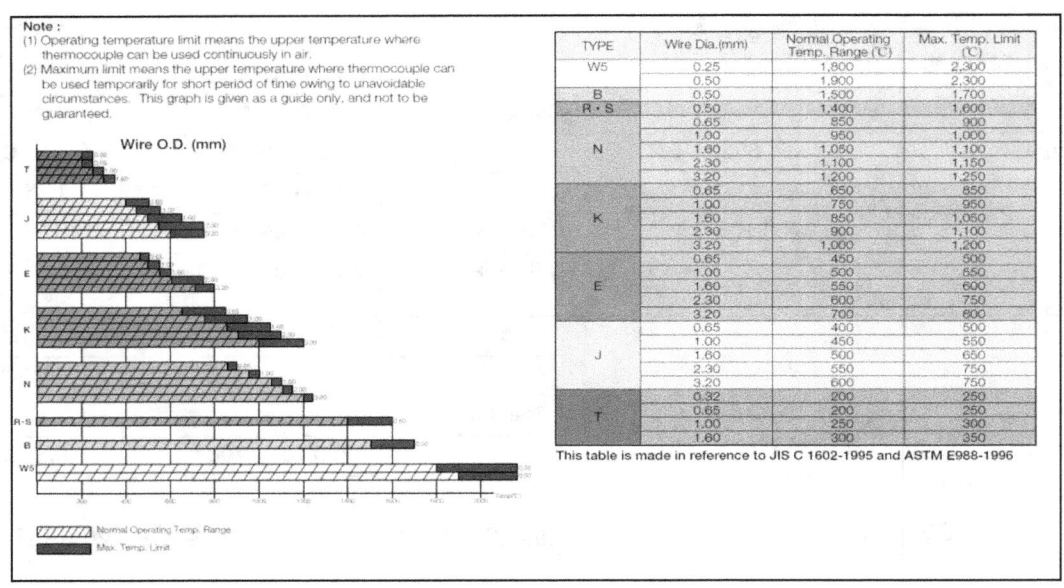

Fig.11 Operating and Maximum Temperature Limits to Conductor Diameter (mm)

2.14 Thermocouple Response Time

The response time for a thermocouple is usually defined as the time taken for the thermal voltage (output) to reach 63% of maximum for the step change temperature in question. It is dependent on several parameters including the thermocouple dimension, construction, tip configuration and the nature of the medium in which the sensor is located.

Sheath Outside Diameter	Types of Measuring Junction	Response Time in Seconds					
		100° C	250° C	350° C	430° C	700° C	850° C
6.0 mm	Insulated	3.2	4.0	4.7	5.0	6.4	16.0
6.0 mm	Grounded	1.6	2.0	2.3	2.5	3.15	8
3.0 mm	Insulated	1.0	1.1	1.25	1.4	1.6	4.5
3.0 mm	Grounded	0.4	0.46	0.50	0.56	0.65	1.8
1.5 mm	Insulated	0.25	0.37	0.43	0.50	0.72	1.0
1.5 mm	Grounded	0.14	0.17	0.185	0.195	0.22	0.8
1.0 mm	Insulated	0.16	0.18	0.19	0.21	0.24	0.73
1.0 mm	Grounded	0.07	0.09	0.11	0.12	0.16	0.6

Table 8. Thermocouple Response Time w.r.t Sheath Diameter

2.15 Mineral Insulated RTD and Thermocouple

Mostly used insulation materials are:

a)   **Alumina (Al2O3):** The maximum operating temperature is 1900 deg. Celsius. High purity aluminum oxide is the standard insulation material. Alumina offers high thermal conductivity and high electrical resistivity.

b)   **Magnesium Oxide (MgO):** Used primarily with compacted sheathed RTD and Thermocouple. Maximum operating temperature is 1370 deg. Celsius. After 1300 degree Celsius MgO becomes conductive. Normally MgO is 99.8 pure, but after 1300 degree Celsius it becomes impure (98%).

MINERAL INSULATED RTD: Mineral insulated RTD are equipped with Pt 100. The inner Copper conductors are embedded in a closely compacted, inert mineral powder (MgO and AL_2O_3), the measuring resistor will be connected to the inner conductors, is also embedded and is surrounded by the metal sheath to form a hermetically sealed assembly.

MINERAL INSULATED THERMOCOUPLES: Mineral insulated Thermocouples consist of thermocouple wire embedded in a densely packed refractory oxide powder insulate all enclosed in a seamless, drawn metal sheath (usually stainless steel). Effectively the thermoelement, insulation and sheath are combined as a flexible cable, which is available in different diameters, usually from 0.75mm to 8mm. At one end cores and sheath are welded and from a "hot " junction. At the other end, the thermocouple is connected to a "transition" of extension wires, connecting head or connector.

Advantages of mineral insulated thermocouple are: -

a) Small over all dimension and high flexibility, which enable temperature measurement in location with poor accessibility.

b) Good mechanical strength

c) Protection of the thermo element wires against oxidation, corrosion and contamination.

d) Fast thermal response

Cross Section View of MI Thermocouple:

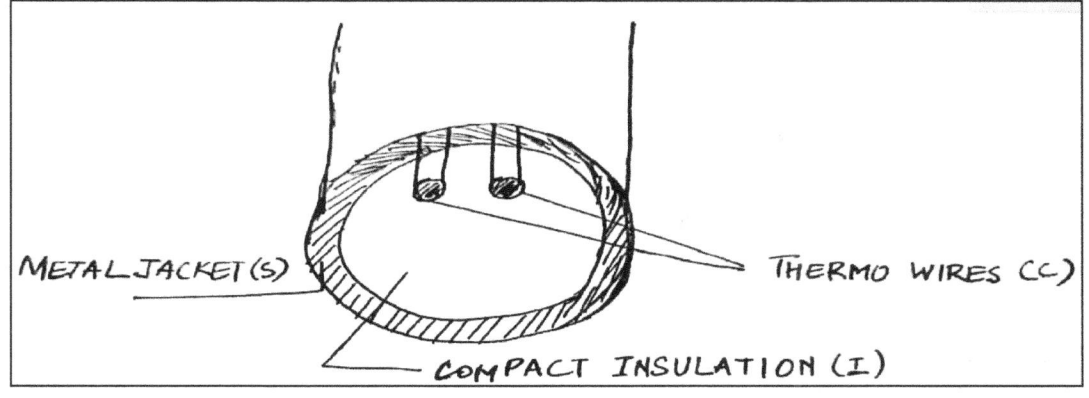

Fig.12 Cross Section of Shell Lines

In Fig.12, S, I, C represents the terms sheath thickness, insulation, conductor respectively. These values depend upon the outer diameter (D) of M.I thermocouple.

Sheath Thickness (S)= 0.10 *D
Conductor © = 0.15 *D

Insulation = 0.08* D

Why MgO used?

Magnesium oxide makes an excellent electrical insulation material because it resists oxidation and ionizing radiation, and it is both chemically and physically stable at high temperatures. The mineral fillings in MI cables provide excellent non-reactive insulation, preventing the thermocouple wires from contact with each other or with caustic substances such as oils, solvents or water. This helps to ensure that the thermocouple probes remain accurate, critical in applications where the quality of the product could be adversely affected by temperature fluctuations.

The insulation in Mineral insulated cables does not burn, which makes it ideal for applications where fire could be catastrophic. Medical devices, power plants and oil rigs are examples of places where installation of MI cables makes operations safer.

Unlike most electrical insulations, magnesium oxide has a relatively high thermal conductivity. This enables the heat to be quickly conducted from the outside sheath and dissipated to the -surrounding air. This conductivity increases when magnesium oxide is compacted.

2.16 Non-Mineral Insulated Thermocouples

In Non- Mineral Insulated thermocouple, thermocouple wires are insulted with ceramic beads covered by a metal sheath. In this type of construction thermocouple wires are protected from the process environment when sheath protection is provided. The construction does not provide flexibility and not found in small sizes.

2.17 Thermowell /Protection Tube

Thermowell are used to protect temperature sensors used to monitor industrial processes while permitting accurate measurement.

The depth of immersion in the fluid is an important consideration. One method of checking for adequacy of immersion is to increase the depth of immersion of the thermocouple well assembly in a constant temperature bath until the thermocouple output becomes constant. A minimum immersion depth of ten times the well outside diameter is a rule of thumb often used.

Types of Thermowell:

a)   Solid Bar Stock Thermowell

b)   Tubular Thermowell

Solid Bar Stock Thermowell:

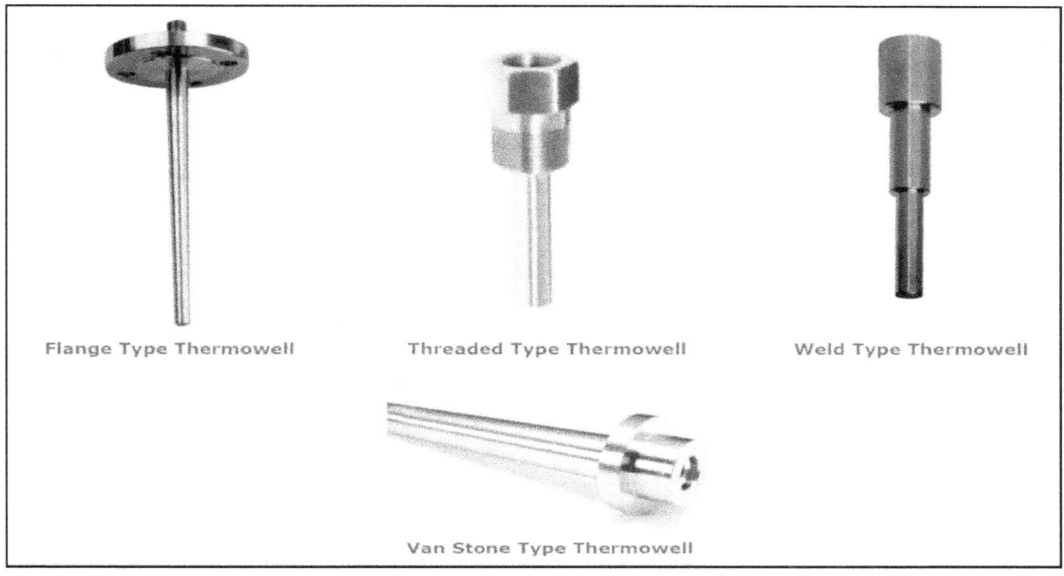

Fig.13 Types of Solid Bar Stock Thermowells

Tubular Thermowell: Intended for light and medium application

a)   Manufactured from tube or pipe materials.

b)   Applications that do not require compliance to ASME PTC 19.3 Calculation

c)   Suitable for low velocity or stagnant media application eg. Tank.

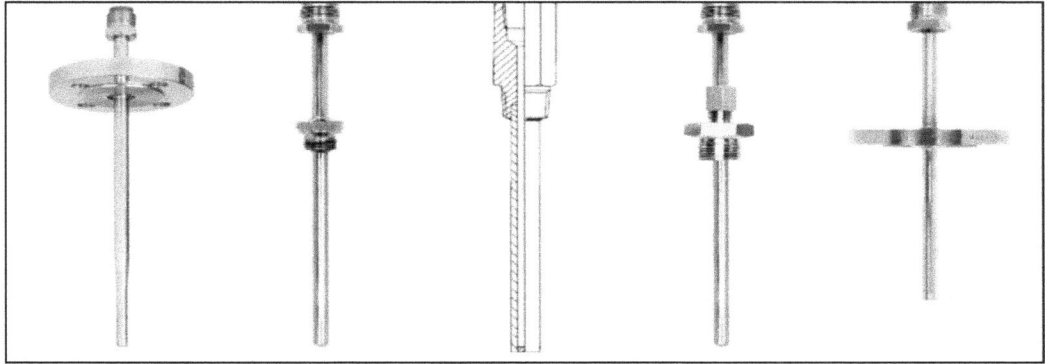

Fig.14 Types of Tubular Thermowells

2.18 Solid Bar Stock Thermowell Terminology

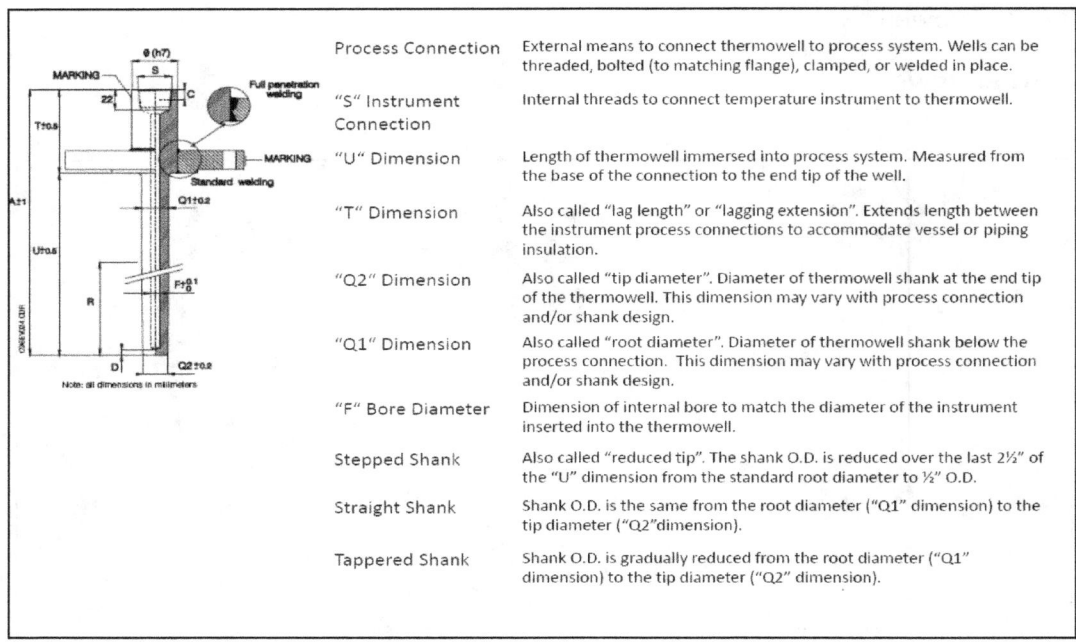

Process Connection	External means to connect thermowell to process system. Wells can be threaded, bolted (to matching flange), clamped, or welded in place.
"S" Instrument Connection	Internal threads to connect temperature instrument to thermowell.
"U" Dimension	Length of thermowell immersed into process system. Measured from the base of the connection to the end tip of the well.
"T" Dimension	Also called "lag length" or "lagging extension". Extends length between the instrument process connections to accommodate vessel or piping insulation.
"Q2" Dimension	Also called "tip diameter". Diameter of thermowell shank at the end tip of the thermowell. This dimension may vary with process connection and/or shank design.
"Q1" Dimension	Also called "root diameter". Diameter of thermowell shank below the process connection. This dimension may vary with process connection and/or shank design.
"F" Bore Diameter	Dimension of internal bore to match the diameter of the instrument inserted into the thermowell.
Stepped Shank	Also called "reduced tip". The shank O.D. is reduced over the last 2½" of the "U" dimension from the standard root diameter to ½" O.D.
Straight Shank	Shank O.D. is the same from the root diameter ("Q1" dimension) to the tip diameter ("Q2"dimension).
Tappered Shank	Shank O.D. is gradually reduced from the root diameter ("Q1" dimension) to the tip diameter ("Q2" dimension).

Fig.15 Thermowell Terminology

Thermowell Shank Styles:

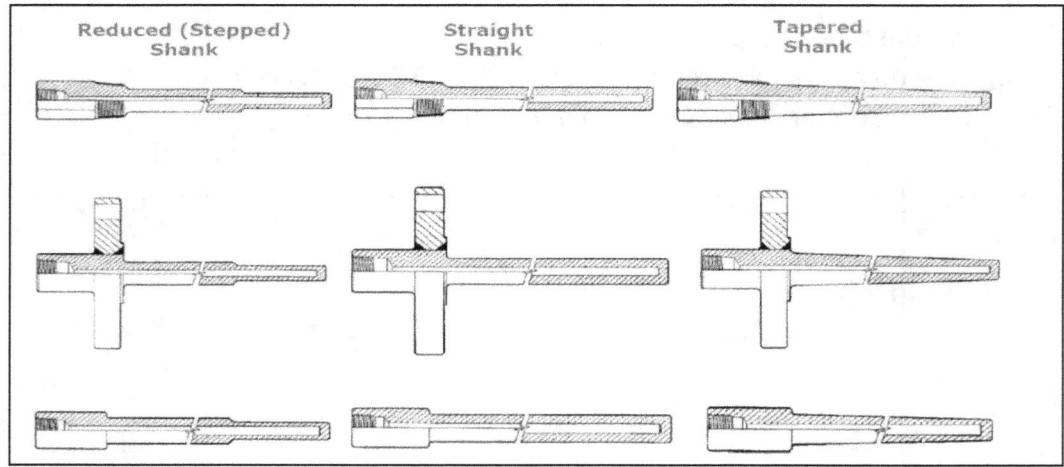

Fig.16 Thermowell Shank Styles

Tapered shank wells provide greater stiffness. The higher strength to weight ratio gives these wells higher natural frequency than for the equivalent length straight shank wells, thus permitting operation at higher fluid velocity.

Thermowell Process Connection:

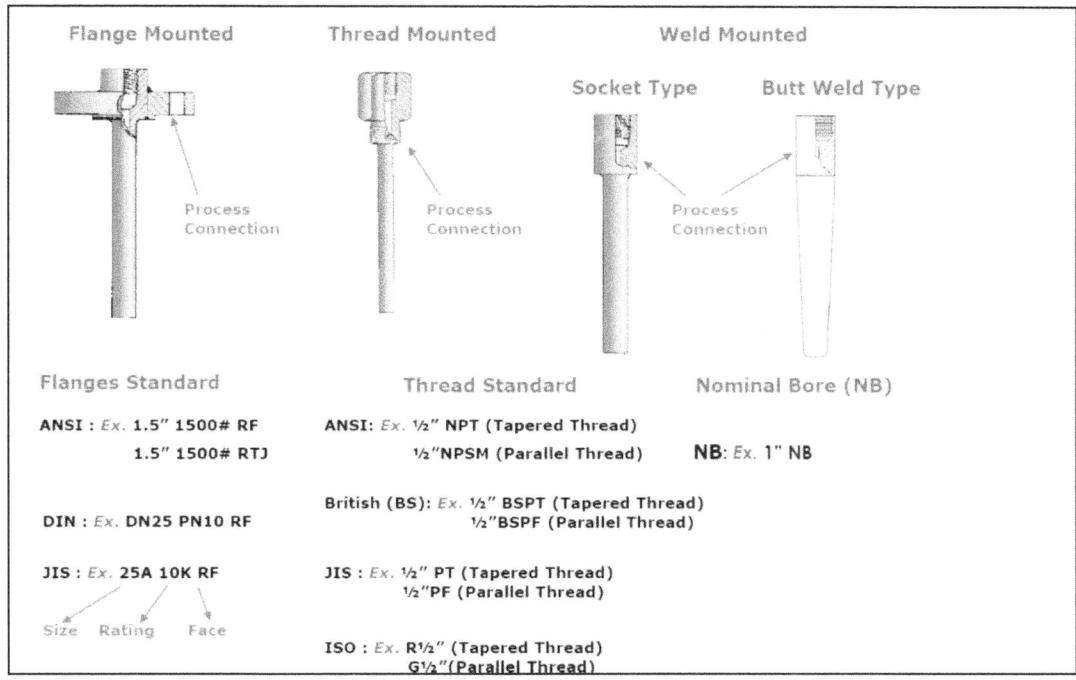

Fig.17 Thermowell Process Connection

Thermowell Process Connection- Threads identification

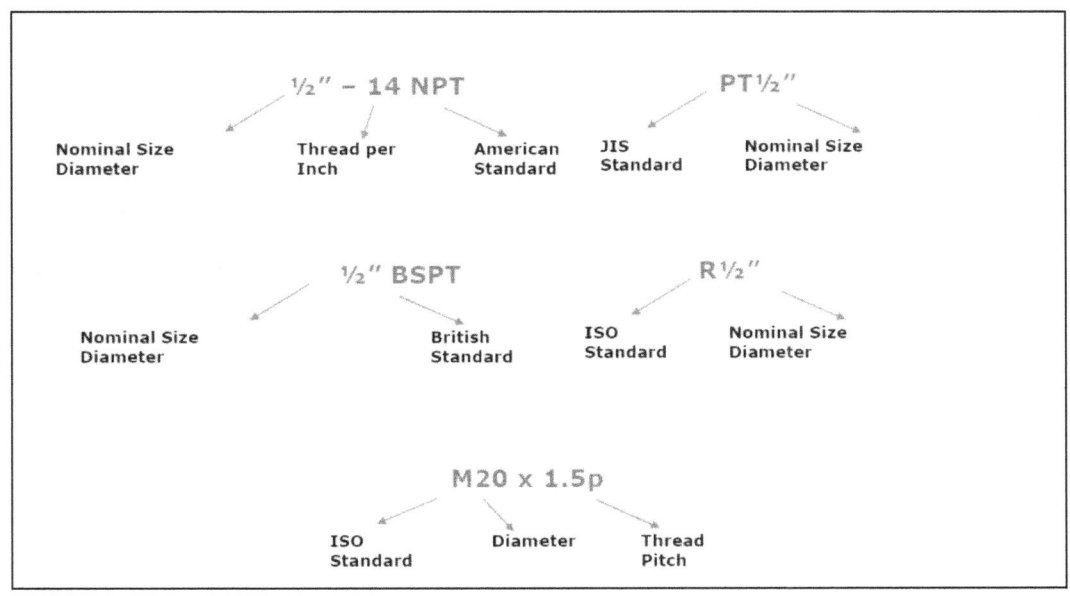

Fig.18 Thermowell Process Connection- Threads Identification

2.19 Tubular Thermowell Terminology

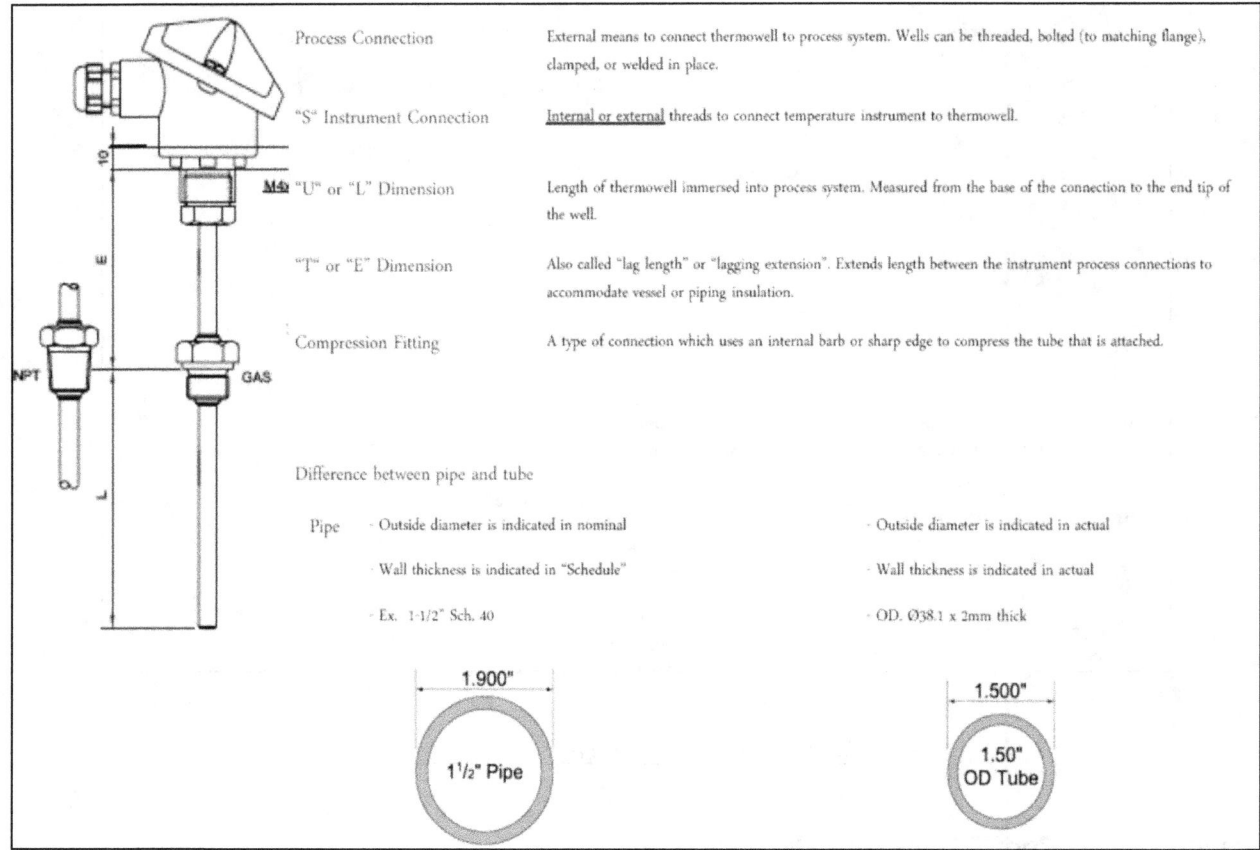

Fig.19 Tubular Thermowell Terminology

2.20 Thermowell Failures

Types of Failure	Reasons
Mechanical Failure	Occurs when excessive force was applied to the thermowell that is greater than its yield strength.
Corrosion Failure	Caused by chemical and/or elevated temperature
Erosion Failure	Resulting from high speed particle impingement on the thermowell
Vibrational fatigue Failure	Occurs due to Von Karman trail vortex frequency

Table 9. Probable Causes for Thermowell Failure

2.21 Thermowell; Quality Pass Criteria

There are four criteria for a thermowell to pass evaluation to PTC 19.3 TW-2010

a)   Frequency Limit: The resonance frequency of the thermowell shall be sufficiently high so that destructive oscillation is not excited by the fluid flow.

b)   Dynamic Stress Limit: The maximum primary dynamic stress shall not exceed the allowable fatigue stress limit.

c)   Static Stress Limit: The maximum steady static stress on the thermowell shall not exceed the allowable stress, determined by the Von Mises Criteria.

d)   Hydrostatic Pressure Limit: The external pressure shall not exceed the pressure rating of the thermowell tip, shank and flange.

2.22 Ceramic Tube

Ceramic Tubes are usually used at high temperature beyond the ranges of metal tube although they are sometimes used at lower temperatures in atmosphere harmful to metal tubes.

The ceramic tube most widely used has a mullite base with certain additives to improve the mechanical and thermal shock properties. The upper temperature limit is 1650 ⁰ C.

Silicon carbides tubes are used as secondary protecting tubes. It is not impermeable to gases.

Fused alumina tubes can be used as primary or secondary protecting tubes or both where temperature to 1900 ⁰ C are expected and a gas tight tube is essential. Fused alumina tubes and insulators should be used with Platinum- Rhodium/ Platinum Thermocouples above 1200 ⁰ C in order to ensure long life and attain maximum accuracy.

2.23 Calibration of Thermocouple

The calibration of a thermocouple by comparison with a reference thermometer is sufficiently accurate for most purposes and can be done conveniently in most industrial and technical laboratories. The success of this method usually depends upon the ability of the observer to bring the measuring junction of the thermocouple to the same temperature as the actuating element of the reference thermometer, such as the measuring junction of a Type S thermocouple, the bulb of a liquid-in glass thermometer, or the resistor of a resistance thermometer. The accuracy obtained is further limited by the accuracy of the reference thermometer. The method of bringing the measuring junction of the thermocouple to the same temperature as that of the actuating element of the reference thermometer depends upon the type of thermocouple, type of reference thermometer, and the method of heating.

Laboratory Furnaces:

The calibration procedure consists of measuring the emf of the thermocouple being calibrated at selected calibration points, the temperature of each point being measured with a reference thermometer. The number and choice of calibration points will depend on the type of thermocouple, the temperature range covered, and the accuracy required.

a)   Noble-Metal Thermocouples: Such thermocouples usually may be calibrated at temperatures from ambient up to 1200°C by comparison with either a Type S or Type R reference thermocouple in electrically heated furnaces. Above 1200°C (2192°F) the Type B thermocouple is a preferred reference thermometer because of its greater stability at high temperatures. This thermocouple may be used to 1700°C (3092°F) or higher.

An automated method of making comparison calibrations of such thermocouples is based upon the simultaneous reading of the emf, the reference thermocouple, and that of the test thermocouple without waiting for the furnace to stabilize completely at any given temperature. The values of emf are measured and recorded with an automatic digital data acquisition system that includes a microcomputer, a switch scanner, and two digital voltmeters. The test thermocouples are connected to one digital voltmeter by means of the switch scanner, and the reference thermocouple is connected to the other digital voltmeter so that measurements of the test and reference thermocouples may be made simultaneously. To reduce the time required to calibrate by this method, the furnace should be so constructed that it may be heated or cooled rapidly.

b)   Base-Metal Thermocouples: The methods of testing base metal thermocouples above room temperature are generally the same as those just described for testing noble-metal thermocouples.

2.24 Data Sheet of RTD

	RESPONSIBLE ORGANIZATION	RTD ASSEMBLY w/wo THERMOWELL Device Specification		SPECIFICATION IDENTIFICATIONS
1			6	
2	(ISA)		7	Document no
3			8	Latest revision Date
4			9	Issue status
5			10	

	PROTECTIVE SHEATH AND FITTING			THERMOWELL OR PROTECTING TUBE
11			50	
12	Housing type		61	Construction type
13	Pad/Collar type		62	Shank style
14	Fitting conn nominal size Style		63	Process conn nominal size Rating
15	Mounting fitting type		64	Process conn termn type Style
16	Sheath alignment		65	Internal conn nom size Style
17	Sheath outside diameter Length		66	Bore diameter
18	Spring loading		67	Outside dia at support
19	Sheath material		68	Outside dia at tip
20	Fitting material		69	Insertion length "U"
21			70	Lagging extension lg "T"
22	SENSING ELEMENT		71	Thermowell/Tube material
23	Sensor type		72	Sheath material-thickness
24	Sensor quantity		73	
25	Temperature coefficient		74	
26	Tolerance class		75	PERFORMANCE CHARACTERISTICS
27	Nominal resistance At temp		76	Max press at design temp At
28	Configuration-no of wires		77	Min working temperature Max
29	Sensor material		78	Max fluid velocity limit At temp
30	Insulator material		79	Temp Lower Range-Limit URL
31			80	Min insertion length
32	LEAD WIRE AND HEAD EXTENSION		81	Min ambient working temp Max
33	Extension type		82	
34	Ext wire nom size-type		83	
35	Extension/Lead length		84	
36	Nipple-union nom size Rating		85	ACCESSORIES
37	Nipple pipe schedule no		86	Moisture seal style
38	Transition type		87	Bayonet adapter size
39	Termination style		88	
40	Connecting wire length		89	
41	Shield - ground wire		90	
42	Nipple material		91	
43	Union/Coupling material		92	SPECIAL REQUIREMENTS
44	Coating-armor material		93	Custom tag
45	Extension wire material		94	Reference specification
46	Ext wire insulation matl		95	Compliance standard
47			96	Construction code
48			97	Calculation report
49	CONNECTION HEAD		98	Calibration report
50	Housing type		99	
51	Cover style		100	
52	Element conn nominal size Style		101	
53	Signal conn nominal size Style		102	PHYSICAL DATA
54	Enclosure type no/class		103	Estimated weight
55	Grounding terminal lct		104	Overall length
56	Enclosure material		105	Removal clearance
57	Terminal block material		106	Mfr reference dwg
58	Terminal material		107	
59			108	

	CALIBRATIONS AND TEST		INPUT OR TEST		OUTPUT	
110						
111	TAG NO/FUNCTIONAL IDENT	MEAS/SIGNAL/TEST	LRV	URV	LRV	URV
112		Temp-Output signal				
113		Test pressure				
114						
115						
116						
117						

	COMPONENT IDENTIFICATIONS		
118			
119	COMPONENT TYPE	MANUFACTURER	MODEL NUMBER
120			
121			
122			
123			
124			
125			

Rev	Date	Revision Description	By	Appv1	Appv2	Appv3	REMARKS

Form: 20T2201 Rev 0

© 2001 ISA

1. Data Sheet of RTD by ISA

2.25 Data Sheet of Thermocouple

	RESPONSIBLE ORGANIZATION	THERMOCOUPLE ASSEMBLY w/wo THERMOWELL Device Specification		SPECIFICATION IDENTIFICATIONS
1			6	
2			7	Document no
3	(ISA)		8	Latest revision · Date
4			9	Issue status
5			10	

	PROTECTIVE SHEATH AND FITTING			THERMOWELL OR PROTECTING TUBE	
11			60		
12	Housing type		61	Construction type	
13	Pad/Collar type		62	Shank style	
14	Fitting conn nominal size · Style		63	Process conn nominal size · Rating	
15	Mounting fitting type		64	Process conn termn type · Style	
16	Sheath alignment		65	Internal conn nom size · Style	
17	Sheath outside diameter · Length		66	Bore diameter	
18	Spring loading		67	Outside dia at support	
19	Sheath/Braid material		68	Outside dia at tip	
20	Fitting material		69	Insertion length "U"	
21			70	Lagging extension lg "T"	
22	SENSING ELEMENT		71	Thermowell/Tube material	
23	Sensor type · Quantity		72	Sheath material-thickness	
24	Wire nominal size		73		
25	Thermocouple type		74		
26	Tolerance class		75	PERFORMANCE CHARACTERISTICS	
27	Measuring junction		76	Max press at design temp · At	
28	Thermocouple wire matl		77	Min working temperature · Max	
29	Insulator material		78	Max fluid velocity limit · At temp	
30			79	Temp Lower Range-Limit · URL	
31			80	Min ambient working temp · Max	
32	LEAD WIRE AND HEAD EXTENSION		81		
33	Extension type		82		
34	Ext wire nom size-type		83		
35	Extension/Lead length		84	ACCESSORIES	
36	Nipple-union nom size · Rating		85	Moisture seal style	
37	Nipple pipe sched no		86	Bayonet adapter size	
38	Transition type		87		
39	Termination style		88		
40	Connecting wire length		89		
41	Shield - ground wire		90		
42	Nipple material		91	SPECIAL REQUIREMENTS	
43	Union/Coupling material		92	Custom tag	
44	Coating-armor material		93	Reference specification	
45	Extension wire material		94	Compliance standard	
46	Ext wire insulation matl		95	Construction code	
47			96	Calculation report	
48			97	Calibration report	
49	CONNECTION HEAD		98		
50	Housing type		99		
51	Cover style		100		
52	Element conn nominal size · Style		101	PHYSICAL DATA	
53	Signal conn nominal size · Style		102	Estimated weight	
54	Enclosure type no/class		103	Overall length	
55	Grounding terminal lct		104	Removal clearance	
56	Enclosure material		105	Mfr reference dwg	
57	Terminal block material		106		
58	Terminal material		107		
59			108		

	CALIBRATIONS AND TEST		INPUT OR TEST		OUTPUT	
110	TAG NO/FUNCTIONAL IDENT	MEAS/SIGNAL/TEST	LRV	URV	LRV	URV
111						
112		Temp-Output signal				
113		Test pressure				
114						
115						
116						
117						

	COMPONENT IDENTIFICATIONS		
118	COMPONENT TYPE	MANUFACTURER	
119			
120			
121			
122			
123			
124			
125			

Rev	Date	Revision Description	By	Appv1	Appv2	Appv3	REMARKS

Form: 20T2301 Rev 0

© 2001 ISA

2. Data Sheet of Thermocouple by ISA

2.26 Colour Codes for Thermocouple Wires and Connector

ANSI Code	ANSI MC 96.1 Color Coding		Alloy Combination		Comments Environment Bare Wire	Maximum T/C Grade Temp. Range	EMF (mV) Over Max. Temp. Range	IEC 584-3 Color Coding		IEC Code
	Thermocouple Grade	Extension Grade	+ Lead	– Lead				Thermocouple Grade	Intrinsically Safe	
J			IRON Fe (magnetic)	CONSTANTAN COPPER-NICKEL Cu-Ni	Reducing, Vacuum, Inert. Limited Use in Oxidizing at High Temperatures. Not Recommended for Low Temperatures.	−210 to 1200°C −346 to 2193°F	−8.095 to 69.553			J
K			CHROMEGA® NICKEL-CHROMIUM Ni-Cr	ALOMEGA® NICKEL-ALUMINUM Ni-Al (magnetic)	Clean Oxidizing and Inert. Limited Use in Vacuum or Reducing. Wide Temperature Range. Most Popular Calibration	−270 to 1372°C −454 to 2501°F	−6.458 to 54.886			K
T			COPPER Cu	CONSTANTAN COPPER-NICKEL Cu–Ni	Mild Oxidizing, Reducing Vacuum or Inert. Good Where Moisture Is Present. Low Temperature & Cryogenic Applications	−270 to 400°C −454 to 752°F	−6.258 to 20.872			T
E			CHROMEGA® NICKEL-CHROMIUM Ni-Cr	CONSTANTAN COPPER-NICKEL Cu–Ni	Oxidizing or Inert. Limited Use in Vacuum or Reducing. Highest EMF Change Per Degree	−270 to 1000°C −454 to 1832°F	−9.835 to 76.373			E
N			OMEGA-P® NICROSIL Ni-Cr-Si	OMEGA-N® NISIL Ni-Si-Mg	Alternative to Type K. More Stable at High Temps	−270 to 1300°C −450 to 2372°F	−4.345 to 47.513			N
R	NONE ESTABLISHED		PLATINUM-13% RHODIUM Pt-13% Rh	PLATINUM Pt	Oxidizing or Inert. Do Not Insert in Metal Tubes. Beware of Contamination. High Temperature	−50 to 1768°C −58 to 3214°F	−0.226 to 21.101			R
S	NONE ESTABLISHED		PLATINUM-10% RHODIUM Pt-10% Rh	PLATINUM Pt	Oxidizing or Inert. Do Not Insert in Metal Tubes. Beware of Contamination. High Temperature	−50 to 1768°C −58 to 3214°F	−0.236 to 18.693			S
U	NONE ESTABLISHED		COPPER Cu	COPPER-LOW NICKEL Cu-Ni	Extension Grade Connecting Wire for R & S Thermocouples. Also Known as RX & SX Extension Wire.					U
B	NONE ESTABLISHED		PLATINUM-30% RHODIUM Pt-30% Rh	PLATINUM-6% RHODIUM Pt-6% Rh	Oxidizing or Inert. Do Not Insert in Metal Tubes. Beware of Contamination. High Temp. Common Use in Glass Industry	0 to 1820°C 32 to 3308°F	0 to 13.820			B
G* (W)	NONE ESTABLISHED		TUNGSTEN W	TUNGSTEN-26% RHENIUM W-26% Re	Vacuum, Inert, Hydrogen. Beware of Embrittlement. Not Practical Below 399°C (750°F). Not for Oxidizing Atmosphere	0 to 2320°C 32 to 4208°F	0 to 38.564	NO STANDARD USE ANSI COLOR CODE		G (W)
C* (W5)	NONE ESTABLISHED		TUNGSTEN-5% RHENIUM W-5% Re	TUNGSTEN-26% RHENIUM W-26% Re	Vacuum, Inert, Hydrogen. Beware of Embrittlement. Not Practical Below 399°C (750°F). Not for Oxidizing Atmosphere	0 to 2320°C 32 to 4208°F	0 to 37.066	NO STANDARD USE ANSI COLOR CODE		C (W5)
D* (W3)	NONE ESTABLISHED		TUNGSTEN-3% RHENIUM W-3% Re	TUNGSTEN-25% RHENIUM W-25% Re	Vacuum, Inert, Hydrogen. Beware of Embrittlement. Not Practical Below 399°C (750°F)--Not for Oxidizing Atmosphere	0 to 2320°C 32 to 4208°F	0 to 39.506	NO STANDARD USE ANSI COLOR CODE		D (W3)

Fig.20 ANSI and IEC Color Code for Thermocouple, Wires and Connector

2.27 Temperature Transmitter

What is Temperature Transmitter?

Temperature transmitters convert the input signal from a wide range of sensors, such as resistance sensors and thermocouples, into a standardized output signal (e.g. 0 ... 10 V or 4 ... 20 mA).

Why use Temperature Transmitter?

The temperature of a remote process must be monitored. Common temperature sensing devices such as thermocouples and RTD's produce very small "signals." These sensors can be connected to a two-wire transmitter that will amplify and condition the small signal. Once conditioned to a usable level, this signal can be transmitted through ordinary copper wire and used to drive other equipment such as meters, dataloggers, chart recorders, computers or controllers.

How to use Temperature Transmitter?

A temperature transmitter draws current from a remote dc power supply in proportion to its sensor input. The actual signal is transmitted as a change in the power supply current. Specifically, a thermocouple input transmitter will draw 4 mA of current from a dc power supply when measuring the lowest temperature of the process. Then, as the temperature rises, the thermocouple transmitter will draw proportionally more current, until it reaches 20 mA. This 20-mA signal corresponds to the thermocouple's highest sensed temperature. The transmitter's internal signal-conditioning circuitry (powered by a portion of the 4-20 mA current) determines the temperature range that the output current signal will represent.

Advantages of temperature transmitters:

a) A.C power is not needed at the remote location to operate a two-wire transmitter. Since transmitters are powered by a low level 4-20 mA output current signal, no additional power must be supplied at the remote location. In addition, the usual 24 Vdc signal necessary for operation is standard in plants that have large amounts of instrumentation.

b) Electrical noise and signal degradation are not a problem for two-wire transmitter users. The transmitter's current output signal lends itself to a high immunity when it comes to ambient electrical noise. Any noise that does appear in the output current is usually eliminated by the common-mode rejection of the receiving device. In addition, the current output signal will not change (diminish) with distance as most voltage signals do.

c) Wire costs drop significantly when using transmitters. Low voltage signals produced by thermocouples almost always require the use of shielded cable when they are sent any significant distance. Ambient electrical noise from arcing electrical relays, motors and ac power lines can raise havoc with these signals that are transmitted in an unshielded cable.

Standard Specifications of Temperature Transmitter:

Input: Universal Input Thermocouple (J, K, E, T, R, S, B, N) , RTD (Pt 100, Pt 200, Pt 500, JPt 100, Cu, Ni120 , 2, 3,4 wire), DC Voltage (mv), Resistance (2/3 wire).

Output: Two wire 4 to 20 mA DC

Output Range: 3.68 mA to 20.8 mA

Power Supply: 10.5 Vdc to 42 Vdc

Communication Line Conditions:
Load Resistance: 250 to 600 Ω (including cable resistance); Maximum Line Length: 2 Km when CEV cable is used; Spacing to power line: 15 cm

Accuracy:
A/D conversion accuracy + D/A conversion accuracy or +/- 0.1 % whichever is greater.

Stability:

RTD: +/- 0.1 % of reading or +/- 0.1 $^{\circ}$ C per 2 years, whichever is greater at 23 +/- 2° C.

Thermocouple: +/- 0.1 % of reading or +/- 0.1 $^{\circ}$C per year, whichever is greater at 23 +/- 2 $^{\circ}$C.

2.28 Various Units of Temperature Measurement

°C – degrees Celsius (or Centigrade)
°F – degrees Fahrenheit
K – Kelvin
R – Rankine

Relationship between different units
°C = (°F - 32)/1.8
°F = 1.8 x °C + 32
 K = °C + 273.15
 R = °F + 459.67
Conversion tables or software can be utilized to facilitate with converting between these units.

2.29 Temperature Sensor Selection Guideline

Temperature Sensor Selection Guide		
	RTD	Thermocouple
Temperature Range	-328°F to 1562°F	-310°F to 3308°F
Accuracy	±0.001°F to 0.1°F	±1°F to 10°F
Response Time	Moderate	Fast
Stability	Stable over long periods	Not as stable
	<0.1% error / 5 yr.	1°F error / 1yr.
Linearity	Best	Moderate
Sensitivity	High	Low
Vibration applications	Poor	Good

Table 10. Selection Guideline for RTD and Thermocouple

2.30 Data Sheet of Temperature Transmitter

	RESPONSIBLE ORGANIZATION	RTD/THERMOCOUPLE TEMPERATURE TRANSMITTER OR SWITCH Device Specification		SPECIFICATION IDENTIFICATIONS	
1			6		
2			7	Document no	
3	(ISA)		8	Latest revision	Date
4			9	Issue status	
5			10		

	TRANSMITTER OR SWITCH			PERFORMANCE CHARACTERISTICS	
11			52		
12	Housing type		53	Accuracy rating	
13	Input sensor type		54	Measurement LRL	URL
14	Output signal type		55	Min ambient working temp	Max
15	Min measurement span	Max	56	Contacts ac rating	At max
16	Temp coef/Tolerance cl		57	Contacts dc rating	At max
17	Isolation type		58		
18	Enclosure type no/class		59		
19	Adjustment type		60		
20	Characteristic curve		61		
21	Digital communication std		62		
22	Signal power source		63		
23	Measurement type		64		
24	Configuration-no of wires		65		
25	Contacts arrangement	Quantity	66		
26	Failsafe style		67		
27	Transient protection		68	ACCESSORIES	
28	Integral indicator style		69	Remote indicator style	
29	Signal termination type		70	Indicator enclosure	
30	Cert/Approval type		71	Air set filter style	
31	Mounting type		72	Air set gauges	
32	Failure/Diagnostic action		73		
33	Dead band type		74		
34	Switch time delay		75	SPECIAL REQUIREMENTS	
35	Temp compensation type		76	Custom tag	
36	Enclosure material		77	Reference specification	
37	Mounting kit material		78	Compliance standard	
38			79	Calibration report	
39			80	Software configuration	
40			81		
41			82		
42			83		
43			84	PHYSICAL DATA	
44			85	Estimated weight	
45			86	Overall height	
46			87	Removal clearance	
47			88	Signal conn nominal size	Style
48			89	Mfr reference dwg	
49			90		
50			91		
51			92		

	CALIBRATIONS AND TEST		INPUT OR SETPOINT			OUTPUT OR SCALE	
110							
111	TAG NO/FUNCTIONAL IDENT	MEAS/SIGNAL/TEST	LRV	URV	ACTION	LRV	URV
112		Temp- Analog output 1					
113		Temp-Analog output 2					
114		Temp-Scale					
115		Temp diff-Digital output					
116		Temp-Digital output					
117		Termn temp-Digital output					
118							
119		Temp setpoint 1-Output					
120		Temp setpoint 2-Output					
121		Temp setpoint 3-Output					
122		Temp setpoint 4-Output					
123							
124							
125							

	COMPONENT IDENTIFICATIONS			
126				
127	COMPONENT TYPE	MANUFACTURER	MODEL NUMBER	
128				
129				
130				
131				
132				
133				

Rev	Date	Revision Description	By	Appv1	Appv2	Appv3	REMARKS

Form: 20T2221 Rev 0

© 2001 ISA

3. Data Sheet of Temperature Transmitter by ISA

2.31 Non- Contact Temperature Measurement

Infrared:

This type of instrument measures the heat – infrared energy – radiated from an object. By focusing this infrared energy through an optical system onto a detector gives temperature measurement using signal processor.

All object emits infrared energy, the hotter the object the more active its molecules are, hence more infrared energy emits.

Since infrared thermometers can determine temperature of a target without any physical contact, the measurement system does not contaminate, damage or interfere with the process being monitored so it has many advantages over contact type measurement.

Thermal Radiation:

Any object whose temperature is above absolute zero, will possess thermal energy, and will radiate a portion of this as infrared energy. An object may also become heated by absorbing infrared waves in fact a human body is capable of emitting and absorbing infrared energy. If an object is hotter than its surroundings, it will emit more radiation than it absorbs and will tend to cool down, conversely if an object is cooler than its surroundings, it will absorb more radiation than it emits. Usually, the object will come to thermal equilibrium with its surrounding, at this point the rate of absorption and radiation of infrared energy will be equal.

Thermal radiation is an electromagnetic radiation emitted from the hot surface. Electromagnetic radiation is a self-propagating wave with electric and magnetic components. Electromagnetic radiation is classified into types based on the frequency of the wave.

Radiation thermometers are designed to respond to wavelength within infrared region of the electromagnetic spectrum which is 0.5 μm to 15 μm. The human eye is responsive to infrared emission within the visible region which is generally from 0.4 micrometer to 0.8 μm. It is this response which give us the ability to human to observe the temperature of an object being heated.

Most infrared emissions are outside the range of a human eye and therefore cannot be observed.

Thermal radiation depends not only on temperature but the composition and surface condition of an object. Different material will emit radiation at different rates an unoxidized stainless steel alloy at a temperature of 1700 degree Celsius will emit infrared energy at a rate of 32W cm-2 while polished aluminum plate at the same temperature will emit infrared energy at a rate of 5 W cm-2. In both cases, the rate can be increased by roughening the surface.

2.32 Absorption, Transmission and Reflection

When an infrared energy radiated by an object reached to another body, a portion of that energy is absorbed, a portion of that energy is reflected and a portion will be transmitted through that body if partially transparent.

As shown in fig.21, we can say that if a, r, and t are the body's fractional value then

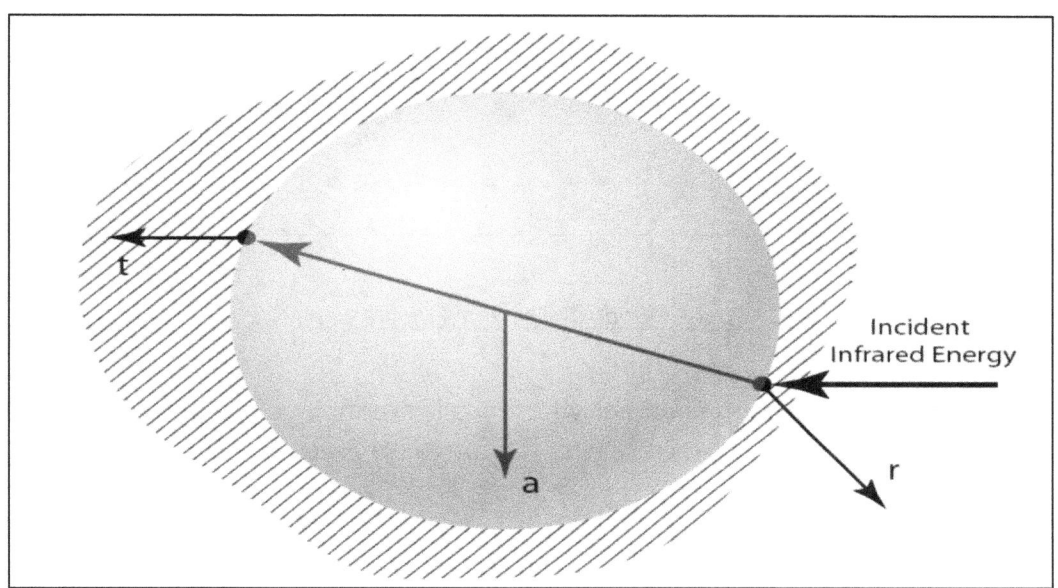

Fig.21 Absorption, Transmission and Reflection

a + r + t=1

It then follows that if a large fraction of radiant energy is transmitted through an object, the material can be regarded as transparent. If reflectivity r is very high, the object is said to be look like mirror. Other material types which have poor transmission and poor reflection characteristics generally absorb much of the radiant energy incident on their surface. The absorption of any non-transparent material generally increases as the surface roughness.

2.33 Black Body Radiation

Kirchoff's Law of Thermal Equilibrium states that "at thermal equilibrium the energy radiated by an object must be equal to the energy absorbed by that object.

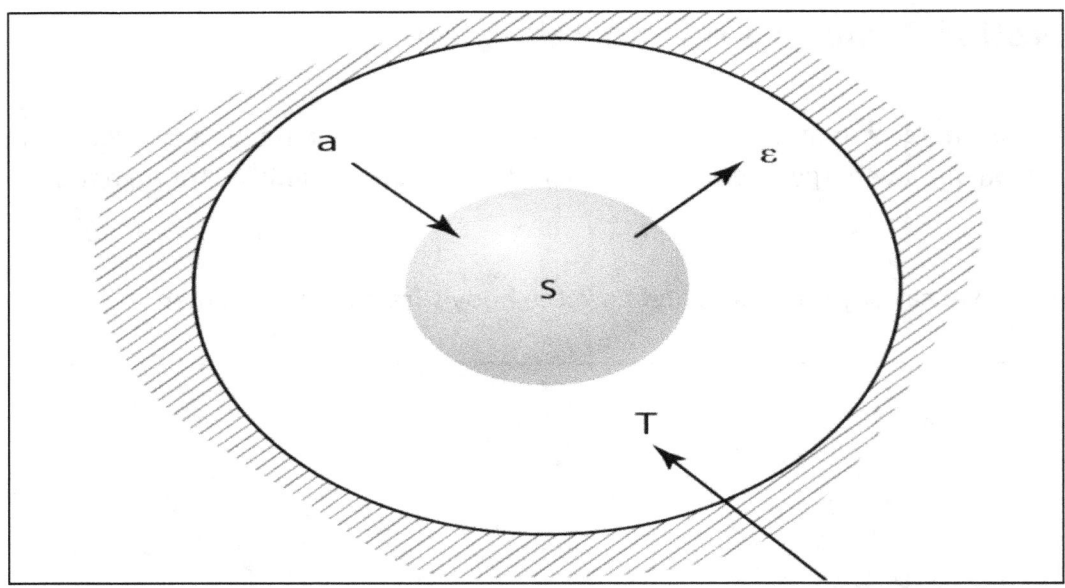

Fig.22 A Body at Thermal Equillibrium

If we consider an object S in a vacuum chamber whose walls are maintained at a uniform temperature T, then at thermal equilibrium, the object must reach the same temperature as the chamber walls.

Kirchoff's Law expressed as:

A=e

Where e is emissivity and in between 0 to 1.

The Blackbody is an ideal surface having the following properties:

a) A Black body absorbs all incident radiation, regardless of wavelength and direction.
b) No Surface can emit more thermal radiation than the blackbody.
c) The radiation emitted by a black body is independent of direction.

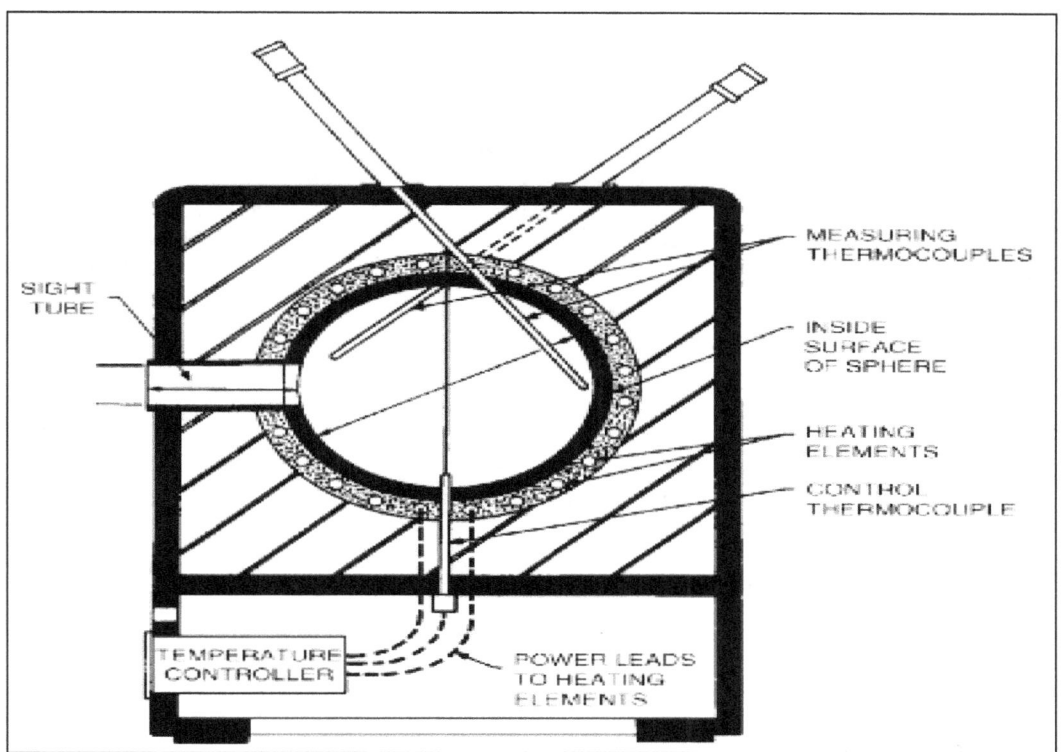

Fig.23 Schematic view of a commercial blackbody furnace

Fig. 23 illustrates schematic view of a commercial blackbody furnace. Entrance sight tube is shown to the left of the hollow sphere. One thermocouple is used to determine the sphere's temperature, another to determine temperature gradients.

2.34 Emissivity

Emissivity is defined as the ratio of the energy radiated by the material to the energy radiated by the black body. It is a measure of a material ability to absorb and radiate energy. A true black body have an emissivity e=1 while any real object would have emissivity <1. Emissivity is a numerical value and does not have units.

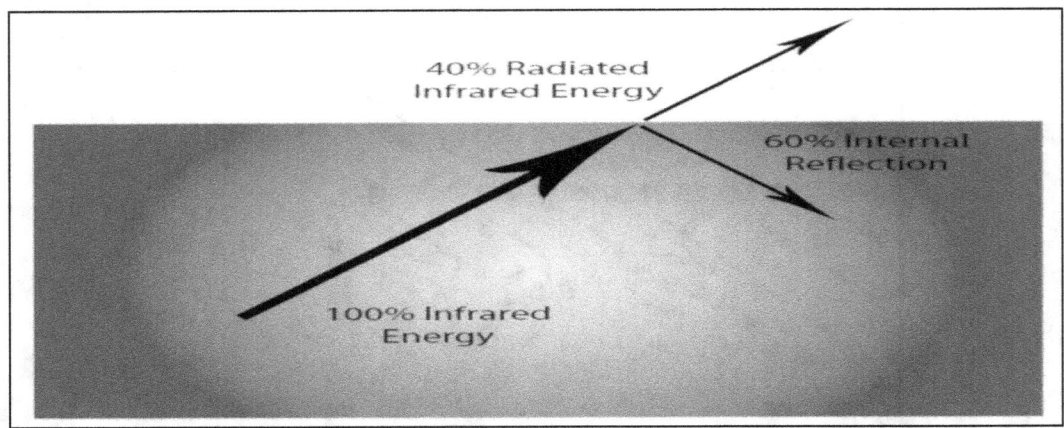

Fig.24 Radiated Energy

The ability of a material to radiate infrared energy depends upon several factors: the type of material, the surface condition. The value of emissivity for an object is an expression of its ability to radiate infrared energy.

Effect of Wavelength:

Emissivity will normally vary with wavelength – for example, the emissivity of polished metals tends to decrease as wavelength becomes longer.

A body at higher temperatures emits electromagnetic radiation. The rate at which energy is emitted depends on surface temperature and surface conditions. The thermal radiation from a body is composed of wavelengths forming an energy distribution. The total emissive power of a black body e_b at a temperature is

$$e_b = \int_0^\lambda e_{b\lambda} \, d\lambda$$

In which λ is wavelength and $e_{b\lambda}$ is monochromatic emissive power. Planck's distribution law relates $e_{b\lambda}$ to the wavelength and temperature:

$$e_{b\lambda} = \frac{2\,\pi\,h\,a^2\,\lambda^{-5}}{\exp\left[\dfrac{ch}{K_B\,\lambda T}\right]-1}$$

In which h is Planck's constant, a is velocity of light, λ is wavelength. is absolute temperature and K_B is Boltzmann constant.

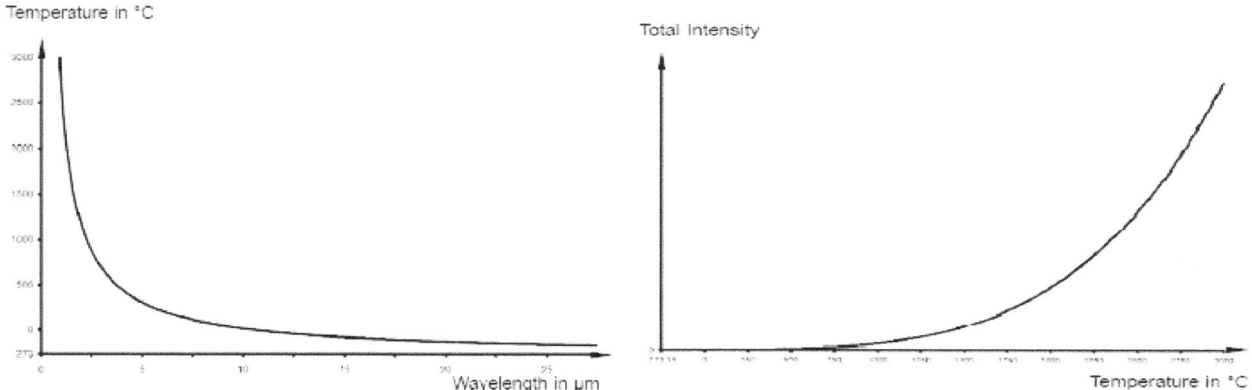

Total emissive power of a black body is

$$e_b = \sigma\, T^4$$

In which is Stefan's Boltzmann constant and its value is 5.67 * 10 $^{-8}$ W m^{-2} K $^{-4}$.

Fig. 25 shows the typical emissivity curve for iron along with grey body object which has constant emissivity across the whole of the infrared spectrum.

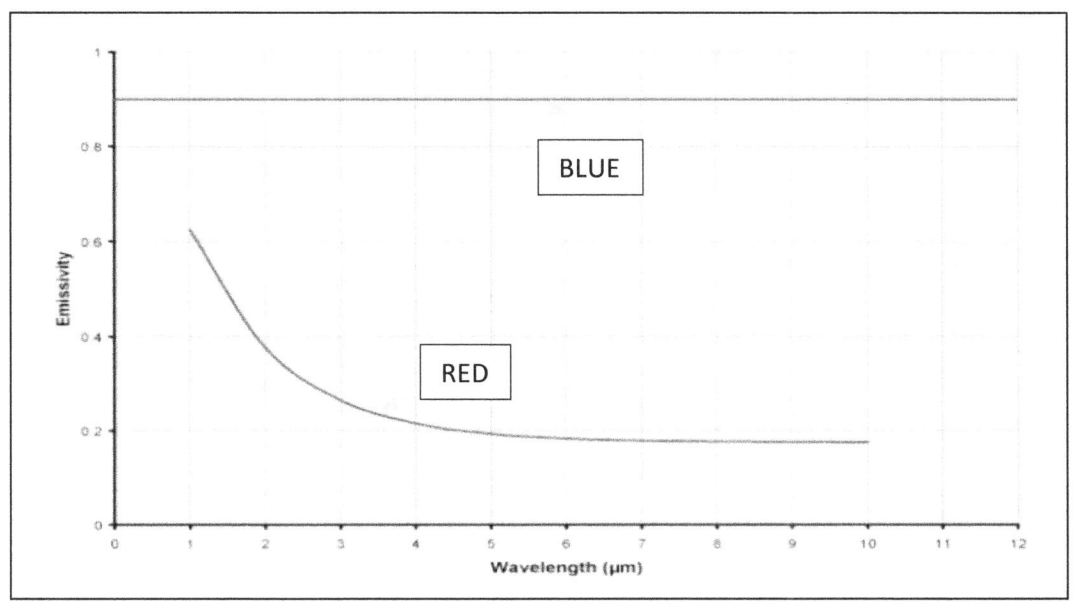

Fig.25 Emissivity vs. Wavelength

Effect of Surface Condition:

In the case of metallic materials, emissivity will decrease with polishing and increases with surface roughness.

Effect if Viewing Angle:

The maximum recommended angle for mounting an infrared thermometer is 45 degrees.

Effect of Temperature:

Emissivity will usually change with temperature if the surface properties of the material change, for example if coatings becomes tarnished or degraded.

Use of Shortest Possible Wavelength Thermometer:

The energy emitted by a hot target changes very rapidly at short wavelengths, but more slowly at long wavelengths. As a result, thermometers which operates at short wavelength minimise the error which occur with change in target emissivity.

Fig. 26 and 27, shows a comparison between the error expected from a short wavelength thermometer (red) and one operates at long wavelength (blue) with changes in emissivity.

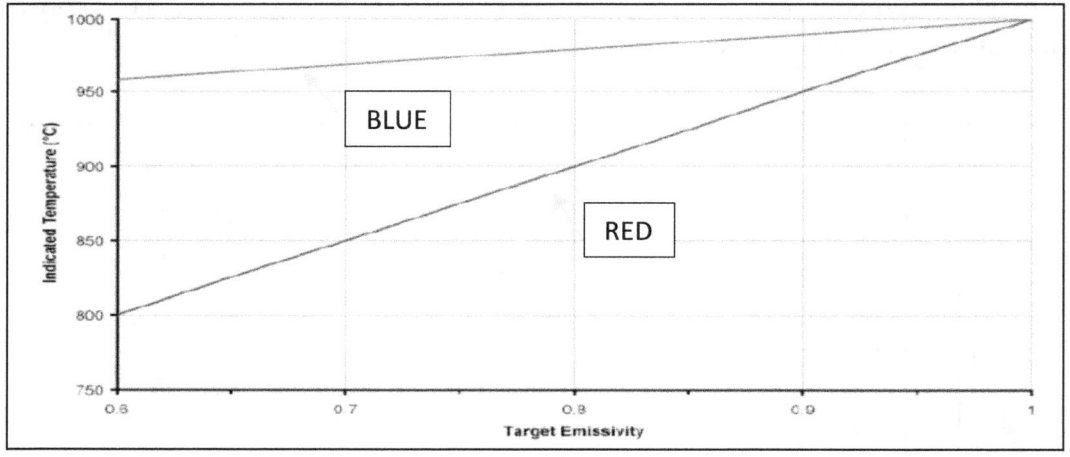

Fig.26 Comparison of measurement errors

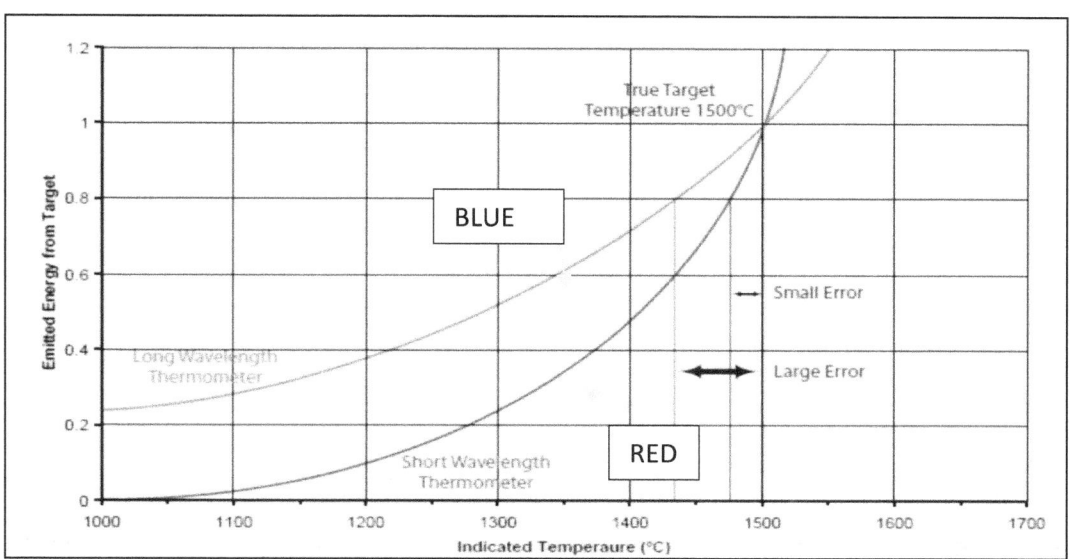

Fig.27 Comparsion of measurement errors

It can be seen that the error from the short wavelength thermometer is about 10 degrees Celsius for a 10 percent change in emissivity for a target at 1000 degree Celsius. The longer wavelength thermometer gives much larger errors with similar changes in target emissivity.

2.35 Radiation Pyrometer

Temperature measurement is based on the measurement of radiation either directly by a sensor or by comparing with the radiation of a body of known temperature. The radiation pyrometer is a non-contact type of temperature measurement.

The wavelength region having high intensity is between 0.1 to about 10µm. In this region, 0.1 to 0.4 µm is the ultraviolet region, 0.4 to 0.7 µm is the visible region and 0.7 onwards is the infrared region. With the increase in temperature, radiation intensity is stronger toward shorter wavelengths. The temperature measurement by radiation pyrometer is limited within 0.5 to 8µm wave length region.

Total radiation pyrometer:

A radiation pyrometer consists of optical component to collect the radiation energy emitted by the object, a radiation detector that converts radiant energy into an electrical signal, and an indicator to read the measurements. Total radiation pyrometers are used to measure temperature in the range 700°C to 2000°C.

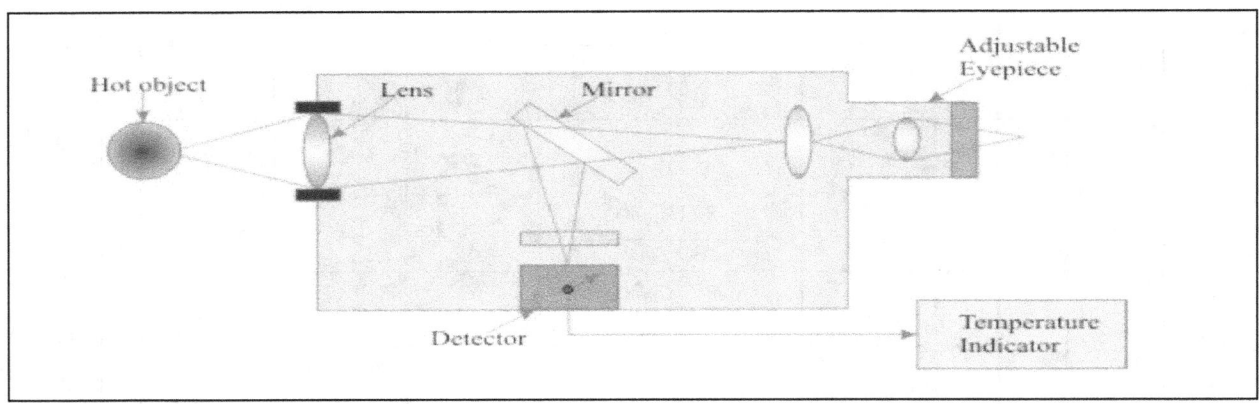

Fig.28 Total Radition Pyrometer

The optical pyrometer is designed to respond narrow band of wavelengths that fall within the visible range of the electro-magnetic spectrum. Thermal detectors are used as sensors. Their hot junction is the radiation sensing surface. Thermopiles can detect radiation of all wavelengths.

A number of semiconductors are developed to sense the radiation. These are materials of Si, PbS, indium antimonides etc. Their response is though instantaneous but it is selective to wavelength. Silicon is suitable only around 0.8 µm - 0.9 µm and lead sulphide around 1 to 2µm.

It is important that gases like CO_2, H_2O and dust should not obstruct the path of radiation. The dust particles scatter the radiation, whereas CO_2 and water vapor selectivity absorbs radiation.

Any instrument built to sense the radiation must be in an enclosure to avoid dirt, dust and gases present in industrial environment. Normally a window is provided with some optical materials to see the radiating body. The materials should have good transmissivity. All optical materials allow only particular wavelength to pass through it with sufficient intensity. For other wavelengths, they are opaque.

Material for Windows	Transmissivity
Glasses like Quartz, Ruby etc.	Good in ultraviolet ad visible region of wavelength but are opaque to infrared. Glass Windows are useful for wavelengths lower than 2.5µm. Beyond 2.5µm, transmissivity decrease drastically.
Barium Fluoride and Zinc Sulphide	They have 60-80% transmissivity in the infrared and visible region.
Calcium Fluoride	It has very good transmissivity in visible and infrared region

Table 11. Materials used for Windows in Radiation Pyrometer

Detectors:

Detectors used for radiation pyrometer falls into two main groups:
a) Thermal Detectors: are which utilize the temperature rise resulting from the absorption of thermal radiation. Their spectral response may be independent of wavelength over a very wide range.
b) Photon Detector: whose characteristics are determined by the absorption of individual photons. Their response is limited to a relatively narrow wavelength range.

Optical Systems:

The aim of the design of Optical System:
a) to isolate the required band or bands of wavelength.
b) for a given source area, to increase the proportion of radiation incident upon the detector, which is usually small.
c) to reduce the effect on the output due to changes in the distance between the source and the detector.
Aperture Optics: If the detector is exposed to the radiation from an extended but finite source, not only does the signal from it depend on the distance of separation, but the presence of other sources in the vicinity may affect the reading. These effects may be overcome by placing an aperture stop and a field stop between the source and the detector as shown in fig.29. in such a way that all rays arriving at the detector arise from some point on the source.

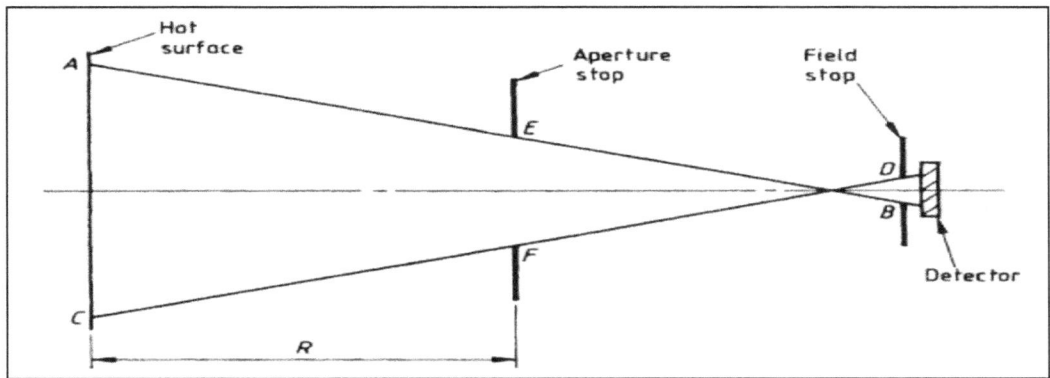

Fig.29 Diagram of an aperture optical system

From fig. 29 the extreme rays AB and CD define the extent of the source AC which is used when the detector has an effective size BD. For a given aperture and field stop geometry, the distance R can increase until the limiting rays move to the edge of the source. If the source is small and the require range of R has to be large, the aperture stop would have to be small, resulting in a low output and loss of accuracy. This difficulty can be overcome by the use of lens to image the source on to the field stop. Fig. 30 shows a small source can be measured without reducing the diameter of the aperture stop. The focal length of the lens should be such that the image should be completely covers the field stop.

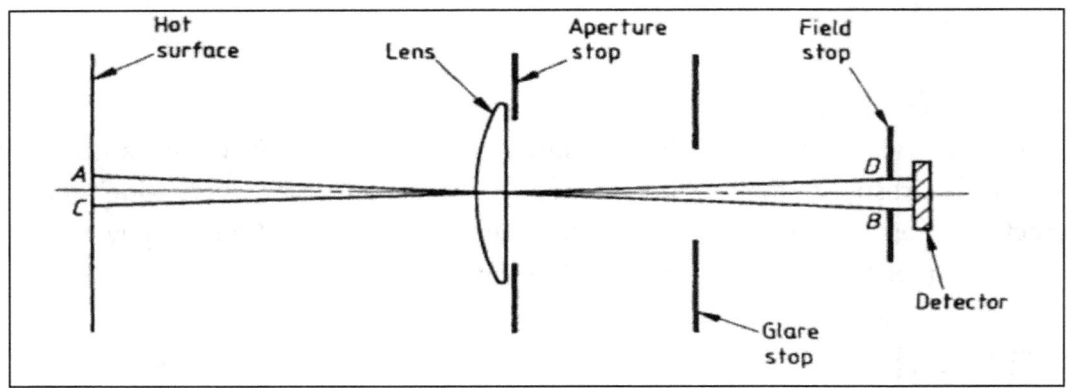

Fig.30 Diagram of lens optical system

Types of Radiation Pyrometer:

The most commonly used thermometers tend to fall into one of three possible classes:

a) Broad Wavelength:

Low temperature thermometers which have a spectral response of 8 to 14 µm will be used on temperature ranges typically of 0 to 250 degree Celsius. It also is a very common wavelength for portable infrared thermometers.

b) Selected Wavelength;

Selected Wavelength Thermometers are usually application specific. For eg. 4.8 to 5.2 µm thermometer for glass surface measurement and 3.43 micrometer thermometer for thin film plastic measurement.

c) Short Wavelength:

These are usually operated below 2.5 µm. A typical example of 1 µm used for temperature range 600 to 1300 degree Celsius. These thermometers are good at minimizing the effect of variable emissivity and are found throughout the steel and other industries. In Steel industries application like coke oven where 1 µm thermometer used, for slag detection generally 3.9 µm thermometer used and for flare stack application 8-14 µm thermometer used. Wavelength bands centered on 0.65 and 2.6 µm are used for metals and can see through quartz windows.

Advantages of Radiation Pyrometer:

a) Since a non-contact method is employed, no part of the instrument need to be brought into the hot zone. As a result, radiation pyrometer is capable of measuring higher temperatures.

b) Good sensitivity can be achieved even at modest temperatures.

c) Moving objects for eg. On a production line can be viewed.

d) Fast response can be achieved hence rapid temperature changes can be monitored.

e) Thermocouple loses its accuracy over a period of time due to impurities in the conductor wire, due to shut error while radiation pyrometer maintains its accuracy over a long period of time.

f) Refractory surface temperature is important when measuring in applications like hot blast in blast furnace, dome of hot stoves in blast furnace as pyrometer measures surface temperature while thermocouple measures atmospheric temperature conditions in same applications.

Limitations of Radiation Pyrometer:

a) Availability of optical materials limit on the wavelengths that can be measured.

b) The surface of the hot object should be clean. It should not be oxidized. Scale formation does not allow to measure the radiation accurately.

c) Emissivity correction is required. Change in emissivity with temperature need to be considered.

Methods for determining the emissivity:

a) The temperature of the object is first determined with a contact thermometer. Then aim the pyrometer to the object. Now adjust the emissivity knob until the same temperature is achieved in both devices. This method can be only be applied for sufficiently large and accessible objects.

b) Coat the material with a special polish whose emissivity approximately equals to 1, accurately know and is stable up to the temperature to be measured. When pyrometer is aimed to the object it first measures the temperature of the coated surface and then it measures the untouched part of the surface. Simultaneously, adjust the emissivity to force the indicator to display the correct temperature.

c) The emissivity of a sample object can be determined by spectrometer analysis.

d)   Standardized emissivity values for most material is available. These can be entered the instrument to estimate the material emissivity value.

2.36 Essential Terms to Know

Thermal Radiation: Electromagnetic radiation emitted by an object by virtue of the thermal motion of the atoms and molecules of which it is composed.

Emissivity: Ratio of the thermally emitted component of the radiant exitance of the radiator to that of a full radiator at the same temperature.

Reflectance: Ratio of the reflected radiant or luminous flux to the incident flux.

Transmittance: Ratio of the transmitted radiant or luminous flux to the incident flux.

Absorptance: Ratio of the absorbed radiant or luminous flux to the incident flux.

Radiation Pyrometer: Instrument measuring the temperature of an object by means of the thermal radiation emitted by the object.

Total radiation Pyrometer: Radiation Pyrometer which measures all the thermal radiation received from the source.

2.37 Data Sheet of Radiation Pyrometer

	RESPONSIBLE ORGANIZATION	THERMAL RADIATION TEMPERATURE SENSOR w/wo MONITOR Device Specification		SPECIFICATION IDENTIFICATIONS
1			6	
2	(ISA)		7	Document no
3			8	Latest revision Date
4			9	Issue status
5			10	

	SENSING ELEMENT			MONITOR Continued
11		60		
12	Sensor type	61	Failure/Diagnostic action	
13	Temperature LRL URL	62	Enclosure material	
14	Spectral response	63		
15	Focus type	64		
16		65	PERFORMANCE CHARACTERISTICS	
17	HEAD AND VIEWER	66	Min head working temp Max	
18		67	Temp accuracy rating	
19	Head type	68	Focus point D:S ratio	
20	Housing style	69	Far point D:S ratio	
21	Enclosure type no/class	70	Min monitor amb working Max	
22	Sighting type	71	Contacts ac rating At max	
23	Lens purge style	72	Contacts dc rating At max	
24	Conn adapter type	73		
25	Cable length	74		
26	Signal termination type	75		
27	Cert/Approval type	76		
28	Mounting type	77		
29	Enclosure material	78		
30		79		
31		80		
32	TRANSMITTER OR MULTIPLEXER	81		
33	Type	82		
34	Signal selector type	83		
35	Cable length	84	ACCESSORIES	
36	Output signal type	85	Blast gate style	
37	Enclosure type no/class	86	Cooling flow regulator	
38	Signal power source	87	Sight tube material	
39	Adjustment type	88	Printer	
40	Number input sensors	89		
41		90		
42	MONITOR	91	SPECIAL REQUIREMENTS	
43	Housing type	92	Custom tag	
44	External input type	93	Reference specification	
45	Output signal type	94	Compliance standard	
46	Zero-span adjustment lct	95	Calibration report	
47	Enclosure type no/class	96	Software configuration	
48	Measurement type	97		
49	Digital communication std	98		
50	Signal power source	99		
51	Dead band type	100	PHYSICAL DATA	
52	Contacts arrangement Quantity	101	Estimated weight	
53	Emissivity adjustment	102	Overall height	
54	Failsafe style	103	Removal clearance	
55	Integral indicator style	104	Signal conn nominal size Style	
56	Signal termination type	105	Mfr reference dwg	
57	Cert/Approval type	106		
58	Mounting type	107		
59	Temperature compensation	108		

	CALIBRATIONS AND TEST		INPUT OR SETPOINT			OUTPUT OR SCALE	
110							
111	TAG NO/FUNCTIONAL IDENT	MEAS/SIGNAL/SCALE	LRV	URV	ACTION	LRV	URV
112		Temp-Analog output 1					
113		Temp-Analog output 2					
114		Temp-Analog output 3					
115		Temperature-Scale					
116		Temp setpoint 1-Output					
117		Temp setpoint 2-Output					

	COMPONENT IDENTIFICATIONS		
118			
119	COMPONENT TYPE	MANUFACTURER	MODEL NUMBER
120			
121			
122			
123			
124			
125			

Rev	Date	Revision Description	By	Appv1	Appv2	Appv3	REMARKS

Form: 20T2401 Rev 0

© 2001 ISA

3. Data Sheet of Pyrometer as per ISA

2.38 Temperature Gauges

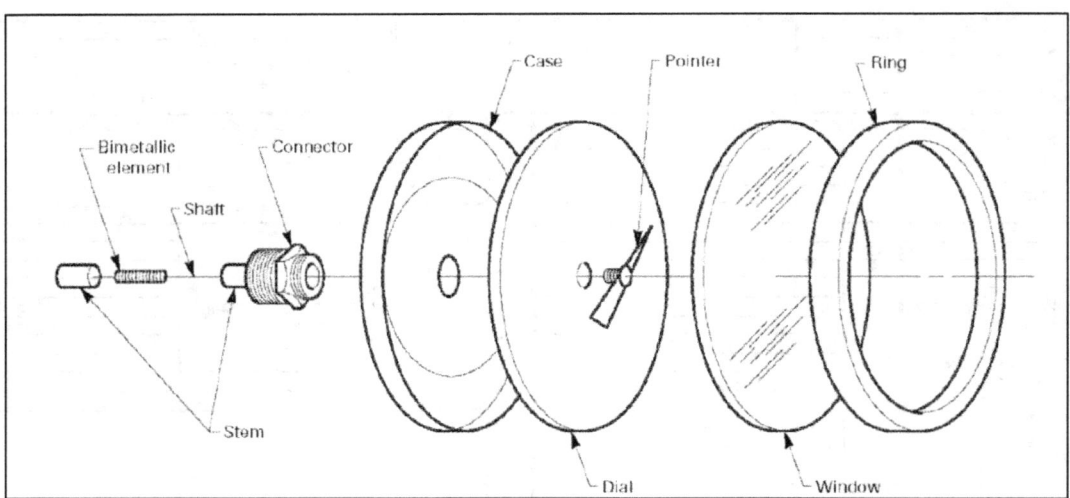

Fig.31 Bimetallic Thermometer-Typical Componenets

Temperature gauges measures the thermal states of a homogenous substance. It operates on three types of principles:

a) Bimetallic Thermometer
b) Gas Expansion Thermometer
c) Liquid Expansion Thermometer

Fig. 31 illustrates, Bimetallic thermometer composed of two or more metals mechanically associated in such a way that relative expansion of the metals, caused by temperature change, produces motion.

Gas Expansion Thermometer: Dial indicating thermometer operating on the principle of change in pressure of a fixed volume of gas in response to temperature change.

Liquid Expansion Thermometer: Liquid expansion thermometer consists of a temperature sensor, a capillary and a Bourdon tube. Dial indicating thermometer operating on the principle of liquid expansion in response to temperature change. Using capillaries from 500 mm to 10000 mm long, measurements can also be taken from remote measuring points. The scale ranges from -40 degrees Celsius to +400 degree Celsius with class 1 and 2 accuracies in accordance with EN13190. The filling fluid is usually an inert hydrocarbon, such as xylene (C_8H_{10}), which has a coefficient of expansion six times that of mercury and makes smaller bulbs possible. Other liquids (even water)
are sometimes used. The criterion is that the pressure inside the system must be greater than the vapor pressure of the liquid to prevent bubbles formation.

In industrial temperature measurements, there are typically two types of dial thermometers are used; Bimetallic Thermometer and Gas Expansion Thermometer.

Bimetallic Thermometer measures the temperature via a spiral tube which consists of two different metals. These metals possess different thermal expansion. If two different straight metal strips are bonded together and heated, the resultant strip will bend toward the side of the metal with the lower expansion rate. Deflection is proportional to the square of the length and the temperature change and inversely proportional to the thickness. Through the mechanical deformation of the bimetal strip in a spiral tube, the rotational movement results caused by temperature changes. This is transferred through a pointer shaft to the pointer of the instrument. Fig. 33 illustrates the working principle of Bimetallic Thermometer. Bimetal Thermometers are generally designed for scale ranges of -70 degree Celsius to +600 degrees Celsius, this only applies for unfilled instruments. When there are vibrations in the process, the operator must often resort to thermometer with liquid filling. The Silicone oil is generally used to dampens the vibrations effects. However, this limits the possible upper temperature range to 250 degrees Celsius.

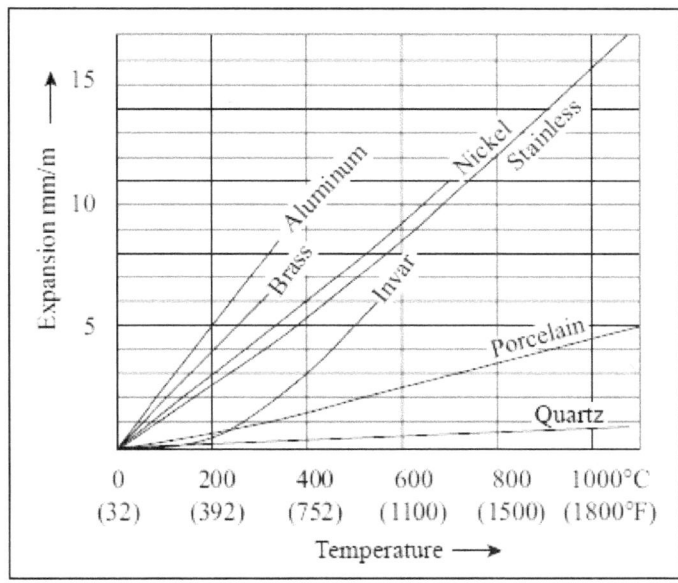

Fig. 32 Thermal expansion of materials

A favorite combination of metals is low-expanding Invar (64% Fe, 36% Ni) against high-expanding nickel-iron alloy is mostly used.

Gas Actuated Thermometer: It operates on completely different principle. The gas preferably helium gas (under pressure condition) expands at elevated temperature and deforms the measuring tube. Gas actuated thermometers generally operate reliably in the range between -200 degrees Celsius to +700 degrees Celsius. However, designer must consider the ambient temperature conditions. With ambient temperature of 23 degrees Celsius +/- 10 degrees Celsius, as per EN 13190 class 1 accuracy can be achieved. For an exact measurement the process temperature, the thermometer must be dimensioned appropriately; the stem diameter should be at least 8 mm. and the active part of the thermometer that is filled with gas, should be at least 100mm long. Fig. 34 illustrates the gas actuated thermometer principle.

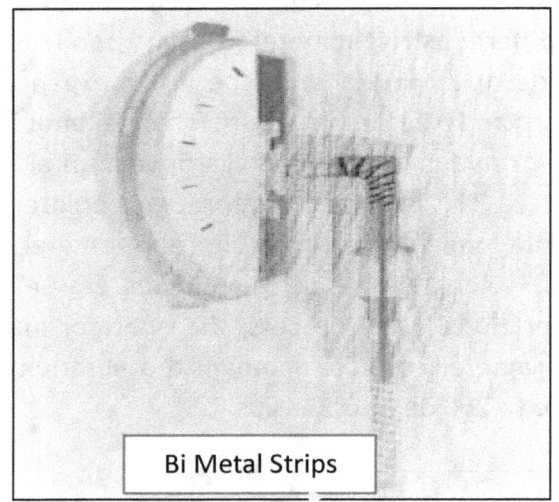

Bi Metal Strips

Fig. 33 Bimetallic Thermometer

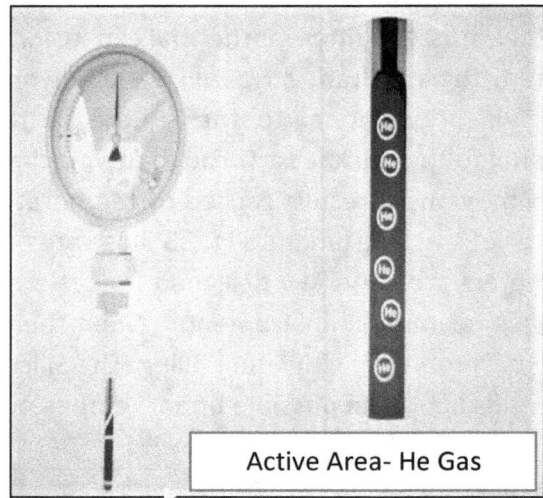

Active Area- He Gas

Fig. 34 Gas Actuated Thermometer

2.39 Selection of Temperature Thermometers

Due to simpler construction, Bimetallic Thermometer are significantly more economical than Gas Actuated Thermometers. Changes in ambient temperature have no influence on the measuring results for bimetallic thermometer. With gas actuated thermometers changes in the ambient temperature affect the measuring result. These can be partially compensated.

Gas Actuated thermometers has its own advantages. Rapid change in pressure in the closed system of gas actuated thermometer, the temperature can be displayed instantaneously with any change, whereas the spiral tube in the bimetal thermometer responds much more delay to the temperature changes. Gas actuated thermometers can be used as much as 60 meters and more away from the measuring point via a capillary. Bimetal thermometers on the other hand can only be read directly at their measuring point. Gas actuated thermometers enables a significantly greater span than bimetal thermometers. Case fillings for gas actuated thermometers are also possible at medium temperatures of over 250 degrees Celsius.

2.40 Thermometer Accuracy w.r.t ASME B40 vs. EN13190

Thermometer accuracy is graded as shown in fig.35. (as per ASME B40.200).

Accuracy is defined as per EN 13190 in terms of class, Class 1 and Class 2. Class 1 may be used with nominal sizes 63 to 160, Class 2 may be used with nominal sizes with 40 to 160.

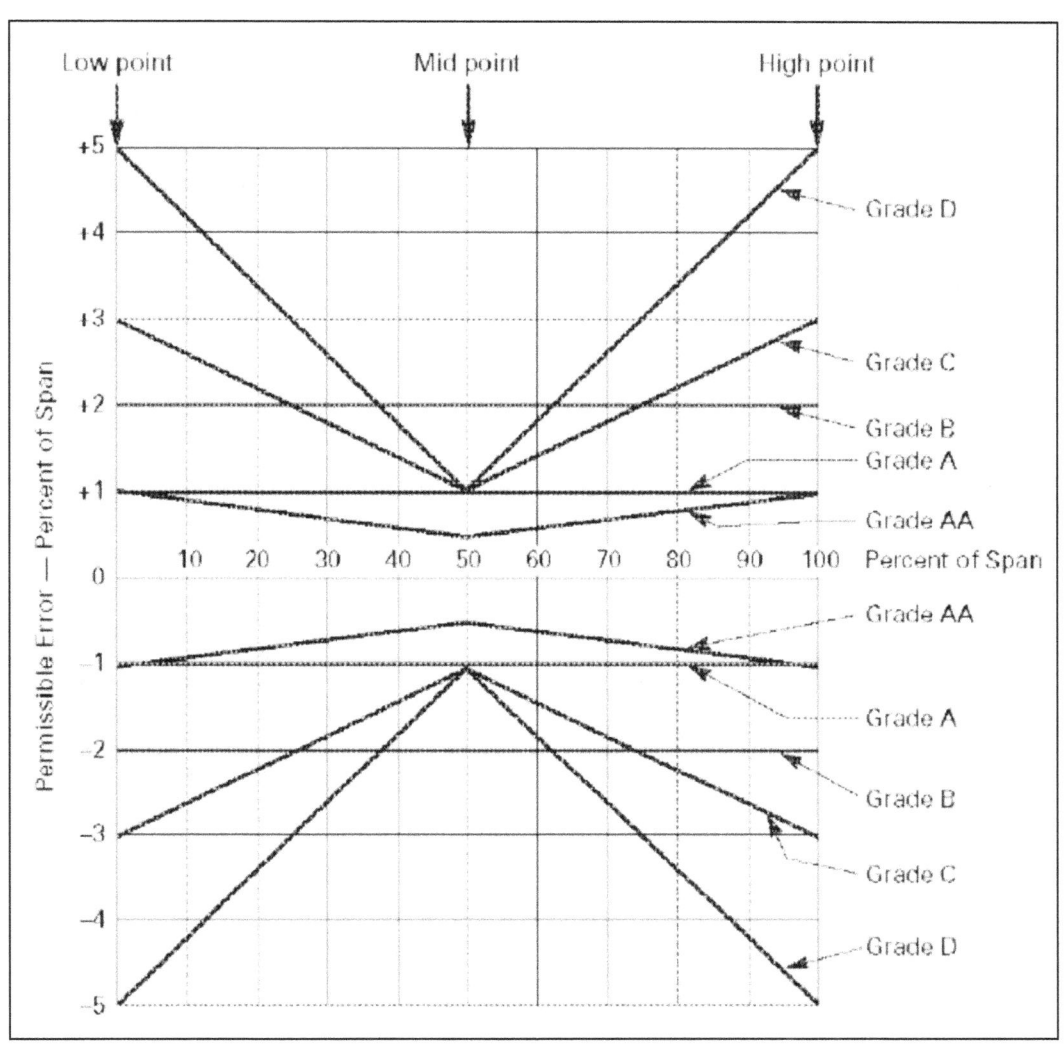

Fig.35 Accuracy Grades as per ASME B40.200

Nominal Range °C	Measuring Range °C	Limits of Error +/- °C	
		Class 1	Class 2
-20 to +40	-10 to +30	1	2
-20 to +60	-10 to +50	1	2
-20 to +120	-10 to +110	2	4
-30 to +30	-20 to +20	1	2
-30 to +50	-20 to +40	1	2
-30 to +70	-20 to +60	1	2
-40 to +40	-30 to +30	1	2
-40 to +60	-30 to +50	1	2
-100 to +60	-80 to +40	2	4
0 to 60	10 to 50	1	2
0 to 80	10 to 70	1	2
0 to 100	10 to 90	1	2

0 to 120	10 to 110	2	4
0 to 160	20 to 140	2	4
0 to 200	20 to 180	2	4
0 to 250	30 to 220	2.5	5
0 to 300	30 to 270	5	10
0 to 400	50 to 350	5	10
0 to 500	50 to 450	5	10
0 to 600	100 to 500	10	15
0 to 700	100 to 600	10	15
50 to 650	150 to 550	10	15
100 to 700	200 to 600	10	15

Table 12. Accuracy as per EN 13190

2.41 Scale Interval

The scale interval shall be chosen from 1 Deg. Cel., 2 Deg. Cel., 5 Deg. Cel. And 10 Deg. Cel.

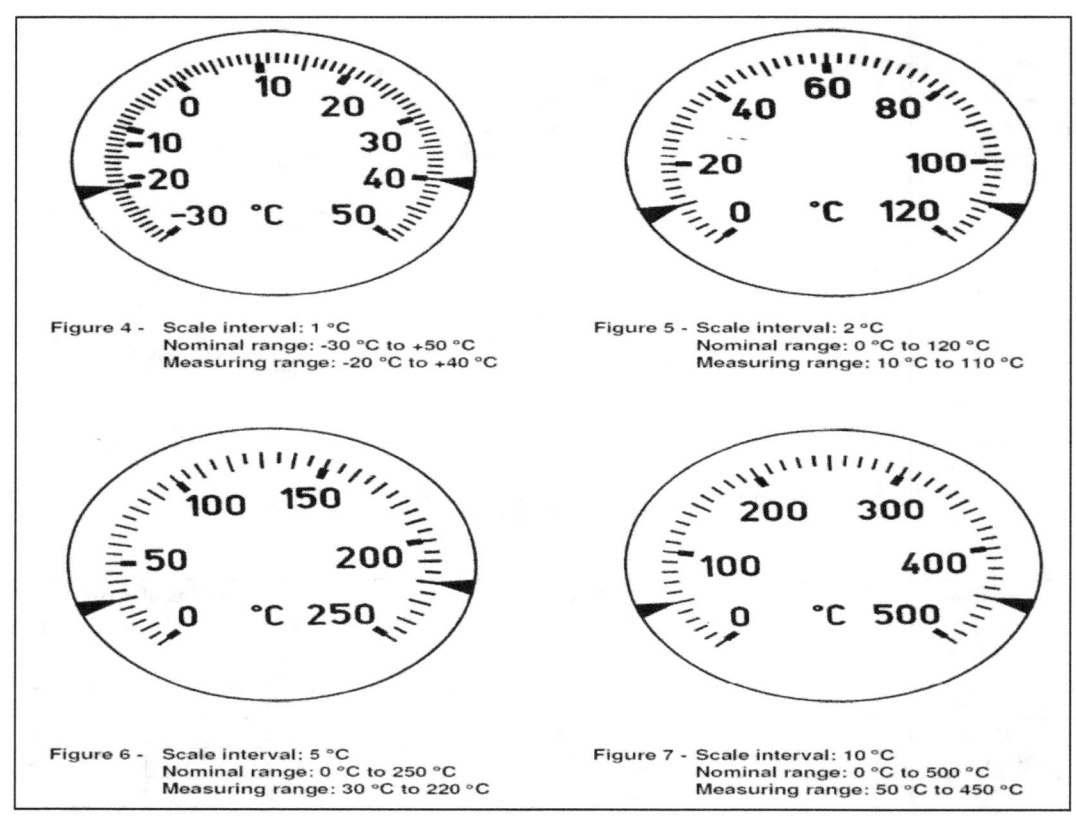

Figure 4 - Scale interval: 1 °C
Nominal range: -30 °C to +50 °C
Measuring range: -20 °C to +40 °C

Figure 5 - Scale interval: 2 °C
Nominal range: 0 °C to 120 °C
Measuring range: 10 °C to 110 °C

Figure 6 - Scale interval: 5 °C
Nominal range: 0 °C to 250 °C
Measuring range: 30 °C to 220 °C

Figure 7 - Scale interval: 10 °C
Nominal range: 0 °C to 500 °C
Measuring range: 50 °C to 450 °C

Fig.36 Nominal Range vs Measuring Range as per EN 13190

2.42 Dial Size

As per EN13190 following dial sizes are available 40, 50, 63, 80, 100, 130, 150 and 160.

As per ASME B40.200 following dial sizes are available 1, 1 ¾, 2 ,3 and 5 inches.

2.43 Filled System Thermometers

Filled system thermometers are separated into classifications of fill media as per following.

Class 1, Class 1A, Class 1B, Class 2, Class 2A, Class 2B, Class 2C, Class 2D, Class 3, Class 3B, Class 4, Class 4A, Class 4B.

Class 1: Liquid actuated, solid filled with a high volumetric expansion organic liquid (not mercury). The elastic element responds to volumetric expansion of liquid in the bulb. Common fill liquids are hydrocarbons and silicone fluids.

Class 1A: A Class 1A thermometer with full compensation to correct the effects of ambient temperature changes on the elastic element and entire length of capillary.

Class 1B: A Class 1B thermometer with full compensation to correct the effects of ambient temperature changes on the elastic element and a specified length of capillary.

Class 2: Vapor actuated, partially filled with a volatile liquid and its vapor at equilibrium.

Class 3: Gas actuated with elastic element, capillary and bulb filled with inert gas. Internal pressure changes occur as a result of temperature changes anywhere on the system.

2.44 Data Sheet for Temperature Gauge

	RESPONSIBLE ORGANIZATION	BIMETALLIC THERMOMETER w/wo THERMOWELL Device Specification		SPECIFICATION IDENTIFICATIONS
1	(ISA)		6	
2			7	Document no
3			8	Latest revision Date
4			9	Issue status
5			10	

	OPERATING PARAMETERS			THERMOWELL OR PROTECTING TUBE	
11			60		
12	Project number	Sub project no	61	Construction type	
13	Project		62	Shank style	
14	Enterprise		63	Process conn nominal size	Rating
15	Site		64	Process conn termn type	Style
16	Area	Cell Unit	65	Internal conn nom size	Style
17	Related equipment		66	Bore diameter	
18	Service		67	Outside dia at support	
19			68	Outside diameter at tip	
20	P&ID/Reference dwg number		69	Insertion length "U"	
21	Material name		70	Lagging extension lg "T"	
22	Maximum pressure		71	Thermowell/Tube material	
23	Minimum temperature		72	Sheath material-thickness	
24	Normal temperature		73		
25	Maximum temperature		74		
26	Material phase		75		
27	Maximum fluid velocity		76	PERFORMANCE CHARACTERISTICS	
28			77	Max press at design temp	At
29			78	Min working temperature	Max
30	PROCESS CONNECTION AND CASE		79	Max fluid velocity limit	At temp
31	Case type		80	Sensitive portion length	
32	Case style		81	Min ambient working temp	Max
33	Case size		82		
34	Connector conn nom size		83		
35	Connector conn type	Style	84		
36	Connection location		85	ACCESSORIES	
37	Ring style		86	Bushing nominal size	
38	Stem outside diameter	Length	87	Bushing material	
39	Stem/Bulb material		88		
40	Case material		89		
41	Ring material		90		
42	Exterior treatment-color		91		
43	Window material		92	SPECIAL REQUIREMENTS	
44	Connector material		93	Custom tag	
45			94	Reference specification	
46			95	Compliance standard	
47	SENSING ELEMENT		96	Construction code	
48	Element type		97	Calculation report	
49	Nominal accuracy grade		98		
50			99		
51			100	PHYSICAL DATA	
52			101	Estimated weight	
53	DIAL AND POINTER		102	Removal clearance	
54	Dial scale type		103	Maximum thickness	
55	Pointer adjustment		104	Max case outside dia	
56	Graduations and color		105	Mfr reference dwg	
57			106		
58			107		
59			108		

	CALIBRATIONS AND TEST			TEST		SCALE	
110							
111	TAG NO/FUNCTIONAL IDENT	MEAS OR TEST		URV		LRV	URV
112		Temperature/Scale 1					
113		Temperature/Scale 2					
114		Test pressure					
115							
116							
117							

	COMPONENT IDENTIFICATIONS		
118			
119	COMPONENT TYPE	MANUFACTURER	MODEL NUMBER
120			
121			
122			
123			
124			
125			

Rev	Date	Revision Description	By	Appv1	Appv2	Appv3	REMARKS

Form: 20T2001 Rev 0

© 2001 ISA

4. Data Sheet of Bimetallic Thermometer as per ISA

2.45 References

A) Temperature Hand Book, Tempsens Instruments
B) Publication No. 1255, Yamari Industries Limited
C) Temperature Measurement Thermocouple, ANSI MC 96.1- 1982
D) Temperature Measurement, BS 1041-3: 1989
E) Standard Specification for Platinum Resistance Thermometer ASTM E 1137
F) Specification forms for process Measurement and Control Instruments: Part 1 General Consideration ISA TR.20.00.01-2001
G) Temperature Measurement, Mark Murphy, PE, Technical Director, Fluro Corp.
H) Thermometers, Direct reading and Remote Reading ASME B40.200-2008.
I) Dial Thermometer EN 13190:2001
J) Temperature Measurement Guide to selection and use of radiation pyrometer, BS 1041-5
K) Land Pyrometer Guide

3

Flow Measurement

After completing this chapter, you should be able to:

Know about Orifice, Venturi Based Flow Meter

Know about ISO Standard for Differential Pressure Based Flow Measurement, Electromagnetic Flow meter, Different Types of Flowmeter in Industrial Application, Calibration of Flow Measuring Devices

3.1 What is Flow Rate

The quantity of fluid flowing through a cross section of a pipe per unit of time.

F= A* V

Where F= Flow rate
A= Area of Cross Section of pipe
V= Velocity of fluid flowing

Flow Rate classified as follows:

a)   Base Flow Rate: The flow rate calculated from flowing conditions to base conditions of pressure and temperature.
b)   Mass Flow Rate q_m: The rate of flow of fluid mass through a cross section of pipe.
c)   Volume Flow Rate q_v: The rate of flow of fluid volume through a cross section of pipe.

3.2 Terminology used for Flow Measurement

a)   **Density (ρ):** A Measure of Mass Per Unit of Volume (lb/ft3 or kg/M3).

b)   **Specific Gravity:** The Ratio of The Density of a Material to The Density Of Water Or Air Depending On Whether It Is A Liquid Or A Gas.

c)   **Base Pressure:** A specified reference pressure to which a fluid volume at flowing conditions is reduced for the purpose of billing and transfer accounting.

d)   **Base Temperature:** A specified reference pressure to which a fluid volume at flowing conditions is reduced for the purpose of billing and transfer accounting.

e)   **Compressible Fluid:** Fluids (Such as Gasses) Where the Volume Changes with Respect to Changes in The Pressure. These Fluids Experience Large Changes in Density Due to Changes in Pressure.

f)   **Non-Compressible Fluid:** Fluids (Generally Liquids) Which Resist Changes in Volume as The Pressure Changes. These Fluids Experience Little Change in Density Due to Pressure Changes.

g)   **Dynamic Pressure:** The increase in pressure above the static pressure that results from the complete isentropic transformation of the kinetic energy of the fluid into potential energy.

h)   **Linear:** Transmitter output is directly proportional to the flow input.

i)   **Square Root:** Flow is proportional to the square root of the measured value.

j)   **Beta Ratio (d/D):** Ratio of a differential pressure flow device bore (d) divided by internal diameter of pipe (D).

k)   **Flashing:** The formation of vapor bubbles in a liquid when the line pressure falls below the vapor pressure of the liquid.

l)   **Pressure Head:** The Pressure At A Given Point In A Liquid Measured In Terms Of The Vertical Height Of A Column Of The Liquid Needed To Produce The Same Pressure.

m) **Laminar Flow:** Flow under condition where forces due to viscosity are more significant than forces due to inertia.

n) **Reynolds Number:** A dimensionless parameter expressing the ratio between inertia and viscous forces.

$Re = Dv\rho/\mu$

Where V= fluid velocity

D= Diameter of the pipe

μ= Dynamic viscosity of the fluid

ρ= Density of the fluid

Irrespective of the pipe diameter, type of fluid or velocity, Reynolds showed that the flow is:

Laminar: Re < 2000

Turbulent: Re> 4000

o) **Turbulent Flow:** Flow under condition where forces due to inertia are more significant than forces due to viscosity.

p) **Repeatability:** The closeness of agreement between successive results obtained with the same method under same condition.

q) **Reproducibility:** The closeness of agreement between results obtained when the conditions of measurement differs.

r) **Maximum Flow Rate:** The highest flow rate at which meter can work satisfactorily as specified by the manufacturer.

s) **Minimum Flow Rate:** The lowest flow rate at which meter can work satisfactorily as specified by the manufacturer.

t) **Rangeability:** is a measure of how much the flow range of an instrument can be adjusted and is defined as the ratio of the maximum flow range (maximum span) and the minimum span.

u) **Turn- down Ratio:** is the ratio of the maximum flow rate to the minimum flow rate for a measuring range that is within a stated accuracy. For eg. The measuring range of a magnetic flow meter might be 0.3 m/s to 12 m/s within an accuracy of 0.3%. This would thus be stated as having a 40:1 turndown ratio (0.3%). In addition, the measuring range might extend from 0.2 m/s to 12m/s with an accuracy of 0.5%, in this case turndown ratio is 60:1(0.5%).

v) **Discharge Coefficient:** Ratio of actual flow rate to theoretical flow rate.

Cd= Qtrue/QTheoretical ;
Cd= Sq.rt [(P1-P2)-PL)/(P1-P2)]
Cd= Sq.rt[1-PL/(P1-P2)]

Where P1- Upstream Pressure before Flow element
P2- Downstream Pressure after Flow element
PL- Pressure loss

w) **Expansion Factor ε:** Dimensionless coefficient given by the formula:

$$\epsilon = \frac{q_m}{\frac{\pi}{4}\alpha d^2 \sqrt{2\Delta p \rho_1}}$$

x) **Flow Coefficient α:** Dimensionless Coefficient given in the case of a flow of fluid considered as not compressible by the formula:

$$\alpha = \frac{q_m}{\frac{\pi}{4} d^2 \sqrt{2\Delta p \rho_1}}$$

Where qm – mass flow rate of the fluid
d- bore dia of the orifice plate or throat dia. of venturi tube
Δp – differential pressure
ρ 1- density of the fluid at upstream

y) **Piezometer Ring:** A pressure equalizing linking together two or more pressure taps installed on one cross section and to which secondary device can be connected.

z) **Velocity Profile:** Fluid will always flow faster along the center of the pipe than along the pipe wall, due to friction along the pipe wall. This is known as velocity profile or flow profile.

aa) **Static weighing:** The method in which the net mass of liquid collected is deduced from tare and gross weighing made respectively before and after the liquid has been diverted for measured time interval into the weighing tank.

bb) **Dynamic weighing:** The method in which the net mass of liquid collected is deduced from weighing made while fluid flow is being delivered into the weighing tank. (A diverter is not required with this method.

cc) **Diverter:** A device which diverts the flow either to the weighing tank or to its by-pass without changing the flow-rate during the measurement interval.

dd) **Flow stabilizer:** A structure forming part of the measuring system, ensuring a stable flow-rate in the conduit being supplied with liquid; for example, a constant level head tank, the level of liquid in which is controlled by a weir of sufficient length.

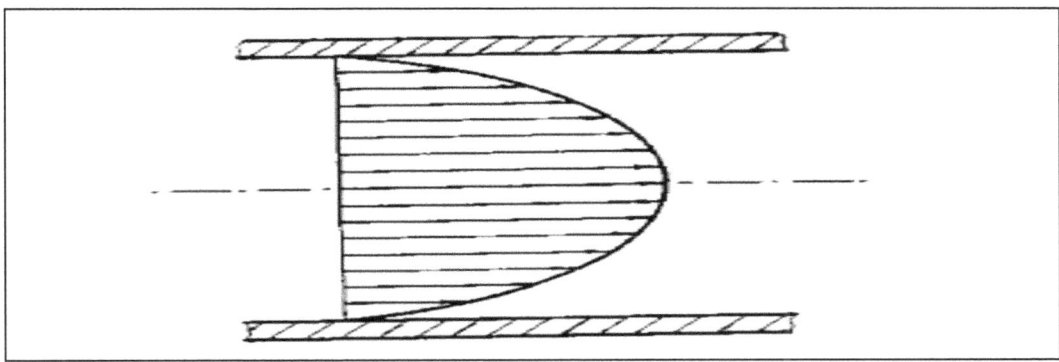

Fig.1 Flow Profile for Laminar Flow

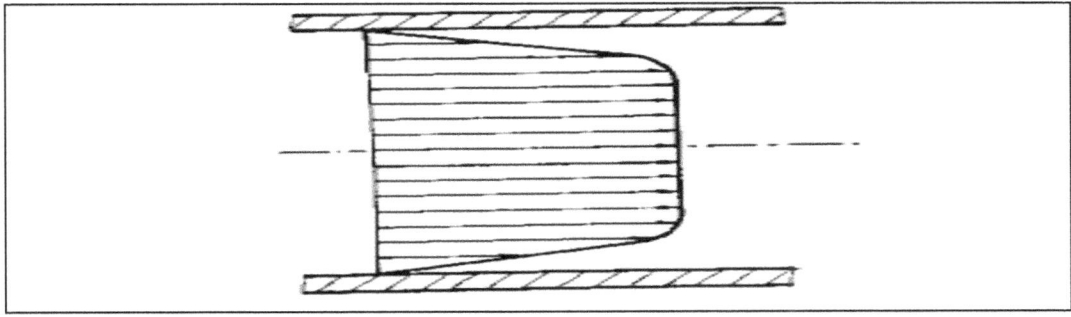

Fig.2 Flow Profile for Turblent Flow

ee) **Meter Factor:** The meter factor is similar to Discharge Coefficient but applies to meters which measures total volume.

ff) **K factor:** The K factor is used to describe the performance of meters the output of which is in the form of electrical pulses.

K= n/ V_t

Where n = No. of pulses per unit time.
V_t= Volume passed in that time.

3.3 Type of Flowmeters

Different type of flowmeters used in industry based on different application: Most Commonly used flowmeter used in industry.

a)   Differential Pressure Based Flowmeter
b)   Electromagnetic Flow meter
c)   Mass Flowmeter
d)   Ultrasonic Flow meter
e)   Variable Flowmeter
f)   Positive Displacement Flowmeter
g)   Vortex Flowmeter

3.4 Electromagnetic Flow Meter

Flowmeter which creates a magnetic field perpendicular to the flow, so enabling the flow-rate
to be deduced from the induced electromotive force (e.m.f.) produced by the motion of a conducting liquid in the magnetic field. The electromagnetic flowmeter consists of a primary device and one or more secondary devices.

Primary Device:

Device containing the following elements:
a)   An electrically insulated meter tube through which the conductive to be metered flow.
b)   One or more pairs of electrode, diametrically opposed, across which the signal generated in the liquid is measured.
c)   An electromagnet for producing a magnetic field in the meter tube.

The primary device produces a signal proportional to the flow rate.

Secondary Device:

Equipment which contains the circuitry which extract the flow signal from the electrode signal and converts it to a standard output signal directly proportional to flow rate. The equipment may be mounted on the primary device.

Calibration Factor of the primary device:

A number which enables the flow signal to be related to the volume flow rate under defined reference conditions for a given value of the reference signal.

Working Principle:

When liquid moves in a magnetic field, voltages (e.m.f.s) are generated in accordance with Faraday's law (see Figure 3). If the field is perpendicular to an electrically-insulated pipe which contains the moving liquid and if the electrical conductivity of the liquid is not too low, a voltage may be measured between two electrodes on the wall of the pipe. This voltage is proportional to the magnetic flux density, the average velocity of the liquid and the distance between the electrodes. Thus, the velocity and hence the flow-rate of the liquid may be measured.

Basic Equation:

In accordance with Faraday's Law of induction, the strength of the induced voltages is given by the simplified expression as,

$$V = B\, D\, U$$

The volume flow meter in the case of the circular pipe is:

$$Q_v = (\pi D^2 /4) * U \quad - (1)$$

Which combines with equation (1) gives

$$Q_v = (\pi D^2 /4) * V/B\, D \quad - (2)$$

$$Q_v = K\, (V/B) - (3)$$

Equation 3 indicates that, in a carefully designed flowmeter, if all other parameters are kept constant, the induced voltage is linearly proportional only to the mean value of the liquid flow.

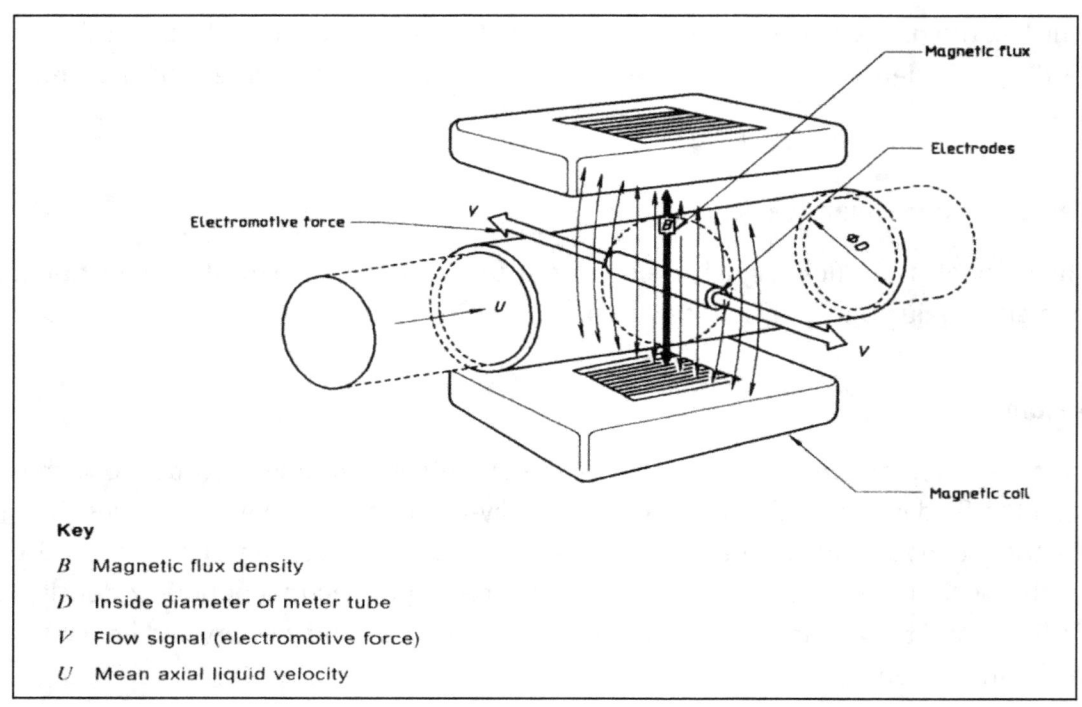

Key

B Magnetic flux density

D Inside diameter of meter tube

V Flow signal (electromotive force)

U Mean axial liquid velocity

Fig.3 Principle of Electromagnetic Flowmeter

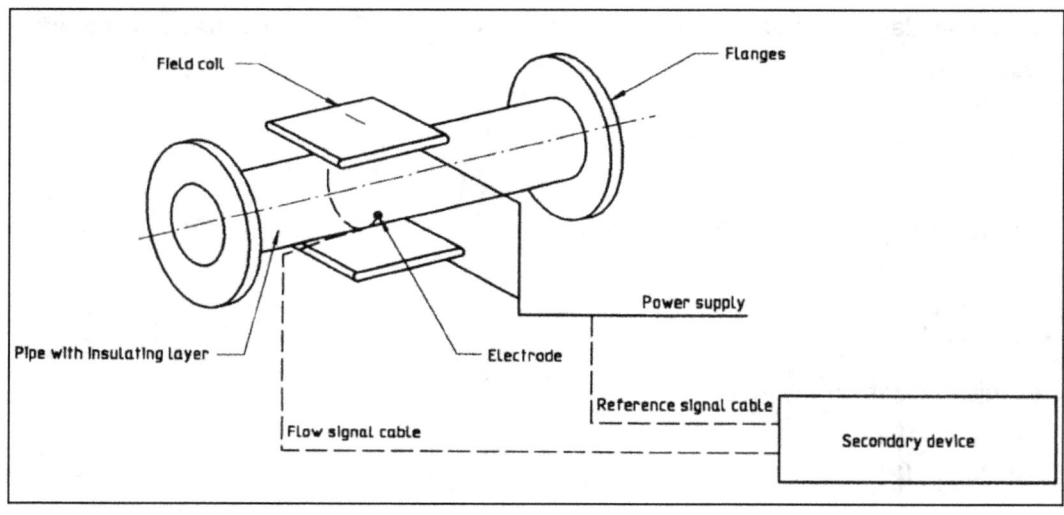

Fig.4 Elements of an Industrial Electromagnetic Flowmeter

3.5 Comparison Details of Coil Excitation

Method	AC POWER	PULSED DC	DUAL FREQUENCY
Waveform	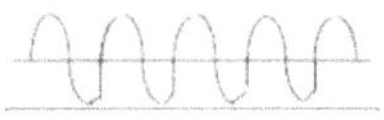		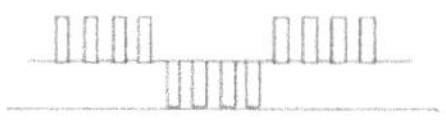
Frequency	**High frequency** **50 Hz**	**Low frequency** **5-10 Hz**	**H& L frequency** **6.25 & 75 Hz**
Zero point	Unstable	Stable	Stable
Response	Fast	Slow	Fast
Slurry noise (Flow noise)	Strong	Weak	Strong

Fig.5 Comparsion Study of Coil Excitation Method for generating Magnetic Flux in tube

Pulsed DC Excitation:

In measuring system with applied pulsed d.c excitation, the magnetic field polarity is alternately reversed. During each magnetic field polarity cycle, the electrode voltage is measured once the magnetic field is considered to be constant. This period is called measuring window.

In a pulsed d.c. system, under ideal or reference conditions, the peak-to-peak value of the electrode signals, $(Vp + Vn)$, is proportional to the flow velocity in the pipeline and Vp is also equal to Vn [see Figure 6)], where Vp = positive voltage and Vn = negative voltage.

In a practical situation, if the zero or "no-flow" signal is offset in the positive direction by an amount Ve then the positive signal is $(Vp + Ve)$ and the negative signal is $(Vn - Ve)$ [Figure 6)]. Hence the overall value of the electrode signal is $(Vp + Vn)$ and the offset zero is eliminated. The same applies if the offset is in the negative direction. The system thus eliminates zero errors automatically at all times and zero adjustment is not usually required, either at start-up/commissioning or at any time during subsequent operation.

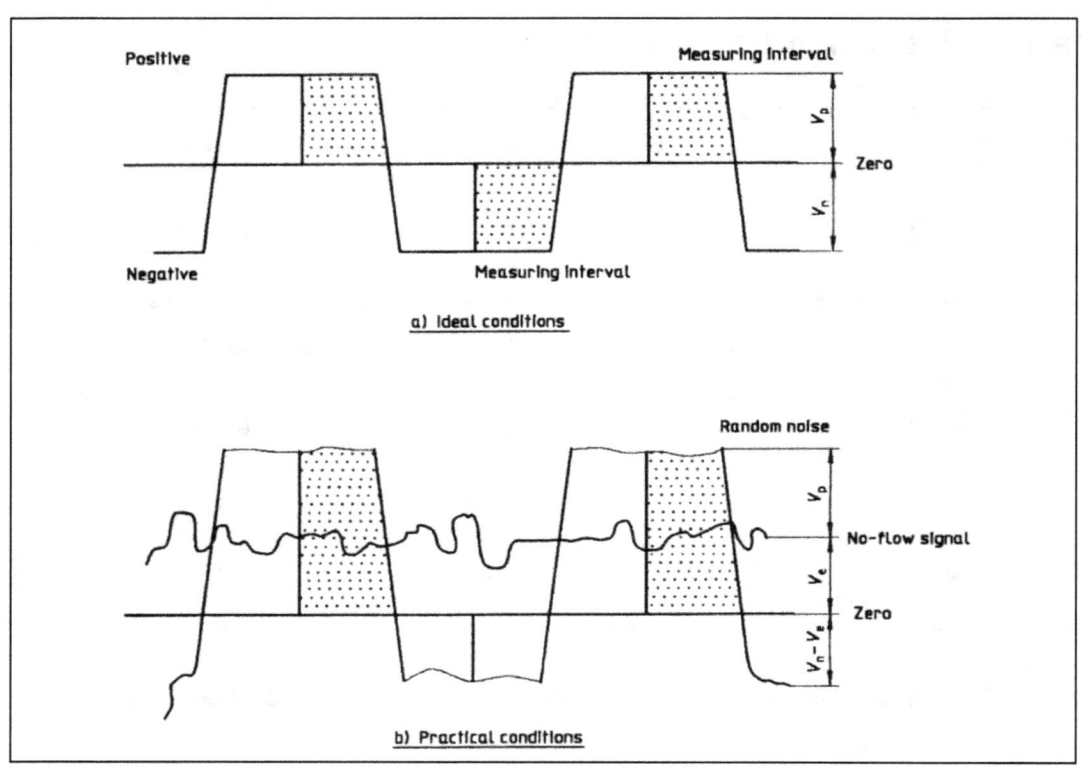

Fig.6 Pulsed DC Excitation Method

The pulsed D.C system is more stable than A.C system in terms of zero stability while A.C system is more superior in applications where process generated noise is present.

Process generated noise is the phenomenon associated with pulse DC system that appears in some slurry processes. Following solution have been developed to solve noise problems.

a) Doubling the coil frequency reduced the noise by half.
b) A noise reduction function was developed for the microprocessor based transmitter.

Why Dual Frequency Excitation is good for Mag flow meter working performance?

Dual frequency method successfully combines the advantages of both high & low excitation methods and provides overwhelming performance. It suppresses process-generated noise securing zero stability without sacrificing response time.

Low signal can secure excellent zero stability by low frequency sampling. LPF works as damping and removes process-generated noise from the low signal. High frequency excitation can keep the high signal stable even in noisy process. For fast response, the noise-free high signal is passed through HPF and added to the low signal filtered through LPF. Therefore, dual frequency method can secure excellent zero stability and output noise-free flow signals with fast response.

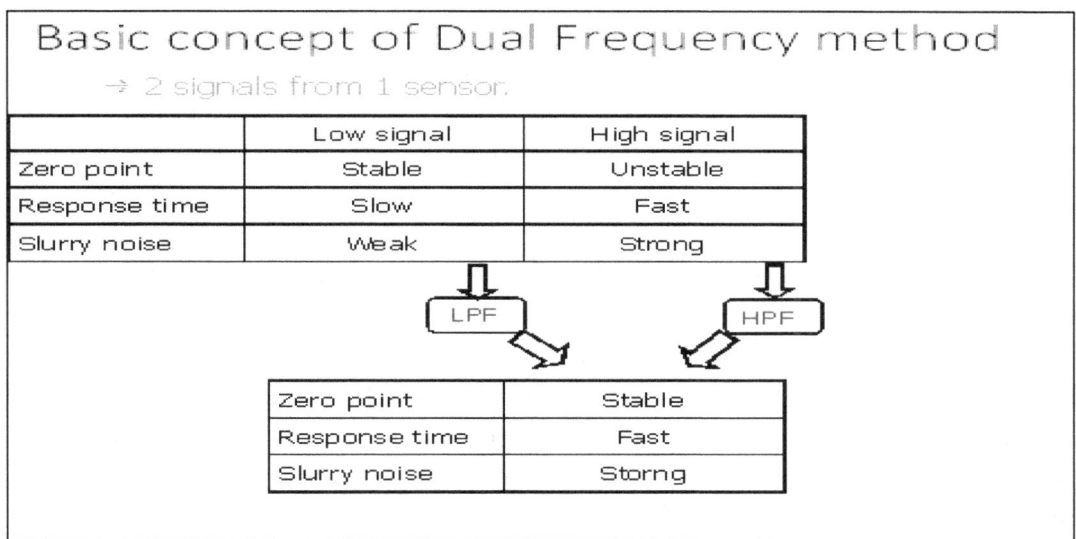

Fig.7 Dual Frequency Method

3.6 Effect of the Liquid Conductivity

If the electrical conductivity of the liquid is uniform in the measuring section of the meter, the electric field distribution is independent of the liquid conductivity and therefore the meter output is generally independent of the liquid conductivity. Minimum operational conductivity requirements should be obtained from the manufacturers. The internal impedance of the primary device obviously depends upon the liquid conductivity, and very large changes in this impedance may produce errors in the output signal. If the conductivity is not uniform throughout the meter, errors may also occur.

3.7 Reynolds Number Effect

In industrial electromagnetic flowmeters, the effect of Reynolds number is usually so small that for practical purposes it can be ignored.

3.8 Layout of Electromagnetic Flow Meter

There is no theoretical restriction on the attitude at which a primary device may be mounted, provided the pipe remains full at all times. Locations close to electrical equipment which may interfere with the flow measurement signal, or locations where currents may be induced in the primary device, should be avoided.

Effect of Layout on velocity distribution:

Ideally, the magnetic field should be so arranged that the calibration factor is always the same, irrespective of the flow pattern. Though this can be done in flowmeters with special electrode arrangements, it cannot be achieved if small electrodes are used. In practice, when a flow velocity profile which is significantly different from that in the original calibration is presented to the electrode plane, an electromagnetic flowmeter may exhibit a shift in calibration. The arrangement of pipe fittings upstream of the primary device is one of the factors which can contribute to the creation of a particular velocity profile.

Precise data on the effects of flow disturbances is not always available, but for most electromagnetic flowmeters it is recommended that any source of flow disturbance, such as a bend, should be at least ten pipe diameters upstream of the electrode plane if the performance is not to be altered by more than 1 %. When the distance is unavoidably less than this, the manufacturer's advice should be sought.

Swirling flow can also alter the calibration factor because, although flow components perpendicular to the pipe axis cannot contribute to the flow-rate, they may contribute to the signal. Furthermore, the amount and distribution of swirl arising from various upstream pipe configurations, such as several bends in different planes, is difficult to predict from the geometry of the pipework. When swirling flow is suspected, it is good practice insert a swirl reducer upstream of the primary device; some types of swirl reducers are described in ISO 7194.

Full pipe requirement:

The primary device shall be mounted in such a position that it will be completely filled with the liquid being metered, otherwise the measurement will not be within the manufacturer's stated accuracy. If the liquid does not make full contact with electrodes the high impedance prevents the current flow hence measurements cannot be taken. Also, If the pipe is not full, even if contact is maintained between the liquid and electrodes, the empty portions of the pipe will lead to miscalculated flow rates.

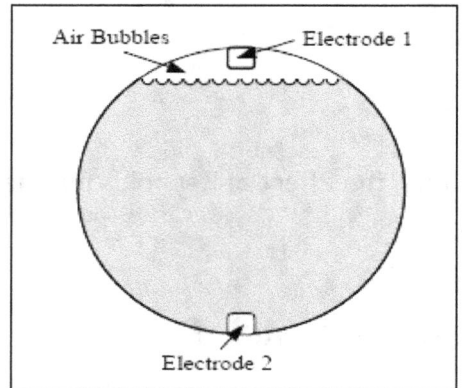

Fig.8 The pipes of electromagnetic flowmeters must be full of liquid at all times for accurate measurements

Installation Guidelines for Electromagnetic Flowmeter:

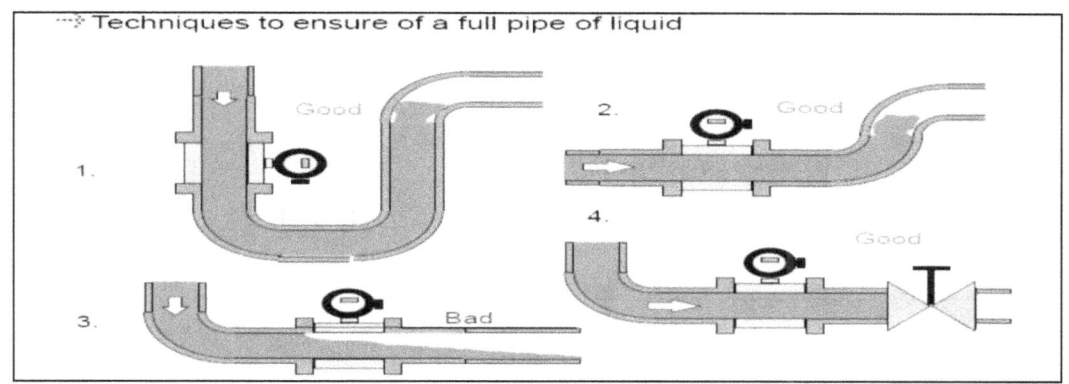

Fig.9 Installation Guidelines for ElectroMagnetic Flowmeter

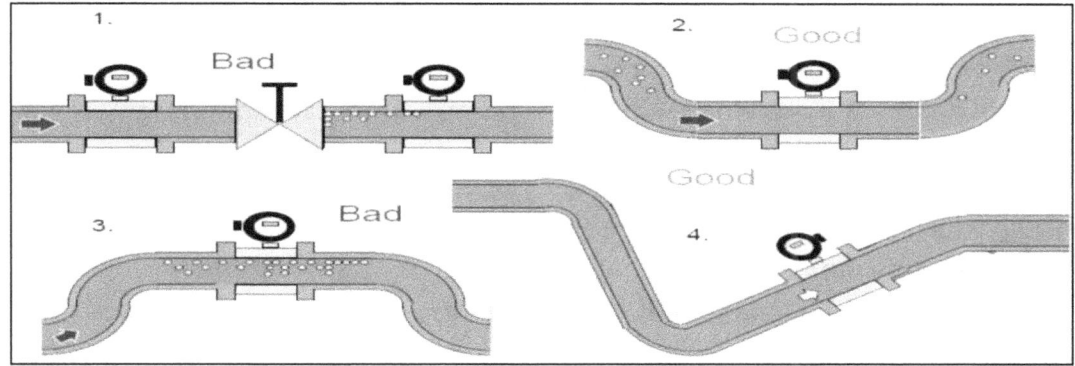

Fig.10 Installation Guidelines for ElectroMagnetic Flowmeter when liquid contains water vapour

Straight inlet run minimum of 5 *DN and outlet run minimum of 3*DN.

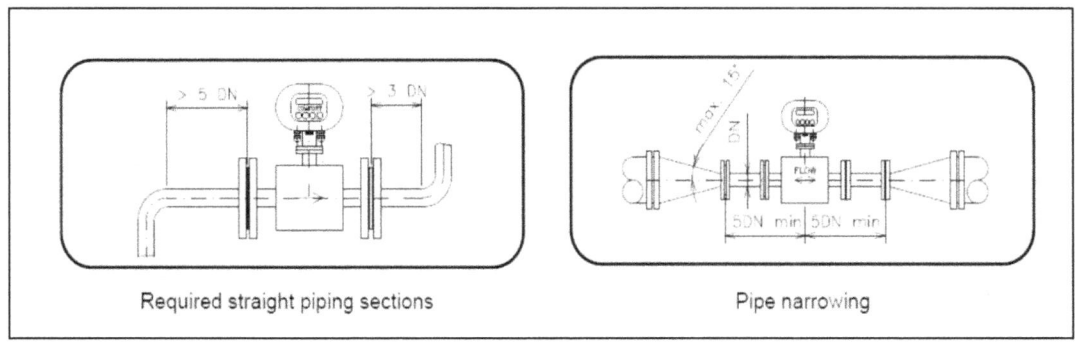

Fig.11 Straight run require for ElectroMagnetic Flowmeter Installation

In the case where the pipeline nominal size is bigger than nominal size of flow meter, it is necessary to use conical reduction with the maximum slope 15° as shown in fig. 11.

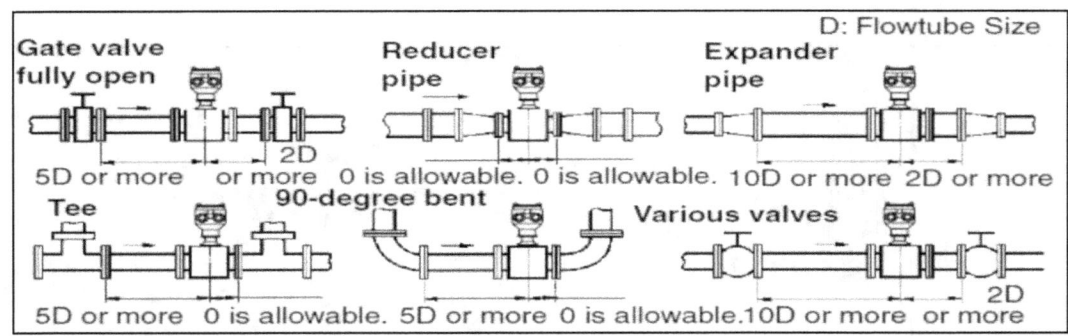

Fig.12 Installation Guidelines as per JIS B7554

3.9 Selection of Meter Sizes

Generally, the size of the primary head is matched to the nominal diameter of the pipeline. Experience has also shown that the optimum flow velocity should be 2-3 m/s. The exact flow velocity can be determined from table 1 for v=1 m/s.

For eg:

Meter Size= DN 80

Desired Measure Range= 55 m3/hr

From the fig.12 at v= 1m/s the flow rate should be 18.10 m3/hr for DN 80.

V= (55 m3/hr * 1m/s) / 18.10 m3/hr

V= 3.04 m/s

Meter Size	Full Scale Range Q_{max} in m3/hr	
DN (MM)	Inch	V= 1m/sec
10	3/8	0.282'7
15	1/2	0.6362
20	3/4	1.131
25	1	1.767
32	-	2.896
40	1 ½	4.524
50	2	7.069
65	-	11.95
80	3	18.10
100	4	28.27
125	-	44.18
150	6	63.62
200	8	113.1
250	10	176.7
300	12	254.5
350	14	346.4
400	16	452.4
500	20	706.9
600	24	1018

700	28	1385
800	32	1810
900	36	2290
1000	40	2827
1200	48	4072
1400	56	5542
1600	64	7238
1800	72	9161
2000	80	11310
2200	88	13685
2400	96	16286
2600	104	19113
2800	112	22167
3000	120	25447

Table 1. Meter size vs. Flow Rate

Overall Length:

For each meter size designation, there is a corresponding fixed overall length L. The length L includes lining if it covers the flange face but excludes accessories such as gaskets, grounding and protection rings.

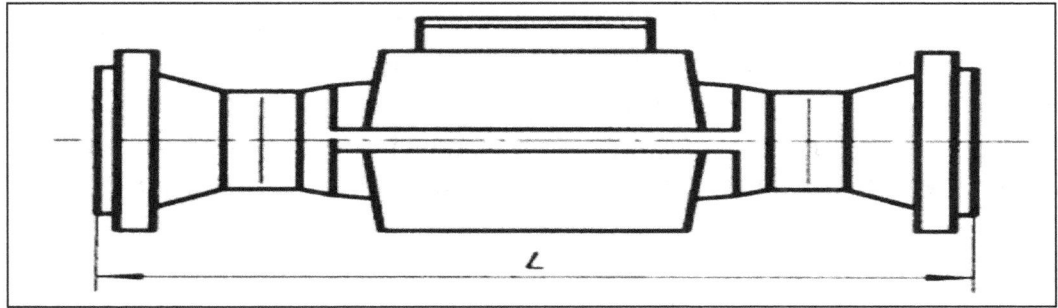

Fig.13 Definition of L

Meter Size	Meter Overall Length	
DN	L (mm)	Tolerance (mm)
15		
20		
25		
32	200	
40		
50		0
65		3
80		
100	250	

125		
150	300	
200	350	
250	450	
300	500	0
350	550	5
400	600	

Table 2. Length Tolerance as per ISO 13359

3.10 Liner

The signal voltage detected by the two sensing electrodes is not electrically short circuited through the tube wall. Consequently, the metering tube must be lined with an insulating material. Such materials have to be selected according to the application.

1) **Elastomers:**
➢ **Hard rubber:** Hard rubber is generally suitable for use within the temperature range 0 °C to 90 °C. It has excellent abrasion resistance against small particles and good chemical resistance, particularly to leaching agents, acid and alkalis.
➢ **Abrasion-resistant rubbers (natural):** Abrasion-resistant natural rubbers are generally suitable for use within the temperature range – 20 °C to + 70 °C. They exhibit excellent
wear resistance and good chemical resistance.
➢ **Neoprene:** Neoprene is generally suitable for use within the temperature range 0 °C to 100 °C. It has good chemical and wear resistance properties, particularly in the presence of oil and greases.

2) **Plastics:**
➢ **Polytetrafluoroethylene (PTFE):** Usually as an extruded sleeve form not bonded to the meter tube, PTFE is generally suitable for use within the temperature range – 50 °C to + 200 °C. It has excellent wear resistance against small particles, and is chemically inert. It may collapse when subjected to sub-atmospheric pressures.
➢ **PFA:** Excellent Chemical Resistance, Excellent Abrasion Resistance as compared to PTFE.

3) **Ceramics:**
➢ This construction material requires no lining, exhibits high form and measuring stability under pressure and temperature variations, and possesses excellent abrasion resistance. Additionally, high chemical resistance to acids and alkaline solutions is characteristic of high-purity Al2O3 ceramics. The service temperature range is from – 60 °C to + 250 °C with full vacuum resistance.

	Abrasion resistance	High temperature (>100 °C)	Chemical resistance	Hygienic design	Drinking water	Vacuum resistance	...
Elastomers[a]							
Hard rubber (ebonite)	B	D	B	D	B	B	
Abrasion-resistant rubbers (natural, soft)	A	D	B	D	D	B	
Neoprene/Chloroprene (CR)	B	D	B	D	D	B	
Polyurethane (PU)	A	D	D	D	B	B	
Plastics							
Polytetrafluoroethylene (PTFE)	C	A	A	B	A	D	
Perfluoroalkoxy-Ploymere (PFA)	C	A	A	A	A	B	
A = Suitable							
B = Usually suitable							
C = Limited suitability							
D = Not suitable							
n.a. = No data available							
[a] All rubber-based materials are attacked by high concentrations of free halogens, aromatics and halogenated hydrocarbons and high concentration of oxidizing chemicals.							

Table 3. Selection of Meter Tube Lining Materials Based on ISO 20456:2017

3.10 Electrode

Following electrode materials used in electromagnetic flowmeter based on application:

1)   **For non-corrosive liquids:** Stainless steel is generally used.

2)   **For corrosive liquids:** The following may be suitable, depending on the chemical properties of the liquid to be metered:

➤   stainless steel
➤   some nickel-based alloys;
➤   platinum
➤   platinum/iridium;
➤   tantalum;

3.11 Accuracy of Electromagnetic Flowmeter

Fig. 14 shows, accuracy vs velocity data. Here A1-10 represents the products available in industry. The figure 14 covers accuracy in terms of % flow rate and % full scale value w.r.t velocity of the fluid.

Velocity m/s	A1	A2	A3	A4	A5	A6	A7	A8	A9	A10
0.1	0.5	0.5	1.25	1.15	1.5	2.2	1.3	1.2	2.4	2.2
0.15	0.35	0.31	0.92	0.82	1.17	1.53	0.97	0.87	1.73	1.53
0.2	0.35	0.28	0.75	0.65	1.00	1.20	0.80	0.70	1.40	1.20
0.3	0.35	0.25	0.58	0.48	0.83	0.87	0.63	0.53	1.07	0.87
0.4	0.35	0.23	0.50	0.40	0.75	0.70	0.55	0.45	0.90	0.70
0.5	0.35	0.22	0.45	0.35	0.70	0.60	0.50	0.40	0.80	0.60
0.75	0.35	0.21	0.38	0.28	0.63	0.47	0.43	0.33	0.67	0.47
1	0.35	0.20	0.35	0.25	0.60	0.40	0.40	0.30	0.60	0.40
1.25	0.35	0.20	0.33	0.23	0.58	0.36	0.38	0.28	0.56	0.36
1.5	0.35	0.20	0.32	0.22	0.57	0.33	0.37	0.27	0.53	0.33
1.6	0.35	0.20	0.31	0.21	0.56	0.33	0.36	0.26	0.53	0.33
1.8	0.35	0.20	0.31	0.21	0.56	0.31	0.36	0.26	0.51	0.31
2	0.35	0.20	0.30	0.20	0.55	0.30	0.35	0.25	0.50	0.30
2.25	0.35	0.20	0.32	0.19	0.54	0.29	0.34	0.24	0.49	0.29
2.5	0.35	0.20	0.31	0.19	0.54	0.28	0.34	0.24	0.48	0.28
3	0.35	0.20	0.30	0.18	0.53	0.27	0.33	0.23	0.47	0.27
4	0.35	0.20	0.29	0.18	0.53	0.25	0.33	0.23	0.45	0.25
5	0.35	0.20	0.28	0.18	0.52	0.24	0.32	0.22	0.44	0.24

Fig.14 Velocity vs Accuracy

Accuracy may be stated in two different ways and specifications often include both. The first is a percent of the meter's full scale or maximum reading. The second is a percent of the meter's flow readng.

Accuracy statement examples	
Type of accuracy specification	**Typical manufacturer's accuracy statements**
Percent of reading	±X % of reading
Percent of full scale	±X % of full scale (or F.S.)
Combination	±X % of reading ±Y mm/s
	±X % of reading ±Y% of full scale

Fig.15 Accuracy Specifications

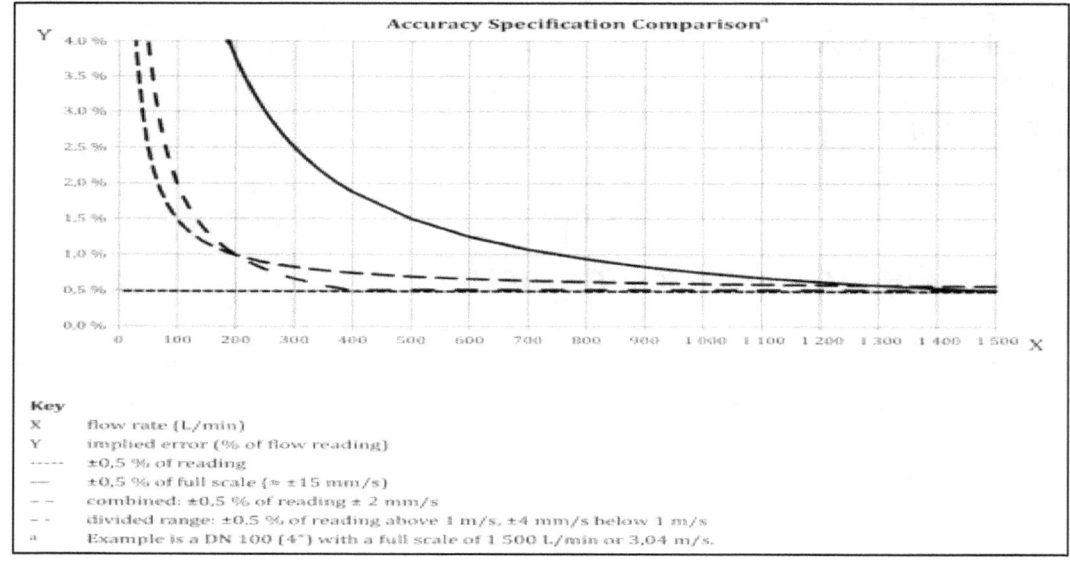

Key
X   flow rate (L/min)
Y   implied error (% of flow reading)
-----   ±0,5 % of reading
——   ±0,5 % of full scale (≈ ±15 mm/s)
– –   combined: ±0,5 % of reading ± 2 mm/s
– –   divided range: ±0,5 % of reading above 1 m/s, ±4 mm/s below 1 m/s
[a]   Example is a DN 100 (4") with a full scale of 1 500 L/min or 3,04 m/s.

Fig.16 Implied error as a function of flowrate

Example 1: Flow rate is 1 500 L/min, 100 % of full scale (3,18 m/s)		Expected error (L/min)	Range of expected readings		Expected error (% of reading)
Accuracy statement	Error calculation		min	max	
±0,5 % of reading	1 500·0,005 =	7,50	1 492,5	1 507,5	0,50 %
±0,5 % of full scale (≈ ±15 mm/s)	1 500·0,005 =	7,50	1 492,5	1 507,5	0,50 %
Combined: ±0,5 % of reading ±2 mm/s	1 500·0,005 + 0,94[a] =	8,44	1 491,6	1 508,4	0,56 %

Example 2: Flow rate is 750 L/min, 50 % of full scale (1,59 m/s)		Expected error (L/min)	Range of expected readings		Expected error (% of reading)
Accuracy statement	Error calculation		min	max	
±0,5 % of reading	750·0,005 =	3,75	746,3	753,8	0,50%
±0,5 % of full scale (≈ ±15 mm/s)	1 500·0,005 =	7,50	742,5	757,5	1,00%
Combined: ±0,5 % of reading ±2 mm/s	750·0,005 + 0,94[a] =	4,69	745,3	754,7	0,63%

Example 3: Flow rate is 150 L/min, 10 % of full scale (0,318 m/s)		Expected error (L/min)	Range of expected readings		Expected error (% of reading)
Accuracy statement	Error calculation		min	max	
±0,5 % of reading	150·0,005 =	0,75	149,3	150,8	0,50%
±0,5 % of full scale (≈ ±15 mm/s)	1 500·0,005 =	7,50	142,5	157,5	5,00%
Combined: ±0,5 % of reading ± 2 mm/s	1 500·0,005 + 0,94[a] =	1,69	148,3	151,7	1,13%

[a] To calculate the volumetric flow error equivalent to the specified velocity error, multiply the velocity error by the cross sectional area of the pipe, and then convert the result to the desired flow units. For example, 7 853,9 mm^2 × 2 mm/s = 15 707 mm^3/s = 0,94 L/min.

Fig.17 Implied error as a function of flowrate

3.12 Empty Pipe Detection

In 3.8 it is mentioned the primary condition for installation of electromagnetic flow meter is that pipe should be always filled with fluid. One of the features of electromagnetic flowmeter is "Empty Pipe Detection". Empty pipe detection is not only used to indicate that the volume reading is incorrect. For eg., in a two-line standby system, one line handles the process and the other is used for standby. Since the standby line does not contain any of the process medium, the flowmeter sensing electrodes are "open circuit" and the amplifier output signal will be subject to random drifting. The resultant falsely generated inputs to any process controllers., recorders etc connected to the system will give rise to false status alarms. Here the "Empty Pipe Detection" system is used to freeze the signal to reference zero.

Another application for "Empty Pipe Detection" is to prevent damage to field coils. Magnetic Flowmeter based on a "pulsed d.c", generates relatively low power to the field coils- typically between 14 to 20 VA. This is usually of little concern regarding heat generation in the field coils. However, flow sensors based on an "a.c generated" magnetic fields, consume power in excess of a few hundred VA. In order to absorb the heat generated in the field coils, a medium is required in the pipe to keep the temperature well within the capability of the field coil insulation. An empty pipe will cause overheating and permanent damage to the field coils and, consequently this type of flowmeter require an "Empty Pipe Detection" system to shut down the power to protect the field coils.

3.13 Grounding and Earthing

To ensure measuring accuracy and avoid corrosion damage to the electrodes of electromagnetic flowmeter, the sensor and the process medium must be at the same electrical potential. This is achieved the primary head as well as the pipeline by any one or more of a number of methods including: earthing straps, ground rings, lining protectors and earthing electrodes.

Improper earthing is one of the most frequent causes of problems in installation. If the earthing is not symmetrical, earth loop currents give rise to interference voltages- producing zero point shifts. Fig. 18 to Fig. 23 shows the most effective earthing configuration:

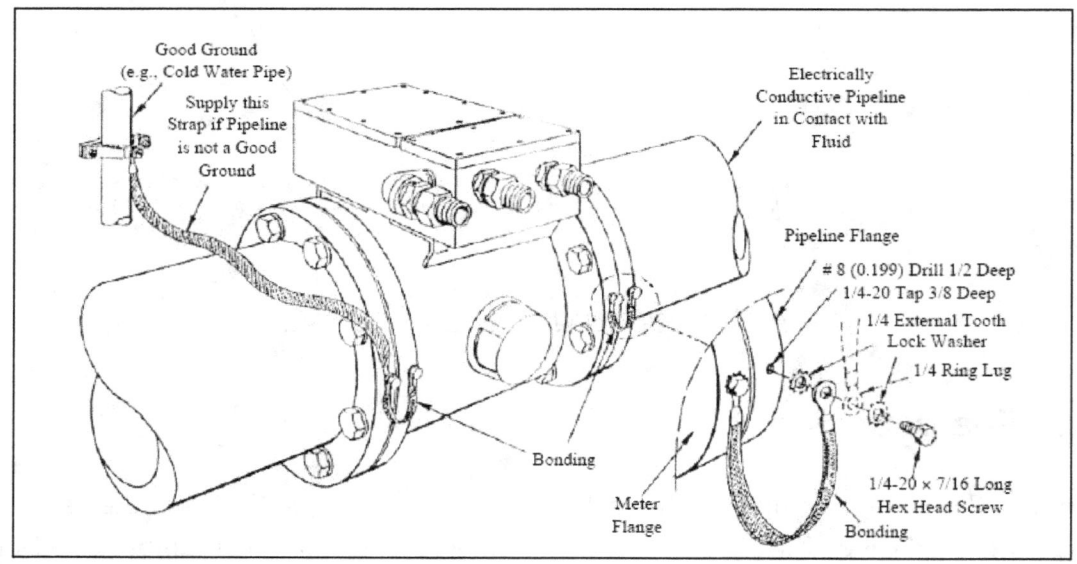

Fig.18 Typical bonding and grounding procedure

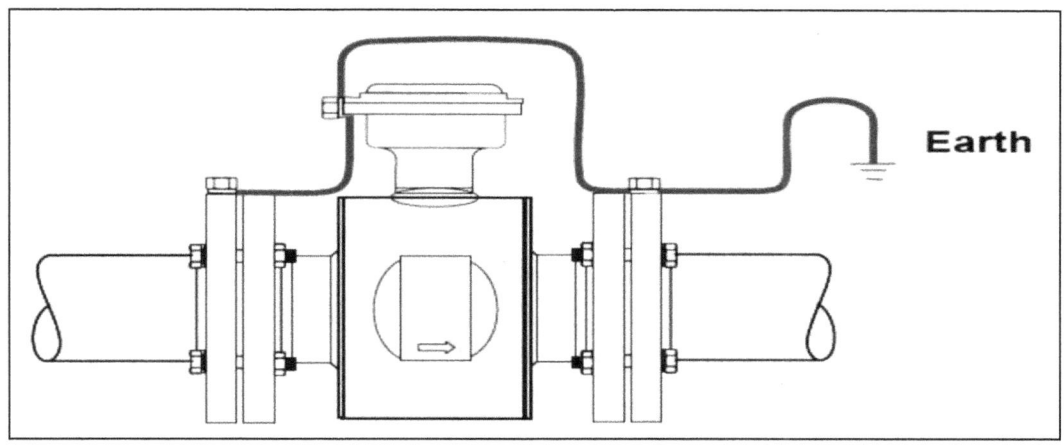

Fig.19 Earthing for conductive unlined pipe and conductive pipe with earthing electrode

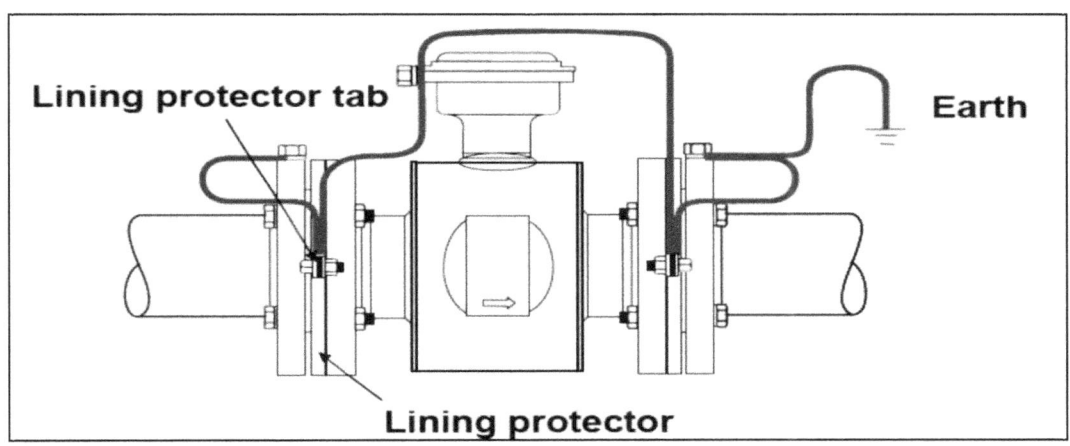

Fig.20 Earthing for conductive unlined and lined pipe with lining protectors

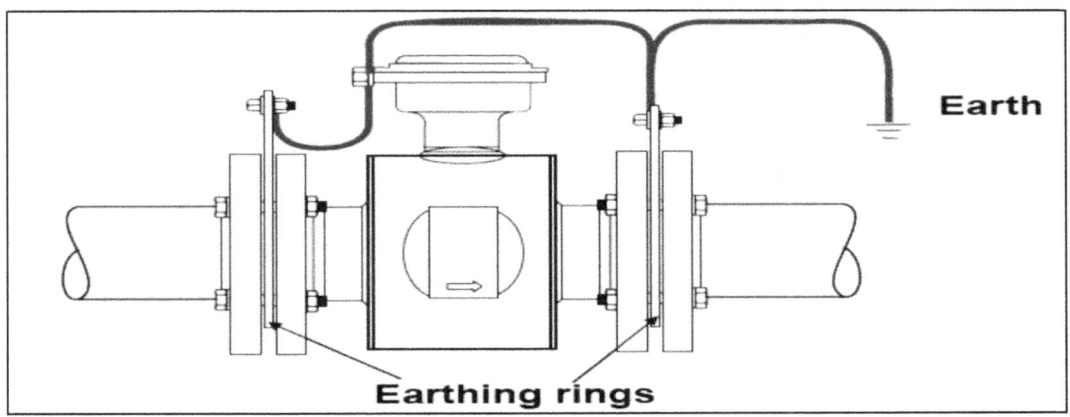

Fig.21 Earthing for non conductive pipe with earthing rings

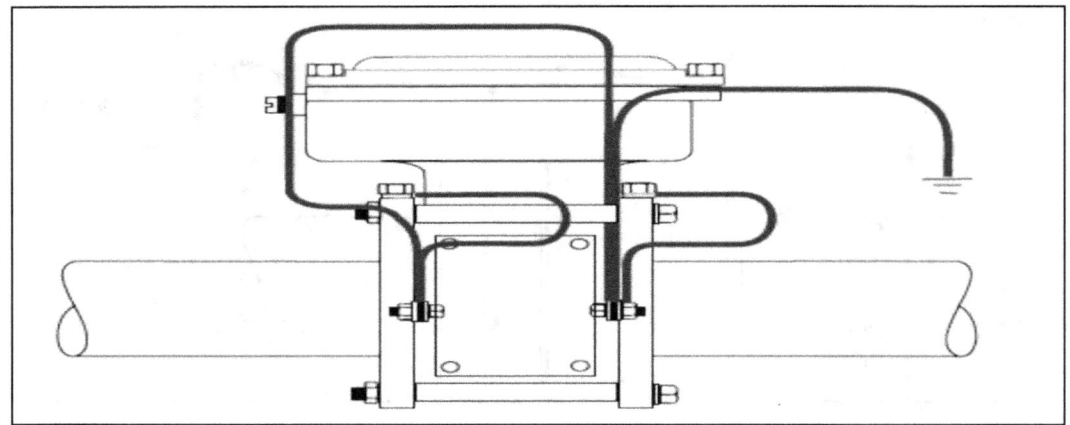

Fig.22 Earthing for conductive lined pipe with earthing rings

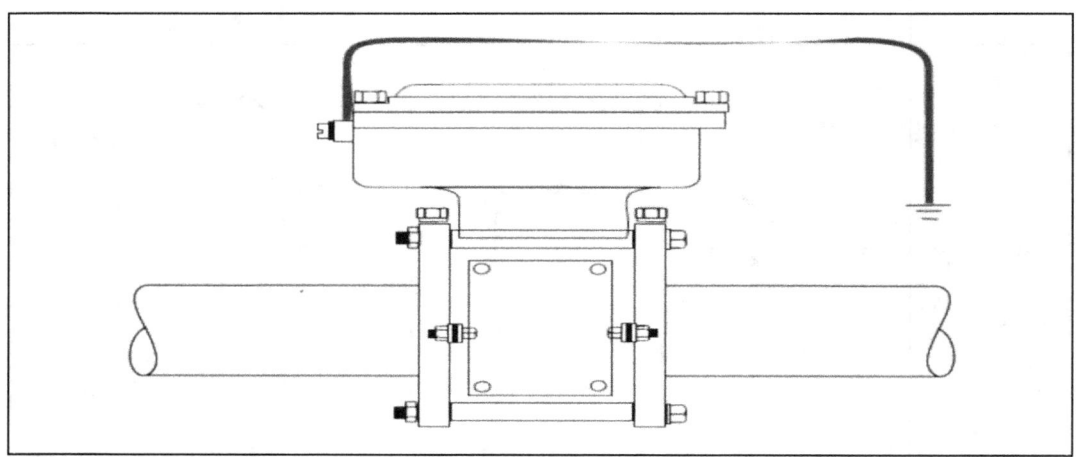

Fig.23 Earthing for non conductive lined pipe with earthing electrodes

3.14 Cleaning and Maintenance of the Primary Device

If insulating materials are likely to be deposited from the conducting liquid onto the electrodes or the walls of the meter tube, provision should be made for mechanical, electrical or chemical cleaning with the following methods:

A)   Withdrawable Electrodes: Withdrawable electrodes can be provided by using mechanical valving and sealing arrangements so that the electrodes can be withdrawn (usually at full pipeline pressure) for external inspection and cleaning.

B)   Mechanical Scraper: In this system, a rotary scraper is fitted to each electrode such that its scraping edge is perpendicular to the electrode face. The scraper is driven by an external electric motor via fluid pressure seals. It may be used continuously or intermittently. This method is becoming less commonly used in modern electromagnetic flowmeters.

C)   Ultrasonic cleaning: In this method, a voltage from the mains supply is connected between the electrodes (the secondary instrument being automatically disconnected during this operation), causing electrolysis on the surface of each electrode. The resultant rapid gas evolution causes removal of deposits. This approach is generally used on oily, greasy and sludge-type coatings.

3.15 Measurement of Liquid Flow in closed circuit – Weighing Method

The method of liquid flowrate measurement in closed conduits by measuring the mass of liquid delivered into a weighing tank in a known time interval.

Statement of the principle:

Static Weighing:

The principle of the flow-rate measurement method by static weighing (for schematic diagrams of typical installations, see figures 24, 25) is:

- to determine the initial mass of the tank plus any residual liquid.

- to divert the flow into the weighing tank (until it is considered to contain a sufficiently quantity to attain desired accuracy) by operation of the diverter, which actuates a timer to measure the filling time;

- to determine the final mass of the tank plus the liquid collected in it.

The flow-rate is then derived from the mass collected, the collection time.

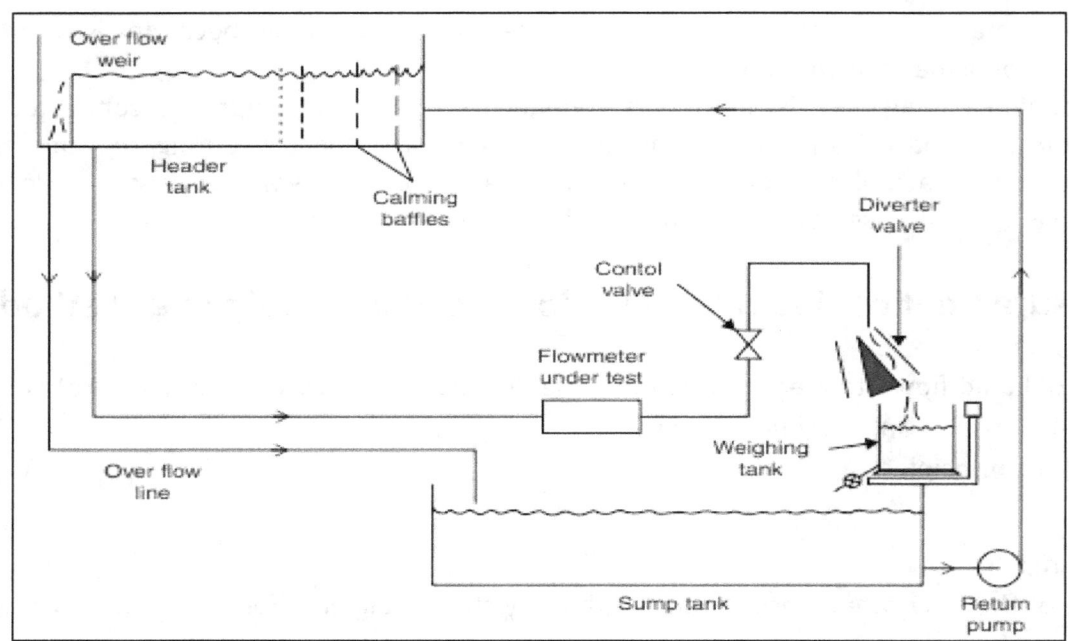

Fig.24 An installation for calibration by weighing (static method, supply by a constant level head tank)

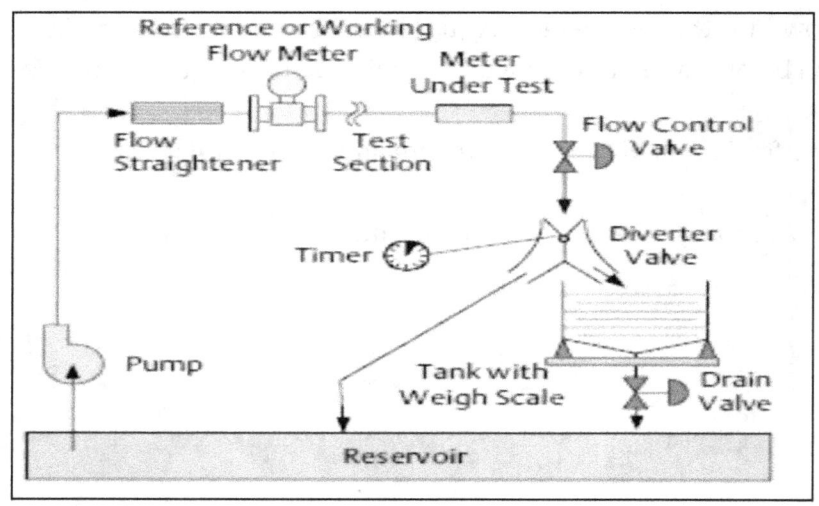

Fig.25 An installation for calibration by weighing (static method, direct pump supply)

3.16 Insertion or Probe Type Magnetic Flowmeter

The probe-type magnetic flowmeter is an "inside out" design in the sense that the excitation coil is on the inside of the probe, as shown in Figure 27. As the process fluid passes through the magnetic field generated by the excitation coil inside the probe, a voltage is detected by the electrodes that are embedded in the probe. The main advantage of this design is its low cost, which is not affected by pipe size, and its retractable nature, which makes it suitable for wet-tap installations. The probe-type magmeter is also suited for the measurement of flow velocities in partially full pipes or in detecting the currents in open waters. When water flow is not constrained by a pipe, flow velocity has to be expressed as a three-dimensional vector. By inserting three magmeter probes parallel with the three axes, one can detect that vector.

The main disadvantage of the magmeter probe is that it detects the flow velocity in only a small segment of the cross-sectional area of the larger pipe. Therefore, if the flowing velocity in that location is not representative of the rest of the cross section, a substantial error can result.

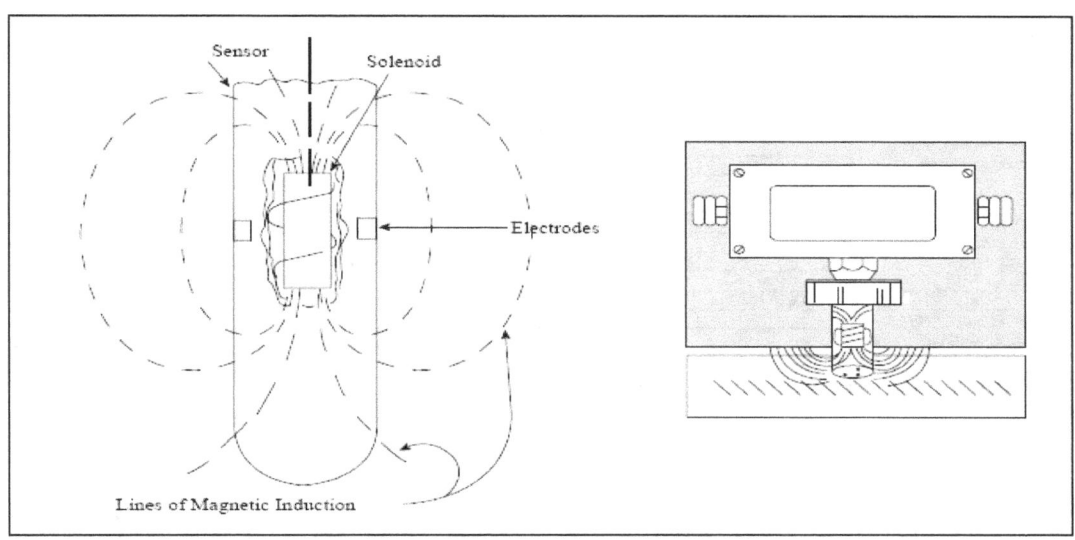

Fig. 26 Insertion Type Mag Flowmeter

3.17 Sizes and Capacities of Commonly Used Electromagnetic Flowmeters

SIZE	FLOW RATE (m3/hr)			STANDARD FLOW RATE	
	0.3m/sec	1 m/sec	10m/sec	m3/hr	m/sec
15	0.191	0.636	6.360	2	3.125
25	0.530	1.770	17.67	6	3.400
40	1.357	4.525	45.25	15	3.315
50	2.120	7.067	70.67	25	3.540
80	5.428	18.10	180.9	60	3.315
100	8.482	28.30	282.7	100	3.540
150	19.10	63.61	636.1	200	3.125
200	33.90	113.1	1131.0	300	2.655

Table 4. Sizes and Capacities of Commonly Used Electromagnetic Flowmeter

3.18 Data Sheet of Electromagnetic Flowmeter

1	RESPONSIBLE ORGANIZATION	MAGNETIC FLOWMETER	6	SPECIFICATION IDENTIFICATIONS	
2		w/wo INTEGRAL TOTALIZER INDICATOR	7	Document no	
3	(ISA)	Device Specification	8	Latest revision	Date
4			9	Issue status	
5			10		

	FLOWMETER BODY			TOTALIZER INDICATOR	
11			60		
12	Body type		61	Totalizer type	
13	Flow tube style		62	Enclosure type no/class	
14	End conn nominal size	Rating	63	Signal power source	
15	End conn termn type	Style	64	Contacts arrangement	Quantity
16	Flow tube diameter	Thickness	65	Totalizer reset style	
17	Hardware mounting kit		66	Integral indicator style	
18	Flow tube material		67	Cert/Approval type	
19	Lining material		68	Mounting location/type	
20	End termination material		69	Enclosure material	
21	Gnd/protective ring matl		70		
22			71	PERFORMANCE CHARACTERISTICS	
23			72	Min press at design temp	At
24	END EXTENSIONS		73	Max press at design temp	
25	End termination type	Style	74	Min working temperature	Max
26	Bolting material		75	Accuracy rating	
27	End termination material		76	Min velocity URL	Max
28	Gasket/O ring material		77	Min liquid conductivity	
29			78	Output signal damping LRL	URL
30	SENSING ELEMENT		79	Min ambient working temp	Max
31	Electrode type		80	Contacts ac rating	At max
32	Insertion length		81	Contacts dc rating	At max
33	Electrode material		82	Max sensor to receiver lg	
34			83		
35	COILS AND HOUSING		84		
36	Housing construction type		85		
37	Coil conn arrangement		86	ACCESSORIES	
38	Enclosure type no/class		87	Connecting cables length	
39	Signal power source		88	Cable Glands	
40	Signal termination type		89	Ultrasonic cleaner style	
41	Cert/Approval type		90	Empty tube detector	
42	Housing material		91	Calibrator adaptor	
43			92	Calibrator/configurator	
44			93		
45	TRANSMITTER OR CONVERTER		94	SPECIAL REQUIREMENTS	
46	Housing type		95	Custom tag	
47	Output signal type		96	Reference specification	
48	Enclosure type no/class		97	Compliance standard	
49	Characteristic curve		98	Calibration report	
50	Digital communication std		99	Software configuration	
51	Signal power source		100		
52	Failsafe style		101	PHYSICAL DATA	
53	Integral indicator style		102	Estimated weight	
54	Signal termination type		103	Face-to-face dimension	
55	Cert/Approval type		104	Overall height	
56	Mounting location/type		105	Removal clearance	
57	Failure/Diagnostic action		106	Signal conn nominal size	Style
58	Enclosure material		107	Mfr reference dwg	
59			108		

	CALIBRATIONS AND TEST		INPUT OR TEST			OUTPUT OR SCALE	
110							
111	TAG NO/FUNCTIONAL IDENT	MEAS/SIGNAL/TEST	LRV	URV	ACTION	LRV	URV
112		Flow rate-Analog output					
113		Flow rate-Digital output					
114		Flow rate-Freq output					
115		Flow rate-Scale					
116		Test pressure					
117							

	COMPONENT IDENTIFICATIONS			
118				
119	COMPONENT TYPE	MANUFACTURER	MODEL NUMBER	
120				
121				
122				
123				
124				
125				

Rev	Date	Revision Description	By	Appv1	Appv2	Appv3	REMARKS

1. Data Sheet of Magnetic Flowmeter as per ISA

3.19 Orifice Plate

Differential pressure flow meters encompass a wide variety of meter types that includes: orifice plates, venturi tubes, nozzles, Dall tubes, target meters, pitot tubes and variable area meters.

The most commonly used flowmeter is the orifice meter. The orifice meter consists of two parts as shown in Figure 27, they are the measurement orifice plate, which is installed in the process line, and the differential pressure transmitter, which measures the pressure developed across the orifice plate.

Orifice Plates are of three types and use according to the industrial application:

a)   Square Edged Concentric Orifice Plate
b)   Segmental Orifice Plate
c)   Eccentric Orifice Plate

The square edged concentric orifice plate is the most frequently used element because of its low cost and adaptability and availability of established coefficients. For most services, orifice plates are made of corrosion resistant materials usually SS 304 and SS 316, other material are used for special services.

Eccentric and Segmental orifice plates should be used for very dirty fluids or slurry or wet gases.

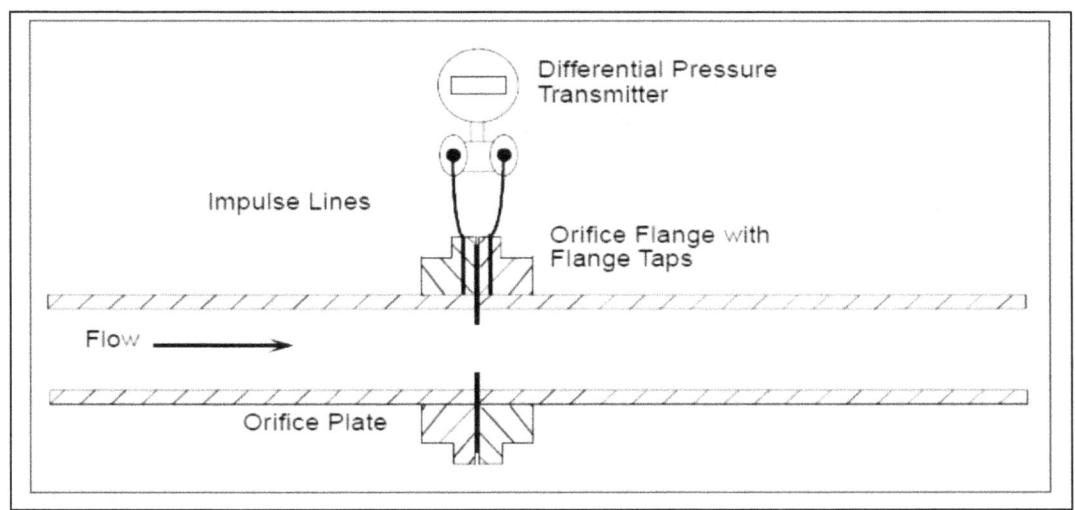

Fig.27 Orifice Plate and Differential Pressure Transmitter

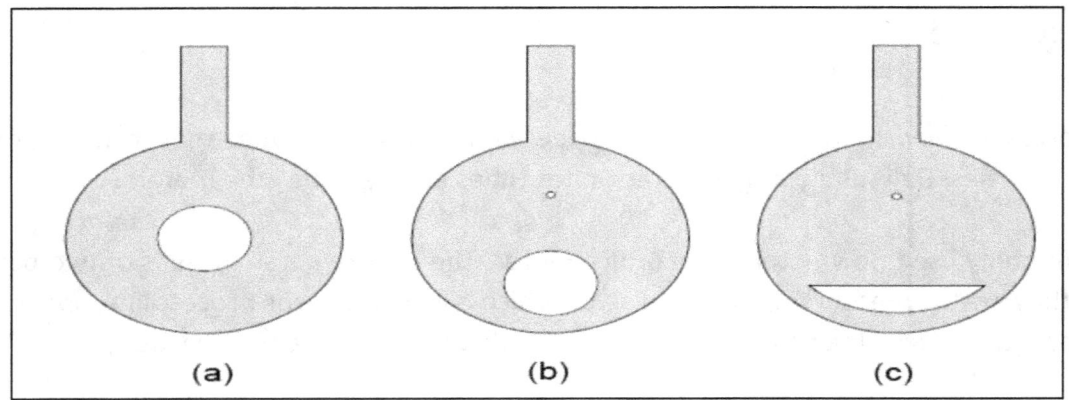

Fig.28 a) Concentric b) Eccentric c) Segmental Orifice Plate

The physical phenomenon of the orifice meter is described by the Continuity equation and the Bernoulli equation. The continuity states that the mass entering a control volume is the same as the mass leaving the control volume. (Control volumes are convenient methods for studying fluid problems). In a numerical form, the Continuity equation is written:

m1 = m2

$A_1V_1/v_1 = A_2V_2/v_2$

To help make this point, imagine the flow of water through a pipe. For this example, we will assume that the liquid is incompressible, that is its volume (v) is constant. This is a legitimate assumption because liquids are virtually incompressible at normally encountered pressures and temperatures.

If v is constant, the continuity equation simplifies to:

$A_1V_1 = A_2V_2$

If the liquid flows through an orifice, the area of the pipe will be reduced in the orifice and the velocity will increase as the fluid passes through the orifice.
The Bernoulli equation relates the increase in velocity to a change in static pressure. In more general terms, Bernoulli's equation relates the energy at one point in the control volume to the energy in a second point in the control volume. Bernoulli's equation written on a per unit weight basis can be represented as:

$V_1^2 \, \rho/2 + Lz_1 + P_1 + u_1 = \rho V_2^2/2 + Lz_2 + P_2 + u_2 + q + w_s$

Where

$\rho V^2/2$ = Kinematic energy

L_z = potential energy due to gravity
P = potential energy due to static
ρ = density
u = internal energy in the fluid
q = change in heat energy of the fluid
w_s= shaft work done by the fluid

For virtually all flow applications, this equation can be simplified. The first simplification is to state that there is no heat transfer in the control volume, so the factor q is 0.0. The next simplification is that there is no shaft work, so w_s is 0.0.

Orifice Equation for Liquids:

Since liquids are incompressible under normal operating conditions, the equations governing orifices can be further simplified for them. The first simplification is made in the continuity equation, specified volume, v, is constant and can be eliminated from the equation yielding:

$V_1A_1 = V_2A_2$ or $V_1 = A_2V_2/A_1$

Bernoulli's equation can be simplified for liquids if these conditions are assumed; the pipe is horizontal and the internal energy is the same in the sections on temperature and fluid properties (i.e. the temperature is the same). Based on these assumptions, Bernoulli's equation becomes:

$P_1 + \rho V_1^2/2 = P_2 + \rho V_2^2/2$

If we substitute V_1 from the continuity equation, we can rewrite Bernoulli's equation as:

$V_2^2/2 \ [\ 1-(A_2/A_1)^2] = 1/\rho \ (P_1-P_2)$ or

$\rho V_2^2 = 1/[\ 1-(A_2/A_1)^2] * [2(P_1-P_2)]$

These are the basic orifice equations. Since it is common to use these measure volumetric flow or mass flow, the square root is usually taken on both sides of the equation to solve for V_2.

$(\rho)^{1/2} V_2 = 1/[\ 1-(A_2/A_1)^2]^{1/2} * [2(P_1-P_2)]^{1/2}$

Since Q is equal to A_2V_2, we multiply both sides of the equation by A_2 and divide both sides by $(\rho)^{1/2}$ yielding:

Also, since the mass flow rate W, is equal to the volumetric flowrate times the density, W can be expressed as:
$W = Q\rho = A_2/[\ 1-(A_2/A_1)^2]^{1/2} * [2(P_1-P_2)]^{1/2}$

The term $1/[1-(A_2/A_1)^2]^{1/2}$ is called the Velocity of Approach factor. Contained within this term is the Beta Ratio.

The Beta Ratio is the ration of the area of the orifice to the area of the pipe, A2/A1. The Beta Ratio is typically depicted by the Greek letter β.

The velocity of approach factor can sometimes be combined with other coefficients. These equations are the theoretical flow equations and are based on the fluid area at the downstream pressure tap. The area of the fluid is no precisely known, and therefore, the bore of the orifice is used for the area. This introduces an error into the calculation.

To correct for this error, a flow coefficient is used to produce what is referred to as the working model of the equation.

To obtain Q actual, we multiply Q theoretical by C, which is the discharge coefficient. By doing this, the equation becomes:

$$Q_{actual} = C\, A_2/[\ 1-(A_2/A_1)^2]^{1/2} * [2(P_1-P_2)]^{1/2} / (\rho)^{1/2}$$

Often the velocity of approach factor is combined with C to yield another discharge coefficient, K.

$$Q_{actual} = K\, A_2\, [2(P_1-P_2)]^{1/2} / (\rho)^{1/2}$$

The common way to get the discharge coefficient is to refer to the standard tables and graphs that plot K versus the Beta ratio. The second method for determining the discharge coefficient is to actually measure it. An assembly that consists of the orifice plate and the inlet and outlet sections of the pipe are wet calibrated to determine the discharge coefficient. This assembly is commonly referred to as a meter run. This approach is usually taken with small pipe sizes and low flow ranges, where installation effects can cause significant errors.

ORIFICE EQUATIONS FOR GASES AND STEAM:

To use the orifice equations for gases and steam, an additional factor must be added to compensate for the fact that gases and steam are compressible. The factor is called the expansion factor and denoted by Y. The expansion factor is determined by taking the ratio of actual flow of a gas or steam through an orifice to the flow that is predicated by the liquid orifice equation. The expansion factor can then be added to the liquid equation to make it suitable for gases and steam. The equation then becomes:

$$Q_{actual} = K\, Y\, A_2\, [2(P_1-P_2)]^{1/2} / (\rho)^{1/2}$$

It is necessary to use the expansion factor when the density of a gas is substantially less downstream than it is upstream of the orifice. Because the expansion factor is affected by the differential pressure across the orifice plate, it will vary throughout the operating range of the meter. To minimize the effect of the variation in Y, it is best to keep the maximum differential pressure less than 2% of the upstream static pressure, P1.

Restriction to High Values of the Reyonlds Number:

Let R_d be the Reynolds number defined by the equation;

$R_d = 4M/ \pi d \eta$

Where η is the viscosity of the fluid

If R_d is large, say $R_d > 200,000$, the value of C found by testing an orifice with water or other liquid, is sensibly independent of the rate of flow, and this shows that the effects of viscosity have become negligible. But if the same orifice is tested with a gas, such as air, the value obtained for C varies with the rate of flow, even though R_d be high enough to make the effects of viscous forces insignificant; for the decrease of density as the pressure falls from p1 to p2, in contradistinction to the constancy of density of a liquid.

3.20 Pressure Taps & Pressure Profile

For each orifice plate, at least one upstream pressure tapping and one downstream pressure tapping shall be installed in one or other of the standard locations, i.e. as D and D/2, flange or corner tappings.

Corner taps are located within the orifice flanges and sense the pressure on the upstream and downstream faces of the orifice plate.

Flange taps are also located in the orifice plates and sense the pressure 1 inch upstream and 1 inch downstream of the orifice plate.

Radius taps or D and D/2 taps or Throat Taps are located 1 diameter upstream and 0.5 diameters downstream of the orifice plate.

Pipe Taps are located 2.5 diameter upstream and 8 diameter downstream of the orifice plate.

The location of the pressure tapping characterizes the type of standard orifice meter.

Orifice plate with *D* and *D*/2 tappings and flange tappings:

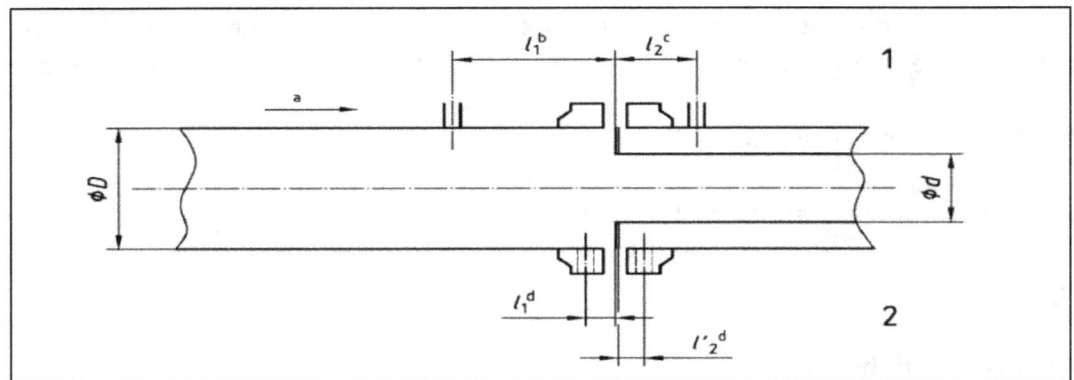

Fig.29 Spacing of Pressure Tapping for orifice plates with D and D/2 tappings and flange tappings

a)   The spacing l of a pressure tapping is the distance between the centerline of the pressure tapping and the plane of a specified face of the orifice plate. When installing the pressure tapping, due account shall be taken of the thickness of the gaskets or sealing materials.

b)   For orifice plate with D and D/2 tapping (see fig.29) the spacing l1 of the upstream pressure tapping is normally equal to D, but may be between 0.9 D and 1.1 D without altering the discharge coefficient.

The spacing l2 of the downstream pressure tapping is normally equal to 0.5 D but may be between the following values without altering the discharge coefficient.

- Between 0.48 D and 0.52 D when β <= 0.6
- Between 0.49 D and 0.51 D when β > 0.6

Both l1 and l2 spacing are measured from the upstream face of the orifice plate.

c) For orifice plates with flange tapping (see fig.29), the spacing l1 of the upstream pressure tapping is nominally 25.4 mm and is measured from the upstream face of the orifice plate.

The spacing l'2 of the downstream pressure tapping is nominally 25.4 mm and is measure   from the downstream face of the orifice plate.

Orifice plate with corner tapping:

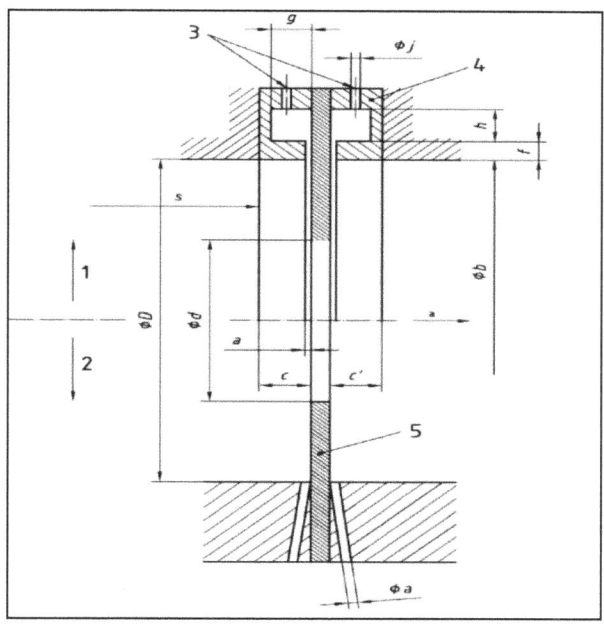

Fig.30 Corner Tapping

a)   The spacing between the centerlines of the tapping and the respective faces of the plate is equal to half the diameter or to half the width of the tapping themselves, so that the tapping holes break through the wall flush with the faces of the plate

Type of Tapping	Pipe Size
Corner Tap	Less than 2 inch pipe size
Flange Tap	More than 2 inch and less than 12 inch
D and D/2 Tap	More than 12 inch

Table 5. Thumb Rule for Tapping Points based on Pipe Size

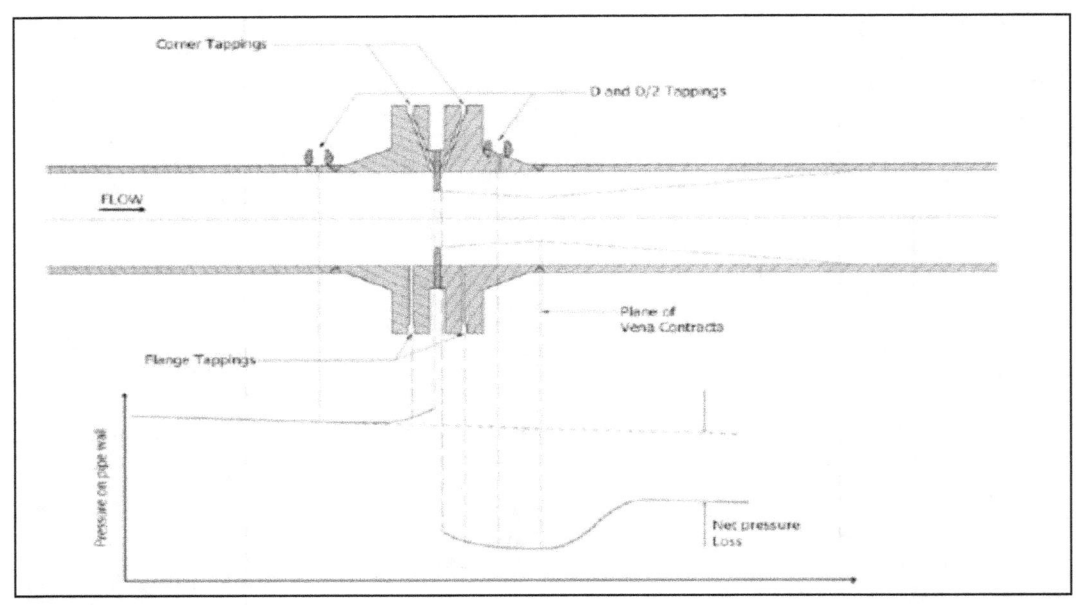

Fig.31 Pressure Profile through an orifice plate with different pressure tappings

General Shape of Orifice Plate:

The axial plane cross – section of a standard orifice plate is shown in fig. 32.

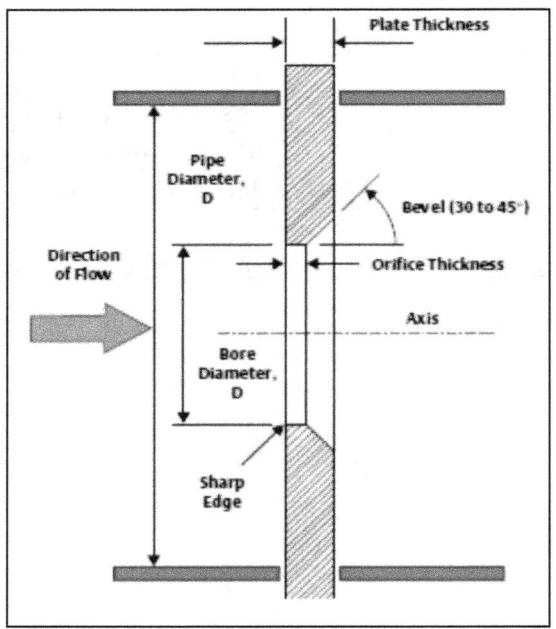

Fig.32 Standard orifice plate

Where
1-Upstream Face

2-Downstream Face
a-Direction of Flow

Thickness E and e:
The thickness e of the orifice plate shall be between 0.005 D and 0.02 D.
The thickness E of the orifice plate shall be between e and 0.05 D.

Angle of Bevel α:
The angle of bevel α shall be 45° +/- 15°.

Diameter of orifice d:
The diameter d shall in all cases be greater than or equal to 12.5 mm. The β ratio shall be always greater than or equal to 0.10 and less than or equal to 0.75. Within these limits, the value of β may be chosen by the user.

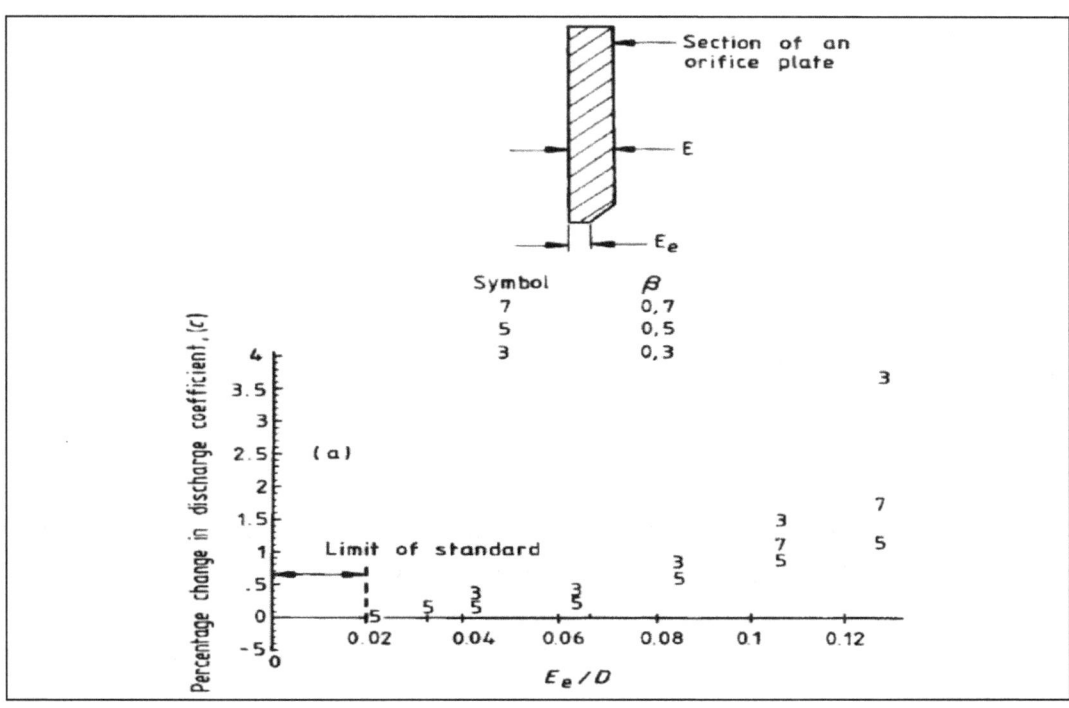

Fig.33 Change in Discharge Coefficient as a function of orifice thickness

3.21 Venturi Tube

Venturi tubes, flow tubes, and flow nozzles, like all differential pressure flow producers, are based upon Bernoulli's theory. They operate on the principal of a high pressure in the low velocity, large diameter inlet section compared to a low pressure in a high velocity, smaller diameter throat section. General performance and calculations are similar to those for orifice plates except for the fact that they operate over a wider dynamic range and are more efficient differential pressure producers and have far less permanent head loss than orifice plates. The meter coefficient (Cd) for these devices is between 0.98 and 0.99 compared to orifice plates that average about 0.62. Thus, almost 60% (98/62) more flow can pass through these elements for the same differential pressure.

The mass flowrate can be determined by the following formula:

$$q_m = \frac{C}{\sqrt{1-\beta^4}}\, \varepsilon \frac{\pi}{4} d^2 \sqrt{2\Delta p \rho_1}$$

For incompressible fluid ε is 1. Expansion factor for venturi tube is also closes to 1.

Similarly, the value of the volume flowrate can be calculated since:

$$q_V = \frac{q_m}{\rho}$$

ρ- fluid density at the temperature and pressure for which volume is stated.

3.22 Classical Venturi Tubes

Three types of standard classical Venturi tube are defined according to the method of manufacture of the internal surface of the entrance cone and the profile at the intersection of the entrance cone and the throat.

a) Classical Venturi tube with an "as cast" convergent section:

This is classical venture tube made by casting in a sand mould, or by some other methods which leave a finish on the surface of the convergent section. The throat is machined.

These classical venture tubes can be used in pipes of diameter between 100 mm and 800 mm and β ratio 0.3 and 0.75 inclusive.

b) Classical Venturi Tube with a machined convergent section:

This is classical venturi tube cast or fabricated as same as (a) but in which convergent section is machined as are the throat.

These classical venturi tubes can be used in pipes of diameter between 50 mm and 250 mm and with β ratio between 0.4 and 0.75 inclusive.

c) Classical Venturi tube with a rough welded sheet iron convergent section:

This is classical venture tube normally fabricated by welding.

These classical venturi tubes can be used in pipes of diameter 200 mm and 1200 mm and with β ratio between 0.4 and 0.7 inclusive.

General Shape of Venturi Tube:

The classical venture tube (fig.34) is made up of an entrance cylinder A connected to a conical convergent section B, a cylindrical throat C and a conical divergent E. The internal surface of the device is cylindrical and concentric with the pipe centerline. The co axiality of the convergent section and the cylindrical throat is assessed by visually.

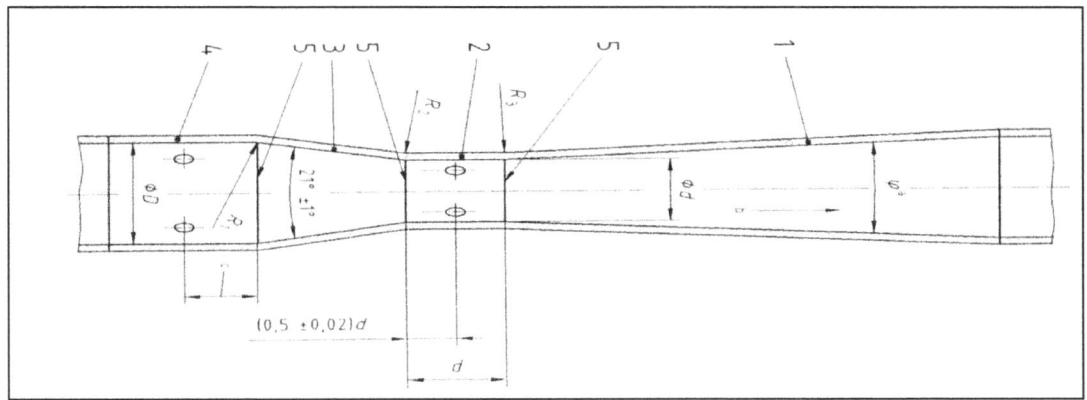

Fig.34 Classical Venturi Tube

Where:
1 Conical Convergent E
2 Cylindrical Throat C
3 Conical Convergent B
4 Entrance Cylinder A
5 Connecting Planes
a 7⁰ <= Ø <= 15 ⁰
b flow direction

a) The convergent section B shall be conical and shall have an included angle of 21⁰ +/- 1⁰ for all types of classical venturi tube.
b) The throat C shall be cylindrical with a diameter d.
c) The divergent section E shall be conical and may have an included angle φ between 7 ⁰ and 15 ⁰.

d) The minimum length of the entrance cylinder A shall be equal to D or 0.25 D (for cast convergent).

e) Radius of Curvature R1 shall be equal to 1.375D +/- 0.275 D (for cast).

f) Radius of Curvature R2 shall be equal to 3.625 d +/- 0.125d (for cast).

g) Radius of Curvature R3 shall be lie between 5d and 15d. (for cast).

h) Minimum length of the entrance cylinder A shall be equal to D (for machined)

i) Radius of Curvature R1 shall be less than 0.25 D and preferably equal to zero.

j) Radius of Curvature R2 shall be less than 0.25 d and preferably equal to zero.

k) Radius of Curvature R3 shall be less than 0.25 d and preferably equal to zero.

3.23 Triple T Arrangement

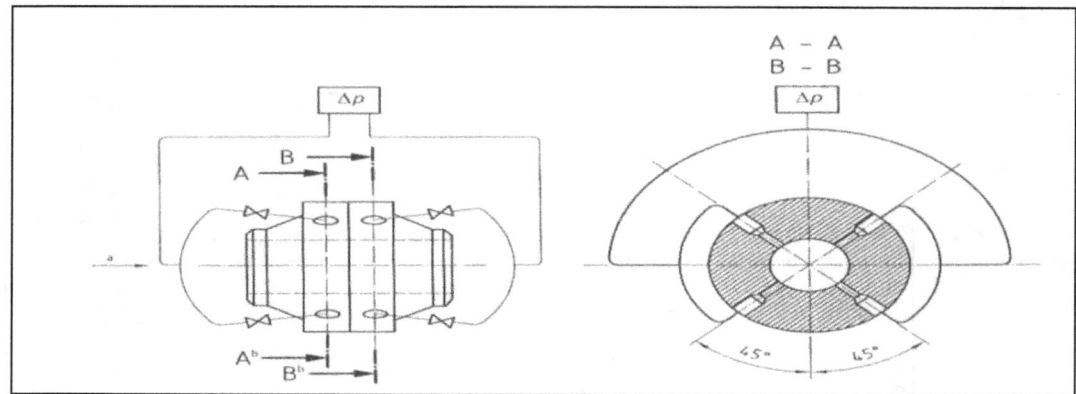

Fig.35 "Triple T" arrangement

The static pressure of the fluid shall be measured by means of an individual pipe wall pressure tapping or several such tapping interconnected, or by means of carrier ring tappings if carrier rings tappings are permitted for the measurement of differential pressure.

Where four pressure tappings are connected together to give the pressure upstream, downstream or in the throat of the primary device, it is best that they should be connected together in a "triple-T" arrangement as shown in fig.35. The "triple-T" arrangement is often used for measurement with Venturi Tube.

Temperature:

The temperature of the fluid shall preferably be measured downstream of the primary device. The thermowell well shall take up as little space as possible. The distance between thermowell and primary device shall be at least equal to 5D (and at most 15 D when the fluid is a gas).

Density of Gas:

It is necessary to know the density of the fluid at the upstream pressure tapping, it can either be measured directly or be calculated from an appropriate equation of state from a knowledge of absolute static pressure, absolute temperature and composition of the fluid at that location.

Density (Result) = 2.7* Specific Gravity (Operating Pressure – Actual Vapour Pressure) / RT + 0.62 Actual Vapour Pressure /T} lbs/ cu.

Density (lb/ft^3) = 2.69883*P(psi)*Specific Gravity/Operating Temperature in Rankine
Density(Kg/m3) = 3.48338*P(kPa)*Specific Gravity/T(K)
Where
R = Gas Law Dev. Coefficient (R in SI unit: 8.31441 KPa*m3/mole*T(K), R in USC unit:10.73151psi*ft3/mole*T(R))
T= Operating Temperature

Specific Gravity of Gas = Molecular Weight of Gas/ Molecular Weight of Air

Density of Water:

Water does not have an absolute density as its density varies with temperature.

At 100 ^{0} C its density is 958.4 kg/m3, at 80 ^{0} C its density is 971.8 kg/m3, at 60 ^{0} C it is 983.2 kg/m3, 40 ^{0} C it is 992.2 kg/m3, 30^{0}C it is 995.65 kg/m3, 25^{0}C it is 997.77kg/m3, 20^{0}C it is 998.2 kg/m3, at 10^{0}C it is 999.70 kg/m3, at 4^{0}C it is 998.97kg/m3.

Specific Gravity of Water: Density of the fluid kg/m3 / Density of the reference (density of water at 4^0 C)

Permanent Pressure Loss for venturi tube:

With exit cone angle of 7^0

$\Delta w = (0.218 – 0.42\beta + 0.38\beta^2) \Delta p$

With exit cone angle of 15^0
$\Delta w = (0.436 – 0.86\beta + 0.59\beta^2) \Delta p$

For Orifice plate, permanent pressure loss:

$$\text{Pressure loss} = \frac{\sqrt{1-\beta^4(1-C^2)} - C\beta^2}{\sqrt{1-\beta^4(1-C^2)} + C\beta^2} \Delta p$$

For Orifice plate, recommended Beta Ratio 0.2 to 0.6 for best accuracy.

3.24 Differential Pressure Transmitter Location Based on the Application

Liquid Service

Tap Locations – The pressure tap location in liquid service orifice meters should be located to prevent accumulation of gas or vapor in the connection between the pipe and the differential pressure instrument. The differential pressure instrument should be close to the pressure taps or connected through downward sloping connecting pipe of sufficient diameter to allow gas bubbles to flow back into the line.

Transmitter Installation – The installation of differential pressure transmitters should be located below the pipe and sloping upwards toward the pipe to prevent the collection of gas bubbles in the impulse tubing.

Vent Holes- are required for venting of any gas in a liquid service. Location of the vent hole in a liquid service is at the top of a pipe, above the center line.

Gas Service

Tap Locations – Pressure tap locations in a gas service must be installed in the top of the line with upward sloping connections towards a pipe. The differential pressure measuring instrument may be close-coupled to the pressure taps in the side of the lines or connected through upward sloping connecting pipe of sufficient diameter to prevent liquid from accumulating in the line.

Transmitter Installation – The installation of differential pressure transmitters should be located above the pipe with the impulse tubing sloping downward towards the pipe so that any condensate drains into the pipe.

Drain Holes – A drain hole is required for draining of any liquid in a gas service. Location of the drain hole is below the center line of the pipe.

Steam Service

Tap Locations - require the use of condensing chambers in steam or vapor applications because condensate can occur at ambient temperatures. Generally, the pressure tap connection has a downward sloping connection from the side of the pipe to the measuring device.

Transmitter Installation – The installation of differential pressure transmitters should be located above the pipe with the impulse tubing sloping downward towards the pipe so that any condensate drains into the pipe.

Drain Holes – A drain hole is required for draining of any condensate liquid in a steam service. The location of a drain hole is below the center line of the pipe.

3.25 Flow and DP Turndown

Flow turndown is defined as the ratio of the maximum flow rate and minimum flow rate. For example, if it is required to measure from 80 m3/hr to 5 m3/hr, the required turndown would be 16:1.

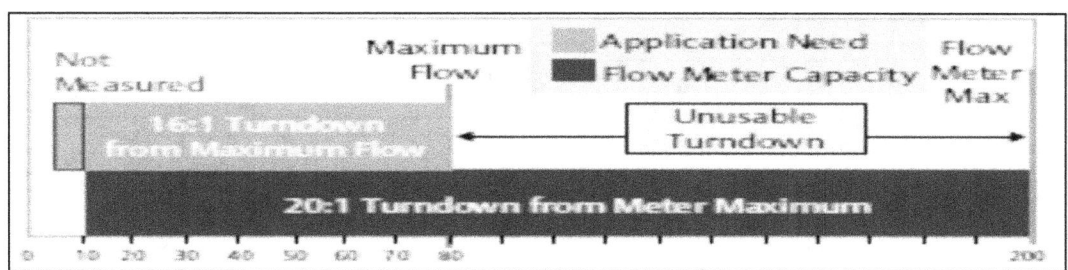

Fig.36 Flow and DP Turndown Relationship

Flow coverage is an important consideration when comparing different flowmeters. The relationship between flow coverage and turndown is illustrated in fig.36.

Since a differential pressure transmitter is required for DP flow, flow turndown is often confused with DP turndown and DP rangedown. DP turndown is defined as the differential pressure at maximum flow rate divided by the differential pressure reading at minimum flow rate. Due to square root relation between differential pressure and flow, an application requiring 10:1 flow turndown will require 100:1 DP turndown.

DP rangedown is defined as the URL of the differential pressure transmitter divided by the minimum span allowed by the device. For Flow applications, differential pressure transmitters are typically calibrated from zero to the differential pressure reading associated with the maximum flow. The DP rangedown is typically not a factor in selecting a DP transmitter for flow application.

Straight Run Pipe Requirement for Orifice Plate:

Traditional Orifice Plate

Upstream to Meter for Orifice Beta	0.20	0.40	0.50	0.65
Single 90° bend or tee	6	16	22	44
Two or more 90° bends in the same plane	10	10	22	44
Two or more 90° bends in different plane	19	44	44	44
Reducer (2 to 1D over 1.5 to 3D in length)	5	5	8	12
Expander (0.5D to D over D to 2D in length)	6	12	20	28
Ball/gate valve, fully open	12	12	12	18
Downstream from meter	4	6	6	7

Table 6. Straight Run Requirement for Orifice Plate

3.26 Data Sheet of Orifice Plate

	RESPONSIBLE ORGANIZATION	ORIFICE PLATE		SPECIFICATION IDENTIFICATIONS
1			6	
2		Device Specification	7	Document no
3	(ISA)		8	Latest revision Date
4			9	Issue status
5			10	

	ORIFICE PLATE AND HOLDER			ACCESSORIES	
11			60		
12	Plate type		61	Orifice holding block	
13	Holder/Seal unit style		62	Gasket set	
14	Nominal size		63		
15	API ring no		64		
16	Orifice flange nom rating		65		
17	Orifice bore type		66	SPECIAL REQUIREMENTS	
18	Orifice inlet edge style		67	Custom tag	
19	Diameter ratio (Beta d/D)		68	Reference specification	
20	Orifice bore diameter		69	Special preparation	
21	Plate outside diameter		70	Compliance standard	
22	Plate thickness		71	Calculation report	
23	Vent/Drain hole size		72		
24	Vent/Drain hole location		73		
25	Stampings required		74	PHYSICAL DATA	
26	Plate material		75	Estimated weight	
27	Seal material		76	Removal clearance	
28	Holder/Ring material		77	Mfr reference dwg	

	CALIBRATIONS AND TEST		INPUT		OUTPUT	
110						
111	TAG NO/FUNCTIONAL IDENT	MEAS/SIGNAL/TEST	LRV	URV	LRV	URV
112		Flow rate-Diff pressure				

	COMPONENT IDENTIFICATIONS			
118				
119	COMPONENT TYPE	MANUFACTURER	MODEL NUMBER	

Rev	Date	Revision Description	By	Appv1	Appv2	Appv3	REMARKS

2. Data Sheet of Orifice Plate as per ISA

3.27 Data Sheet of Venturi Tube

	RESPONSIBLE ORGANIZATION		VENTURI OR FLOW TUBE w/wo METER TUBE Device Specification		6	SPECIFICATION IDENTIFICATIONS	
1					6		
2	(ISA)				7	Document no	
3					8	Latest revision	Date
4					9	Issue status	
5					10		

	BODY OR TUBE				PERFORMANCE CHARACTERISTICS	
11			60			
12	Tube form		61	Max press at design temp	At	
13	Fabrication style		62	Min working temperature	Max	
14	Inlet conn nominal size	Rating	63	Uncalibrated accuracy		
15	Outlet conn nominal size	Rating	64	Perm head loss at URL		
16	End conn termn type	Style	65			
17	Flange facing finish		66			
18	Pressure tap quantity		67			
19	Press tap conn nom size	Rating	68			
20	Press tap conn type	Style	69			
21	Press tap orientation		70			
22	Body/Tube material		71			
23	Bolting material		72			
24	Exterior coating material		73			
25	End termination material		74			
26	Gasket material		75			
27			76			
28			77			
29			78			
30	THROAT		79			
31	Mounting type		80			
32	Diameter ratio (beta d/D)		81			
33	Throat diameter		82			
34	Cone material		83			
35	Throat material		84			
36			85	ACCESSORIES		
37			86	Purge system		
38			87	Cleaner style		
39	METER TUBE		88	Valve style		
40	Bore type		89	Valves and Nipple matl		
41	End conn nominal size	Rating	90			
42	End conn termn type	Style	91			
43	Flange facing finish		92	SPECIAL REQUIREMENTS		
44	Upstream length		93	Custom tag		
45	Pipe schedule no		94	Reference specification		
46	Downstream length		95	Special preparation		
47	Coupling quantity		96	Compliance standard		
48	Coupling nominal size	Rating	97	Construction code		
49	Coupling locations		98	Calibration report		
50	Internal bore diameter		99	Weld radiographs		
51	Flow straightener type		100			
52	Pipe/Tube material		101	PHYSICAL DATA		
53	Bolting material		102	Estimated weight		
54	Exterior coating material		103	Face-to-face dimension		
55	End termination material		104	Removal clearance		
56	Gasket material		105	Mfr reference dwg		
57			106			
58			107			
59			108			

	CALIBRATIONS AND TEST		INPUT OR TEST		OUTPUT	
110						
111	TAG NO/FUNCTIONAL IDENT	MEAS/SIGNAL/TEST	LRV	URV	LRV	URV
112		Flow rate – Diff pressure				
113		Test pressure				
114						
115						
116						
117						

	COMPONENT IDENTIFICATIONS		
118			
119	COMPONENT TYPE	MANUFACTURER	MODEL NUMBER
120			
121			
122			
123			
124			
125			

Rev	Date	Revision Description	By	Appv1	Appv2	Appv3	REMARKS

3. Data Sheet of Venturi Tube as per ISA

3.28 Variable Area Flowmeter

The variable area flowmeter is composed of a body containing the fluid and a "float," which is free to move in the body to a position related to the flow rate. The balance of forces positions the float. Gravity pulls the float downward. The buoyancy of the float plus the velocity related dynamic fluid forces lift the float. The float rises to increase the flow area until the fluid forces lifting the float match the downward force. As annular area between float and tube wall increases, the differential pressure across the float decreases. Every float position corresponds to a particular flow rate for a particular fluid density and viscosity. The meter must be oriented with flow vertically up for the analysis to be correct.

Basic equation for liquid flow through a rotameter:

$$Q = C \, A_a \, [(\rho F - \rho f)/\rho f]^{1/2}$$

Where Q- Volumetric Flow Rate
C- Meter Constant
A_a-Annular area between float and tube
ρF – Density of the float
ρf- Density of the fluid

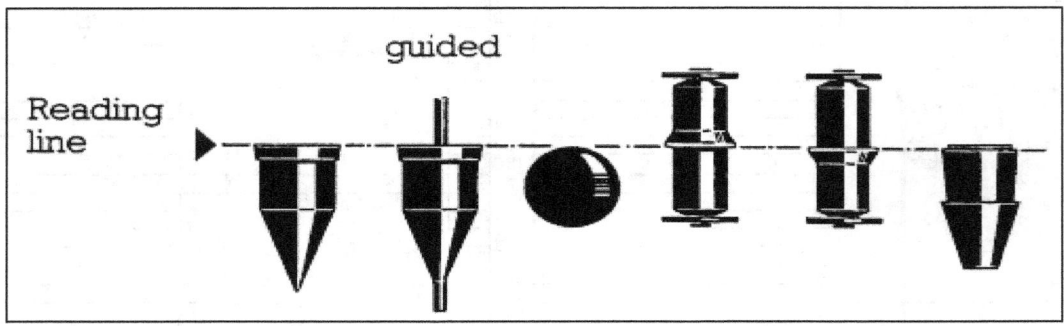

Fig.37 Typical Rotameter bob geometrics

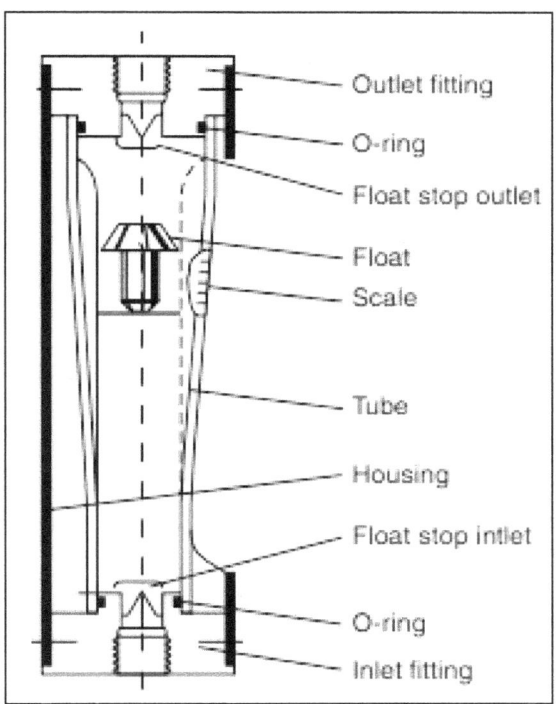

Fig.38 Nomenclatue of Variable Area Flowmeter

Fig. 38 consists of the following:

a) Float
b) Metering Tube
c) Scale
d) Packing and seals
e) Upper Body
f) Lower Body
g) Process Connection
h) Accessories

Float: The float is the body in the flowing fluid that moves in response to fluid flow. It is typically circular in cross section when viewed from the top. From the side, the float geometry may be simply a sphere.

Metering Tube: The tube is that part of the body which surrounds and contains the float. It increases in cross section area from the bottom to the top.

Scale: The scale is that part of the meter which shows the relation between the float position and the flow rate. Some have printed or engraved marks and numbers on a transparent metering tube. For metal tube meters, a magnetically coupled indicator is commonly used. This is coupled to the float, and an electronic or pneumatic device may be attached to develop a signal to be transmitted to another location.

Packing and Seals: For all but the simplest one-piece purge meters (see Fig. 39), some device is required to seal the metering tube to the upper and lower bodies. O-Rings are used in some meters, and packing is common in the larger meters. The selection of packing materials depends on the process fluid properties, including maximum and minimum pressures; and normal, maximum, and minimum temperatures.

Upper Body: The upper body supports the top or outlet of the metering tube. It usually includes a packing or sealing device. It also provides the support for the flow outlet
process connection.

Lower Body: The lower body is at the bottom or inlet of the flow tube. It is similar in function and design to the upper body.

Process Connection: The process connections are used to install the meter to the associated piping system. Standard connections include standard inch and millimeter piping threads and flanges.

Accessories: Accessories include switches controlled by the float position; signal -transmitting devices, check valves to prevent reverse flow, needle valves to control flow, and constant differential relays to stabilize flow.

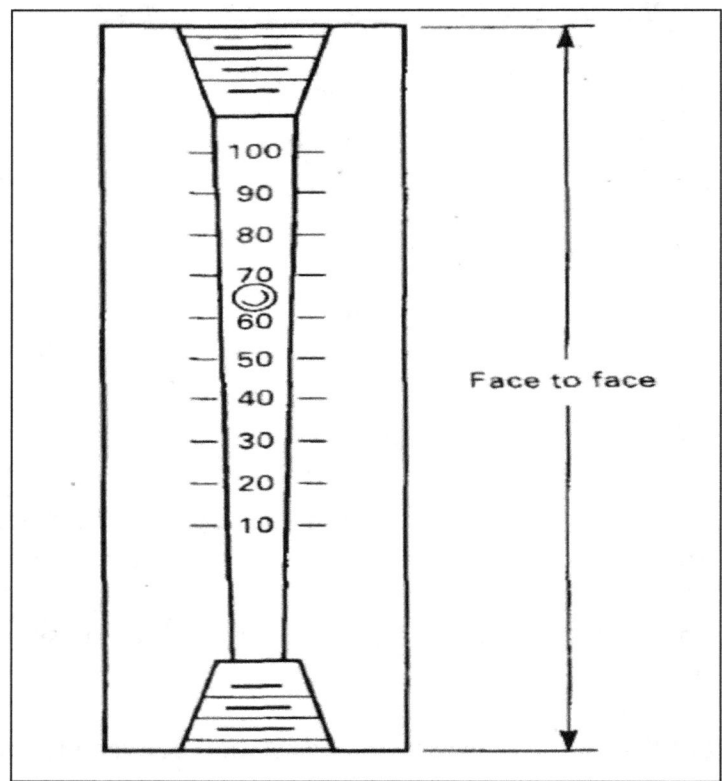

Fig.39 Purge Meter

3.29 Classes of Variable Area Flowmeter

Variable area flowmeters are of three general classes: purge or miniature meters, laboratory meters, and process flowmeters.

Purge meter:

Purge meters (see Fig. 39) are small and typically have 1/4 NPT (6 mm) or smaller connections. Because the applications do not justify it, calibration is unlikely. Catalog claims of 2% repeatability and an uncertainty of 5% of flow rate may not always be realized in practice.

Laboratory Meters:

The laboratory meters are usually longer [300 mm to 600 mm (12 in. to 24 in.)], have longer scales, and include more graduations than other meters of the same connection size and capacity. Repeatability is advertised as ½% and a standard accuracy of 1% is promised, which may be improved to 1/2% with calibration.

Process Meter:

Process meters with 1/2 in. (13 mm), or smaller connections typically have standard calibration uncertainties of 2%. Meters larger than 1/2 in. can often have certified uncertainty of 1 % at the specified conditions if they are calibrated. Tubes are typically between 150 mm and 250 mm long.

3.30 Float Type and Material

For all but the smallest sizes, there may be a choice of float type or style. Most purge type meters use spherical floats. The center line of the sphere is the reference point for reading the flow. Larger meters may have floats which look like a carpenter's plumb bob with various parts added to improve the stability of the float. Most commonly, the point on the float used to read the flow rate is at the point of maximum diameter.

3.31 Accuracy Classes for Variable Area Flowmeters

The accuracy specifications for the Variable Area Flowmeters are defined by the various Accuracy Classes in the VDE/VDI Guidelines.

Flowrate in %	Accuracy Class				
	1	1.6	2.5	4	6
	Total Error in % of Rate				
100	1.000	1.600	2.500	4.000	6.000
90	1.028	1.644	2.569	4.111	6.167
80	1.063	1.700	2.656	4.250	6.375
70	1.107	1.771	2.768	4.429	6.643
60	1.167	1.807	2.917	4.667	7.000
50	1.250	2.000	3.125	5.000	7.500
40	1.375	2.200	3.438	5.500	8.250
30	1.583	2.533	3.958	6.333	9.500
20	2.000	3.200	5.000	8.000	12.000
10	3.250	5.200	8.125	13.000	19.500

3.32 Coriolis Mass Flowmeter

Coriolis meters directly measure mass flow rate, and some can measure the flowing density of the process fluid.

Principle of Operation:

The measuring principle of CMF is Coriolis force, which appears in rotating and oscillating (vibrating) systems. Such a vibrating system is shown in Figure 40, for a straight tube. The tube is excited by an external force F_E. The excitation frequency is kept at the natural frequency of the tube, which minimizes the energy needed for vibration. The general expression for the Coriolis force is $F_C = 2*m*v*\omega$, where $q = m*v$ is mass flow and ω is the rotation vector. When fluid is not flowing within a vibrating tube, the Coriolis force is zero $F_C=0$. When fluid begins to flow, the Coriolis force is no longer zero ($F_C \neq 0$), and the shape of the tube is illustrated by superimposing Figure, panel a and panel b. At the inlet section, the Coriolis force tends to decelerate the movement of the oscillating tube, whereas, for the outlet section, the Coriolis force tends to accelerate the movement. In the middle of the tube, the Coriolis force is always zero, since either ω is zero for straight tubes or q is parallel with ω for curved tubes, bringing the product $q \times \omega$ to zero. As soon as the fluid begins to flow, the Coriolis force induces a phase shift along the tube. This phase shift is proportional to the mass flow. The mass flow can then be determined by measuring the phase shift between two sensor positions, S1 and S2. Since the oscillation is kept at the natural frequency of the system, the frequency changes with changing density of the fluid in the tube; i.e., the natural frequency increases with decreasing density. Therefore, by knowing the actual frequency of the system, the density of the fluid can be calculated directly. Coriolis mass flowmeters have the proven ability to record the total mass flow to better than 0.1% for water at moderate velocities. Each Coriolis instrument gets its own calibration factor that depends only on the geometrical data and material properties of the tube. Thus, the calibration factor is independent of fluid properties.

Consider a straight tube conveying a fluid. We first look at the first eigenmode of this system, which is shown in Figure 40, panel (a). The tube is fixed at both ends, and the velocity of the fluid v shall be zero. The movement of the sensors S1 and S2 is described by the differential equation.

$$M_E \cdot \ddot{y}_E + K_E y_E = F_E$$

Where

M_E is effective mass
F_E is excitation force
y_E is lateral excitation displacement at the sensor
K_E is stiffness of the tube for the excitation mode

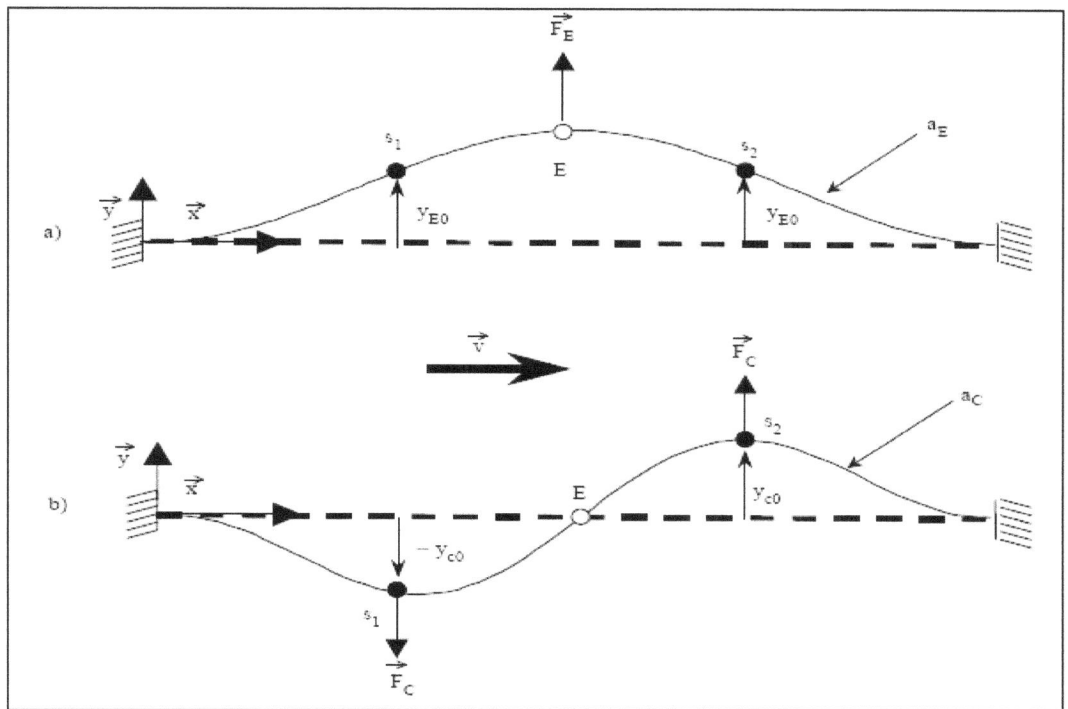

Fig.40 Panel a) describes the movement of a straight tube conveying a fluid, which is oscillating at the excitation frequency. The oscillation is maintained with the excitation force FE at location E. The measuring signal is detected with the two sensors S1 and S2. When the fluid begins to flow, the Coriolis force FC induces an oscillation as shown in panel b). The final lateral displacement is the superposition of both oscillations.

For commercially available instruments the amplitude for varies between 10 μm and 1 mm, and the frequency, $f_E = \omega_E/2\pi$, typically ranges from 80 Hz to 1100 Hz.

When the fluid begins to flow, the second mode is induced by the Coriolis force as shown in Figure 40, panel (b). For the Coriolis mode, the differential equation is

$$M_C \cdot \ddot{y}_C + K_C y_C = F_C$$

where y_C is the lateral Coriolis displacement of the tube at S1 and S2, F_C is the Coriolis force, M_C is the effective mass, and K_C represents the stiffness of the tube for the Coriolis mode.

The final lateral displacement of S1 and S2 is the superposition of excitation mode and Coriolis mode. As seen in Figure 40, the total lateral displacement of S1 is $y_{S1} = y_E - y_C$, and for S2 it is $y_{S2} = y_E + y_C$. The time difference $\Delta\tau$ between the two sensors becomes

$$\Delta\tau = \frac{\Delta\varphi}{\omega_E} \approx \frac{2 \cdot \hat{y}_C}{\omega_E \cdot \hat{y}_E} = \frac{2 \cdot \hat{y}_C}{\hat{v}_E} = \frac{2}{\omega_E} \cdot \frac{(y_{S2} - y_{S1})}{(y_{S2} + y_{S1})}$$

where Δτ is the time lag and Δφ is the phase shift between the two sensors.

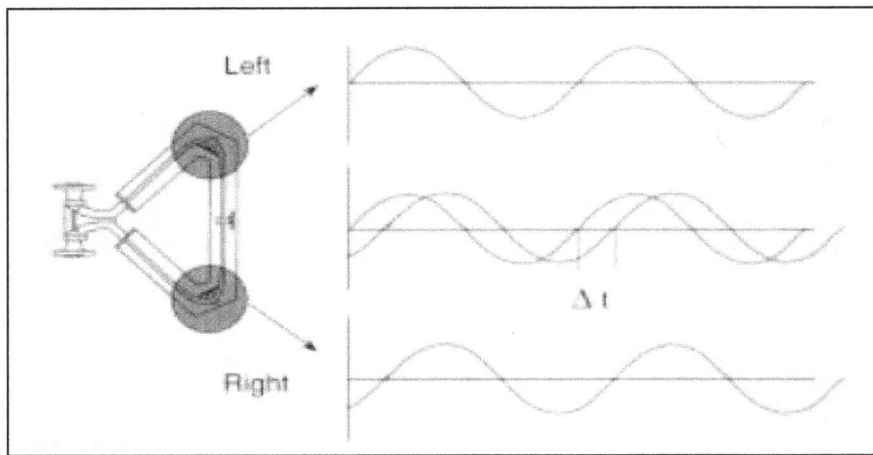

Fig.41 Phase shift between inlet side and outlet side

Coriolis Flow Sensor:

In one class of meters, the flow tube is anchored at two points and oscillated at a position between the two anchors, thus giving rise to opposite oscillatory rotations of the two halves of the tube. In another version, a section of tube is oscillated in a rotational direction and a transverse Coriolis force is generated. Meters can have one or more tubes that can be straight or curved. The movement of the flow tube(s) is measured at various points. When flow is present, Coriolis forces act on the oscillating tube(s), causing a small displacement, deflection, or twist that can be observed as a phase difference between the sensing points.

Coriolis forces (and hence distortion of the tube) only exist when both axial flow and forced oscillation are present. When there is forced oscillation but no flow, or flow with no oscillation, no deflection will occur and the meter will show no output. The sensor is characterized by flow calibration factors that are determined during manufacture and calibration.

The flow is split into two tubes as shown in Figure 43. Sensors are mounted at the inlet and outlet section of the tubes, measuring the phase difference between these two points. The tubes are forced into oscillation by the driver, which is mounted between the two tubes. Thus, the tubes are automatically driven in counter phase which is the preferred type of motion.

To vibrate the flow tubes, all commercially available CMFs use a magnet and a coil as the driving mechanism. Typically, the coil is mounted on one tube, and the magnet is mounted on the opposite tube. To protect the measuring system from any external disturbances, the tubes are fixed into a rigid carrier housing, which is strong enough to isolate the system from the environment. The tubes are vibrated at their natural frequency. As shown before, this frequency requires the least amount of energy to excite the system. Even large meters can be vibrated with only a few milliamps of excitation current. The natural frequency depends mainly on the mass of the system and the elastic properties of the measuring tubes. The total mass of the system includes the mass of the tube itself, the mass of the fluid within the tube, and the mass of any attached items such as driver and sensors. Therefore, since the material properties remain constant, a change in natural frequency directly indicates a change in the density of the fluid.

A key parameter to achieve a precise and stable CMF reading is the decoupling of the internal measuring system from any environmental and external disturbances. If CMFs are not decoupled to near perfection, the oscillations from the measuring tube will be transmitted to the connected process piping, which in turn begins to vibrate as well. Vibrating process piping can then cause the CMF to be excited by undefined vibrations. Depending on the magnitude and the strength of such external excitations, this can lead to a disturbed reading of the CMF. Therefore, it is an important requirement of a CMF to be a balanced system, in which oscillations of the measuring tube are well defined within the meter and are not transmitted to flanges and process piping. This requirement is also a general rule to ensure a good zero-point stability.

CMFs are not suitable for gas applications with low in-line pressure, since low-pressure gases have low densities. To generate enough mass flow to provide a sufficient Coriolis signal, the velocity of the gas must be quite high. This may lead to a large pressure drop across the meter.

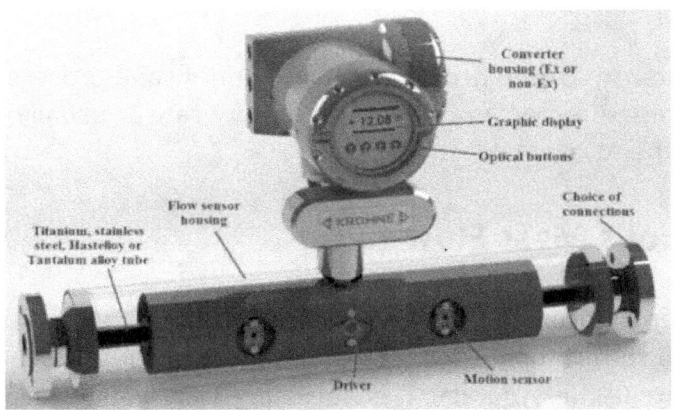

Fig.42 Single Straight Tube Coriolis Meter

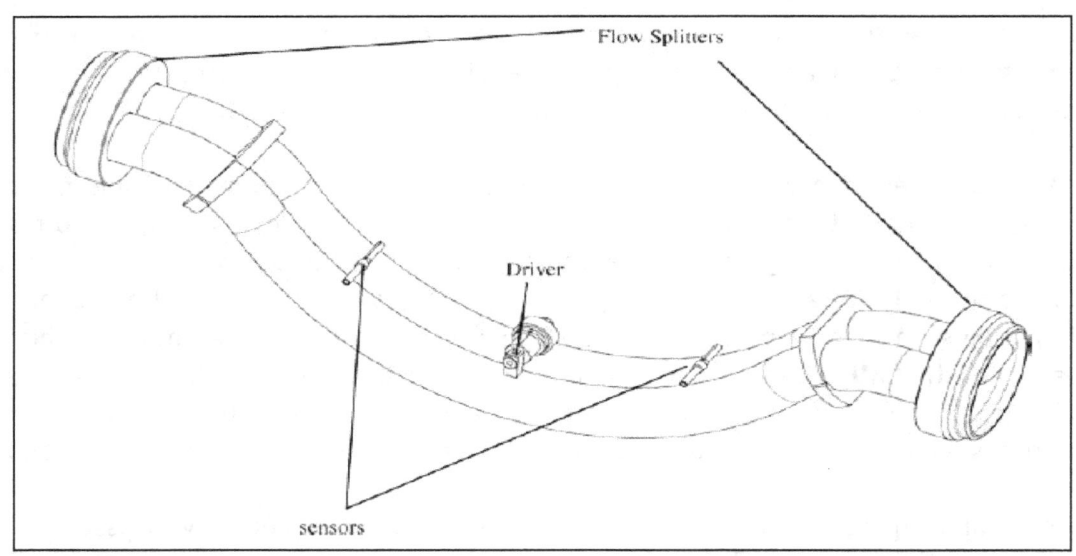

Fig. 43 Tube assembly of a typical Coriolis flowmeter

Coriolis Transmitter:

A Coriolis meter requires a transmitter to provide the drive energy and process the measurement signals to produce a mass flow rate measurement. Also, the mass flow rate is usually integrated over time in the transmitter, thus providing the total mass.

3.33 Accuracy of Coriolis Mass Flowmeter

For Coriolis meters, the term "accuracy" refers to the combined effects of linearity, reproducibility, repeatability, hysteresis, and zero stability. Zero stability is often given as a separate parameter in mass per unit time. In order to determine the accuracy, it is necessary to calculate zero stability as a percentage of the reading at a specified flow rate and add this value to the combined effects of linearity, repeatability, and hysteresis stated in units of percent of reading. A typical equation for accuracy:

$$\text{Accuracy} = \pm(0.20\% \pm (\text{zero stability}/\text{flow rate}) \times 100):$$
$$\text{where the combined effects (without zero}$$
$$\text{stability) are} \pm 0.20\% \text{ of reading}$$

Repeatability can also be given as a separate parameter, expressed as a percentage of the reading. Accuracy and repeatability statements are usually made at reference conditions that are specified by the manufacturer. Zero Stability is a measure of meter sensitivity and it is measured by the manufacture.

3.34 Factors affecting Mass Flow Measurement

Density and Viscosity:

Density and viscosity have a negligible effect on the accuracy of measurements of mass flow with a Coriolis meter.

Temperature:

Temperature changes affect the mechanical structure of the flow sensor, and compensation is necessary. This compensation, based on an integral temperature sensor, is performed by the transmitter. However, large differences in temperature between the oscillating tube(s) and the ambient temperature can cause errors in the temperature compensation. The use of insulation materials can reduce these effects.

Pressure:

Pressure changes can affect the flow calibration factor and compensation may be necessary. Pressure changes can also induce an offset in the meter output at zero flow. This effect can be eliminated by performing a zero adjustment.

Installation:

Stresses exerted on the sensor from the surrounding pipe work can introduce an offset in the meter output at zero flow. This offset should be checked after the initial installation or after any subsequent change in the installation. Zero adjustment should be performed if the offset is unacceptable.

3.35 Density Measurement

Principle of Operation:

Coriolis meters are typically operated at their natural or resonant frequency. For a resonant system, there is a relationship between this frequency and the oscillating mass. The natural frequency of a Coriolis meter viewed as a resonant system can be written as:

$$f_R = \frac{1}{2\pi}\sqrt{\frac{C}{m}}$$

With

$m = m_t + m_{fl}$

and

$$m_{fl} = \rho_{fl} * V_{fl}$$

Where:

C – mechanical stiffness or spring constant of the measuring tube arrangement
V_{fl} – Volume of fluid within the tube
f_R – Natural Frequency
m – Total oscillating mass
m_{fl} – Oscillating mass of fluid within the tube
m_t – Oscillating mass of measuring tube
ρ_{fl} – density of fluid at operated conditions

The mechanical stiffness or spring constant of the measuring tube arrangement depends on the design of the meter and the Young's modulus of elasticity of the tube material.

Equations (3), (4), and (5) can be used to solve for the fluid density, which is given by:

$$\rho_{fl} = C/V_{fl} * (2\pi f_R)^2 - m_t/V_{fl}$$

$$\rho_{fl} = K1 + K2/f_R^2$$

where K1 and K2 are coefficients for the density measurement that are determined during the calibration process. K1 and K2 are temperature and may be automatically compensated for by means of integral temperature measurement.

Factors Affecting Density Measurement:

Temperature:

Temperature changes can affect the density calibration factor of the sensor. Compensation for these changes is necessary and is frequently performed in the transmitter.

Flow Effect:

Density calibration is usually carried out under static conditions (i.e., without any fluid flowing). Operation on a flowing fluid can influence the density measurement. Fluid velocities that give rise to such an effect will vary depending on the sensor size and design.

Corrosion, Erosion, and Coating:

Corrosion, erosion, and coating may affect the mass and stiffness of the measuring tube. These effects will induce errors in the density measurement. In applications where these effects are likely, care should be taken in specifying suitable materials, selecting the most appropriate meter size.

3.36 Data Sheet of Coriolis Mass Flowmeter

#	RESPONSIBLE ORGANIZATION	CORIOLIS MASS FLOWMETER w/wo TOTALIZER INDICATOR Device Specification	#	SPECIFICATION IDENTIFICATIONS
1	(ISA)		6	
2			7	Document no
3			8	Latest revision Date
4			9	Issue status
5			10	

#	FLOWMETER BODY AND HOUSING		#	TOTALIZER INDICATOR	
11	FLOWMETER BODY AND HOUSING		58	TOTALIZER INDICATOR	
12	Body/Housing type		59	Totalizer type	
13	End conn nominal size	Rating	60	Enclosure type no/class	
14	End conn termn type	Style	61	Signal power source	
15	Body wetted material		62	Contacts arrangement	Quantity
16	Housing material		63	Totalizer reset style	
17	End termination material		64	Integral indicator style	
18			65	Cert/Approval type	
19			66	Mounting location/type	
20			67	Enclosure material	
21	FLOWTUBE ASSEMBLY		68		
22	Flowtube type		69	PERFORMANCE CHARACTERISTICS	
23	Flowtube diameter		70	Max press at design temp	At
24	Flowtube material		71	Min working temperature	Max
25			72	Flow rate accuracy rating	
26			73	Density accuracy rating	
27			74	Min flow URL	Max
28	CONNECTION HEAD		75	Density LRL	URL
29	Housing type		76	Sec enclosure press rating	
30	Enclosure type no/class		77	Pressure drop at flow URL	
31	Signal termination type		78	Min ambient working temp	Max
32	Cert/Approval type		79	Contacts ac rating	At max
33	Housing material		80	Contacts dc rating	At max
34			81	Max sensor to receiver lg	
35			82		
36			83	ACCESSORIES	
37	TRANSMITTER OR COMPUTER		84	Connecting cables length	
38	Housing type		85	Heating kit style	
39	Measurement compensation		86	Calibrator/configurator	
40	Output signal type		87		
41	Enclosure type no/class		88		
42	Span-zero adjustment		89	SPECIAL REQUIREMENTS	
43	Characteristic curve		90	Custom tag	
44	Digital communication std		91	Reference specification	
45	Signal power source		92	Special preparation	
46	Failsafe style		93	Compliance standard	
47	Integral indicator style		94	Calibration report	
48	Signal termination type		95	Software configuration	
49	Cert/Approval type		96		
50	Mounting location/type		97	PHYSICAL DATA	
51	Failure/Diagnostic action		98	Estimated weight	
52	Enclosure material		99	Face-to-face dimension	
53			100	Overall height	
54			101	Removal clearance	
55			102	Signal conn nominal size	Style
56			103	Mfr reference dwg	
57			104		

CALIBRATIONS AND TEST

#	TAG NO/FUNCTIONAL IDENT	MEAS/SIGNAL/TEST	INPUT OR TEST LRV	URV	ACTION	OUTPUT OR SCALE LRV	URV
110	CALIBRATIONS AND TEST		INPUT OR TEST			OUTPUT OR SCALE	
111	TAG NO/FUNCTIONAL IDENT	MEAS/SIGNAL/TEST	LRV	URV	ACTION	LRV	URV
112		Mass flow-Analog output					
113		Meas-Analog output 2					
114		Meas-Analog output 3					
115		Meas-Freq output 1					
116		Meas-Freq output 2					
117		Measurement-Scale					
118		Temp-Digital output					
119		Meas-Digital output					
120		Density-Digital output					
121		Test pressure					

COMPONENT IDENTIFICATIONS

#	COMPONENT TYPE	MANUFACTURER	MODEL NUMBER
122	COMPONENT IDENTIFICATIONS		
123	COMPONENT TYPE	MANUFACTURER	MODEL NUMBER
124			
125			
126			
127			

Rev	Date	Revision Description	By	Appv1	Appv2	Appv3	REMARKS

4. Data Sheet of Coriolis Mass Flowmeter as per ISA

3.37 Pitot Tube

Pitot tubes make use of dynamic pressure difference. Orifices in the leading face register total head pressure, dynamic + static, while the hole in the trailing face only conveys static pressure. Pressure difference between the two gives dynamic pressure in pipe, from which flow can be calculated.

A direct way to measure the fluid velocity is by using a pitot tube. A pitot tube measures the pressure due to fluid coming to rest on a fixed point in a flow stream. This pressure is called stagnation pressure and is related to the velocity by:

P0 = D*v2/2 g

Where P0- stagnation pressure, which is the differential pressure across the pitot tube (Pt-Ps)

Pt- Total pressure

Ps- Static Pressure

Fig.44 suggest the velocity of the fluid in front of pitot tube tip is determined by measuring differential pressure and knowing the fluid density. In this case, velocity cannot be seen, so differential pressure, which can be seen is measured in order to determine the velocity. The total pressure is sensed at the pitot tip and the static pressure from holes drilled through the surface of the pitot cylinder at right angle to the flow.

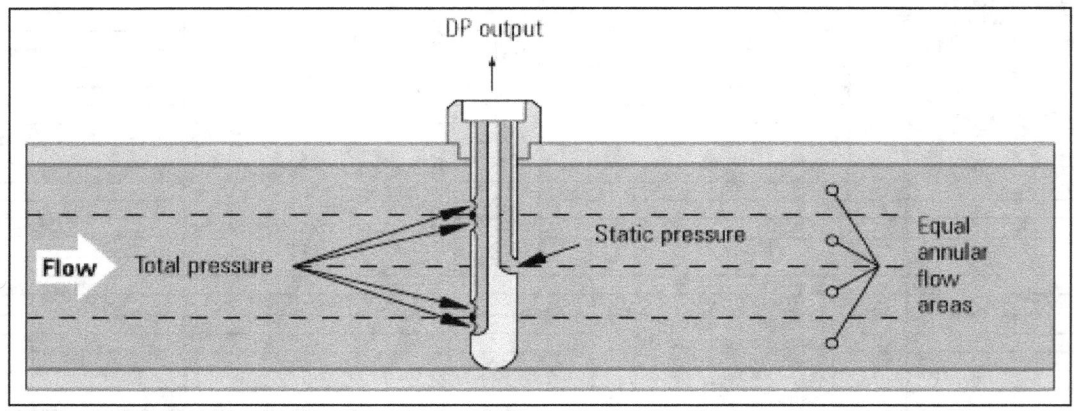

Fig. 44 Tube assembly of a typical Coriolis flowmeter

Averaging Pitot Tube:

Fig.45 shows averaging pitot tube. The averaging pitot tube is a sampling, or insertion meter. This means that is seeing only a portion of the flow field. For a developed flow, this is not an issue, as a sample of any diameter represents the whole. For undeveloped flow, it matters where the averaging pitot tube is inserted into the pipe. This is known as orientation.

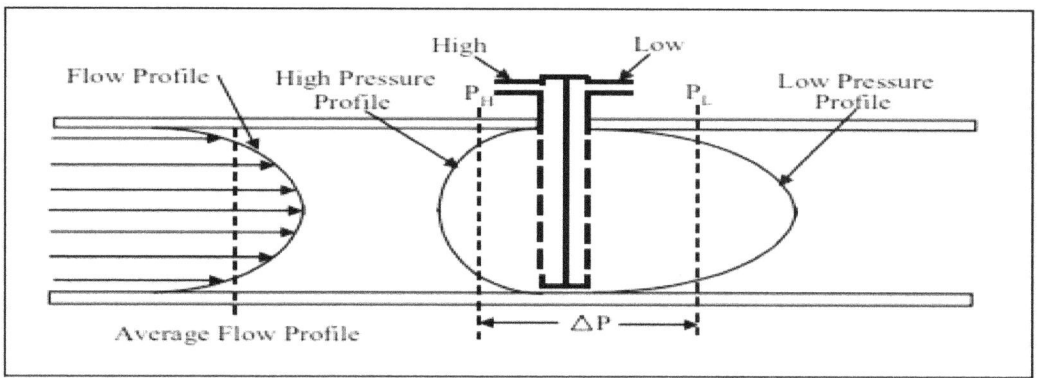

Fig.45 A averaging pitot tube in a pipe

Pitot tube Flow Equation:

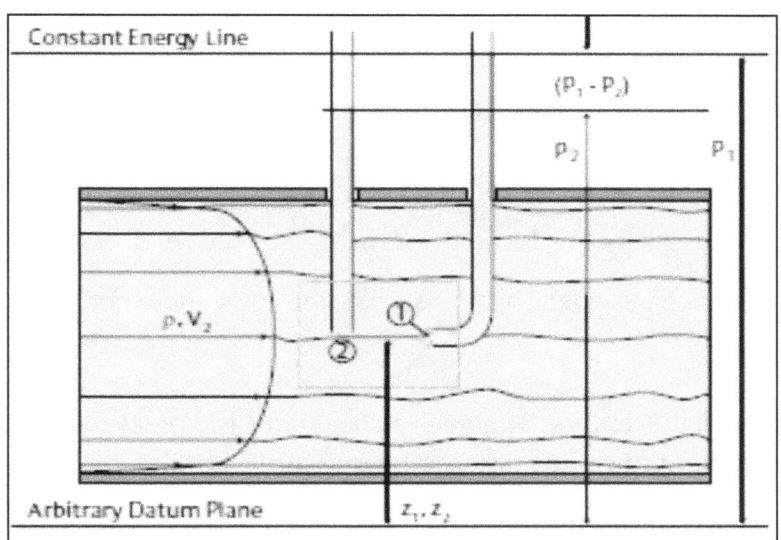

Fig.46 The Pitot Tube DP Energy System

Using Bernoulli's Energy Equation:

$$\left(\frac{V_2^2}{2g_c} - \frac{V_1^2}{2g_c}\right) = \left(\frac{P_1 - P_2}{\rho}\right) = \left(\frac{P_T - P_S}{\rho}\right)$$

It is seen that the value V1=0, or that the fluid is brought to rest, without losses (ie. Isentropic process) at the pitot tip with the value of P1 as the total pressure Pt, and the value of P2 the pipe static pressure Ps, then eq:

$$\frac{V_2^2}{2g_c} = \left(\frac{P_T - P_S}{\rho}\right)$$

Solving for V_2:

$$V_2 = \sqrt{\frac{2g_c(P_T - P_S)}{\rho}}$$

To obtain the incompressible volumetric flow rate Qv= V2*A

So for the pitot tube DP flow meter:

$$Q_v(Theor) = \frac{\pi}{4}D^2\sqrt{\frac{2g_c(P_T - P_S)}{\rho}}$$

Where g_c- inertial force conversion constant

Terminology used in Pitot Tube:

Discharge Velocity: The ratio of the volume flow rate to the area of the measuring cross section.

Static Pressure Tapping: A group of holes for the measurement of fluid static pressure.

Total Pressure Tapping: The pressure produced by bringing the fluid to rest without change in entropy.

Differential Pressure: Difference between the pressure at the total and static pressure taps.

Performance of Measurements:

a) Measurement of Differential Pressure: The device chosen for the measurement of differential pressure shall be capable of measurement of steady differential pressure.

b) Differential Pressure Fluctuations: Differential pressure fluctuations be damped by applying to the measuring apparatus the minimum damping allowing easy reading without long term fluctuations.

Inspection and Maintenance of Pitot tube:

The pitot tube does not require any special maintenance, but it shall be ensured, before and after the measurements, that the tube used complies following points:

a) the pressure sensing holes and their connecting pipes are not blocked.

b) there is no leakage between the chambers inside the Pitot tube which receive the total pressure and the static pressure.

c) the head of the tube is truly perpendicular to the supporting stem.

3.38 Data Sheet of Pitot Tube

1	RESPONSIBLE ORGANIZATION		PITOT TUBE	6	SPECIFICATION IDENTIFICATIONS	
2			w/wo INSERTION ASSEMBLY	7	Document no	
3	(ISA)		Device Specification	8	Latest revision	Date
4				9	Issue status	
5				10		

	BODY OR MOUNTING ASSEMBLY				PERFORMANCE CHARACTERISTICS	
11	BODY OR MOUNTING ASSEMBLY			60	PERFORMANCE CHARACTERISTICS	
12	Body/Assembly type			61	Max press at design temp	At
13	Sensor end support style			62	Min working temperature	Max
14	Process conn nominal size	Rating		63	Uncalibrated accuracy	
15	Process conn termn type	Style		64	Temp accuracy rating	
16	End support conn nom size	Style		65	Flow repeatability	
17	Pipe spool schedule			66	Min Reynolds number	
18	Mounting fitting type			67	Rated flow coefficient K	
19	Body/Assembly material			68		
20	Packing material			69		
21	Mounting fitting mat'l			70		
22				71		
23				72		
24	SENSING ELEMENT			73		
25	Sensor type			74		
26	Nominal line size			75		
27	Sensor nominal size			76		
28	Temperature element type			77		
29	Sensor material			78		
30				79		
31				80		
32	CONNECTION HEAD AND VALVES			81		
33	Connection head type			82	ACCESSORIES	
34	Connection head style			83	ACCESSORIES	
35	Process conn nominal size	Rating		84	Manifold valve style	
36	Process conn termn type	Style		85	Manifold valve material	
37	Press conn nominal size	Style		86		
38	Pressure tap orientation			87		
39	Head orientation			88		
40	Connection valve style			89		
41	Head material			90	SPECIAL REQUIREMENTS	
42	Flange material			91	Custom tag	
43	Connection valve mat'l			92	Reference specification	
44				93	Special preparation	
45				94	Compliance standard	
46				95	Construction code	
47	INSERTION ASSEMBLY			96	Calibration report	
48	Insertion assembly type			97	Special inspections	
49	Isolation valve style			98		
50	Process conn nominal size	Rating		99		
51	Process conn termn type	Style		100	PHYSICAL DATA	
52	Valve nom press rating			101	Estimated weight	
53	Valve body material			102	Face-to-face dimension	
54	Valve seat material			103	Overall height	
55	Wetted material			104	Removal clearance	
56	Packing material			105	Mfr reference dwg	
57				106		
58				107		
59				108		

	CALIBRATIONS AND TEST		INPUT OR TEST		OUTPUT	
110	CALIBRATIONS AND TEST		INPUT OR TEST		OUTPUT	
111	TAG NO/FUNCTIONAL IDENT	MEAS/SIGNAL/SCALE	LRV	URV	LRV	URV
112		Flow rate-Diff pressure				
113		Test pressure				
114						
115						
116						
117						

118	COMPONENT IDENTIFICATIONS			
119	COMPONENT TYPE	MANUFACTURER	MODEL NUMBER	
120				
121				
122				
123				
124				
125				

Rev	Date	Revision Description	By	Appv1	Appv2	Appv3	REMARKS

Form: 20F2071 Rev 0 © 2004 ISA

5. Data Sheet of Pitot Tube as per ISA

3.39 Ultrasonic Flowmeter

What is Sound and Ultrasound?

Sound is a pressure waves that travels across a media. Above 20000 Hz waves human cannot hear and that's called ultrasound.

Producing Ultrasound:

The ultrasonic signals required for the flow measurement are generated and received by transducers. Piezoelectric transducers (barium titanate or lead zirconate-titanate) employ crystals or ceramics, which are set into vibration when an alternating voltage is applied to the piezoelectric element. The vibrating element generates sound waves in the fluid. Since the piezoelectric effect is reversible, the element will become electrically polarized and produce voltages related to the mechanical strain when the crystal is distorted by the action of incident sound waves. Because the acoustic impedance of the gas is much smaller than that of the piezoelectric element, a layer of material is typically used between the gas and the piezoelectric element to maximize the acoustic efficiency.

Transit time ultrasonic flowmeters employs two transducers located upstream and downstream of each other. Each transmits a sound wave to the other, and the time difference between the receipt of the two signals indicates the fluid velocity.

Fluid Velocity Measurement:

Several techniques can be used to obtain average speed of propagation of an acoustic signal in a moving liquid. Two approaches generally utilise in Ultrasonic Flowmeters

a) Transit Time Difference

b) Frequency Difference

Transit Time Difference:

All ultrasonic flowmeter used in gas custody transfer employ transit time measurement. The advantage of transit time approach is that only the time difference is used in the velocity calculations, this greatly reduces the effect of changes in fluid properties such as pressure, temperature and composition. Ultrasonic Flowmeter based on transit time principle shown in fig.47. It is assumed that the transducer element is in direct contact with the liquid and that the acoustic signal propagates normal to the transducer/liquid interface. In most of the cases, it is desirable to protect the transducers in direct contact with liquid by using intervening materials. If such materials used then equation V_m shown in fig.47 changes:

$Q = A*V_A$

$Q = \pi D2/4*K*V_L$

Flow under normalised conditions:

$Q_{comp} = \pi D2/4 * K * V_L * P/P_0 * T_0/T$

Where $K = V_A/V_L$

Where Cu and Cd velocity of sound.

V is gas velocity

l is the distance between upstream and downstream sensors

φ is the angle between the pipe wall and the direction of acoustic propagation

k is conversion coefficient for the average flow velocity

Q-Volumetric Flowrate

V_A - Cross Section Mean Velocity

V_L- Linear Mean Velocity

Q_{comp} – Compensated Volumetric Flow Rate

Due to actual fluid velocity distribution in the pipe cross section, linear mean velocity is not equal to the cross section mean velocity. K is the correction factor between V_A and V_L.

For Ultrasonic meters, the velocity is the function of the time and the geometry of the meter body.

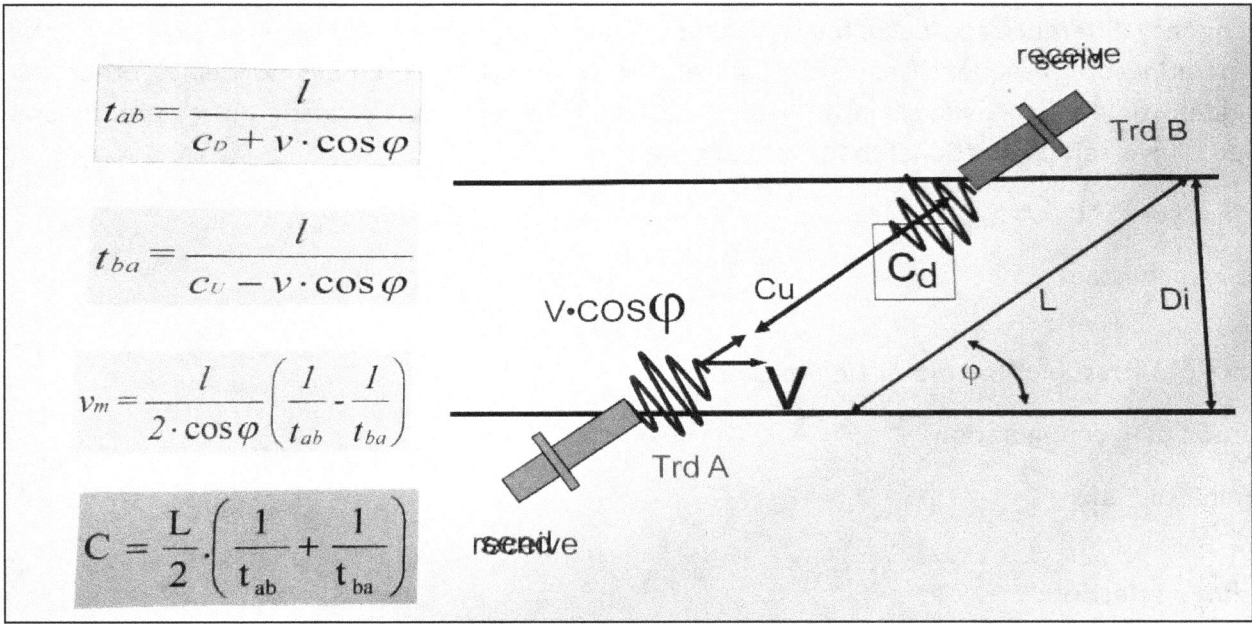

$$t_{ab} = \frac{l}{c_D + v \cdot \cos \varphi}$$

$$t_{ba} = \frac{l}{c_U - v \cdot \cos \varphi}$$

$$v_m = \frac{l}{2 \cdot \cos \varphi} \left(\frac{1}{t_{ab}} - \frac{1}{t_{ba}} \right)$$

$$C = \frac{L}{2} \cdot \left(\frac{1}{t_{ab}} + \frac{1}{t_{ba}} \right)$$

Fig.47 Ultrasonic Flowmeter Measuring Principle on Time Difference (Wetted Configuration)

From fig.48 (UFM with intervening material) eq V_m becomes:

$V_m = l/2.\cos\varphi \, (1/(t_{ab}-t_o) - 1/t_{ba}-t_o)$

Where t_o is the function of temperature, is the transit time of the acoustic signal through the intervening materials.

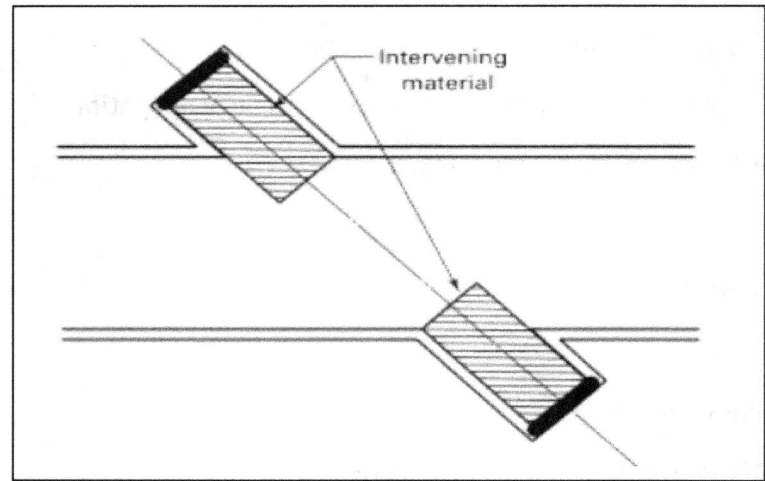

Fig.48 Ultrasonic Flowmeter Measuring Principle on Time Difference (Protected Configuration)

Frequency Difference:

In a frequency difference approach, the reception of an acoustic signal at the receiver is used as a reference for generating a subsequent acoustic signal at the transmitter. Assuming no delays other than the propagation time of acoustic signal in liquid, the frequency at which acoustic pulses are generated or received is inversely proportional to the transit time.

$V_m = (L*k_f/2\cos\varphi) * (f_{down}-f_{up})$

Where k_f is constant

Accuracy of Ultrasonic Flowmeter Depends on:

a) Acoustic path configuration

b) Number of Paths

Transducer Selection:

Key Selection Criteria:

a) Pressure Range

b) Temperature Range

c) Chemical Resistance

d) Acoustic Attenuation

e) Control Valve Noise

Various Types of Transducer Designs and Frequencies:

a) Epoxy Based: Excellent acoustic and chemical properties, used for pressure upto 500 bars

b) Full Titanium: Used upto pressure 150 bar

Types of Ultrasonic flowmeter:

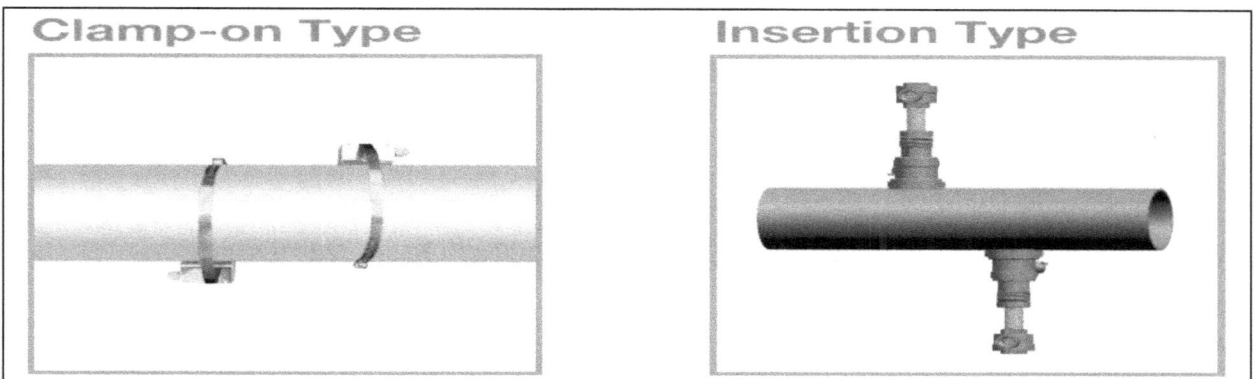

Fig.49 Types of Ultrasonic Flowmeters

Transit Time Ultrasonic Flowmeter (Externally Mounted or Clamp on)

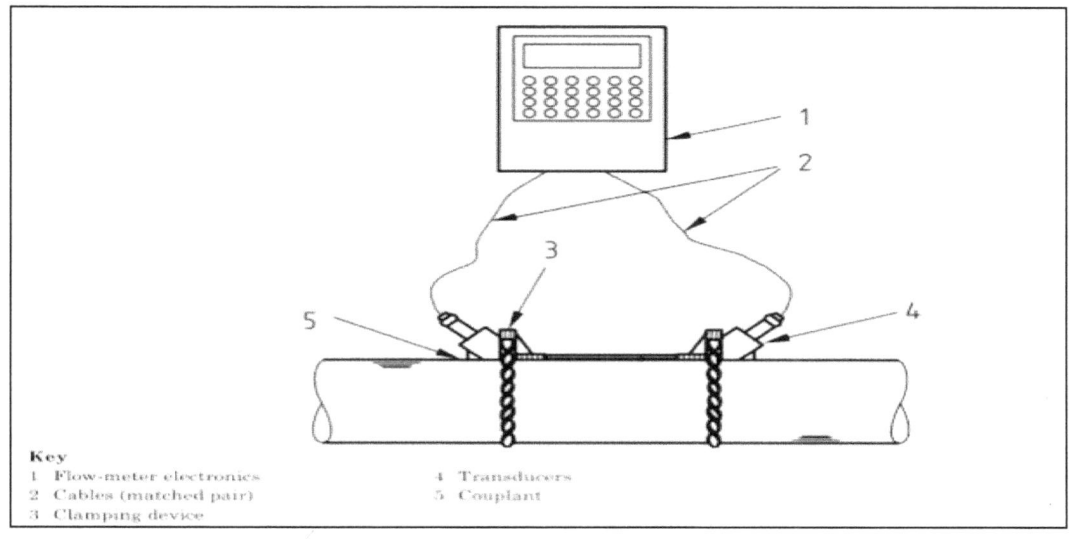

Fig.50 Clamp-On Type Ultrasonic Flowmeter

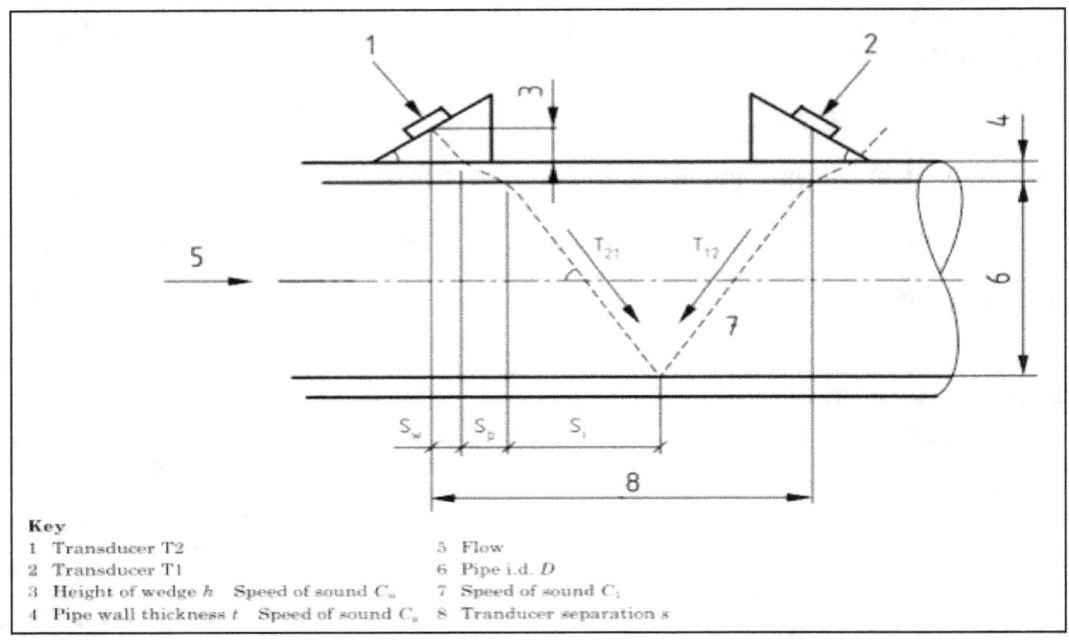

Key
1 Transducer T2
2 Transducer T1
3 Height of wedge h Speed of sound C_w
4 Pipe wall thickness t Speed of sound C_p
5 Flow
6 Pipe i.d. D
7 Speed of sound C_1
8 Tranducer separation s

Fig.51 Schematic of Transit Time Clamp-on Ultrasonic Flowmeter

The flow rate is determined using transit time difference between the time taken for ultrasound to travel from transducer1 to transducer2 and the time taken for ultrasound to travel from transducer2 to transducer1. If the ultrasonic beam in the fluid at an angle φ to the pipe axis and the path orientation as per fig. 51, then volumetric flow rate:

$Q_v = K_h \pi D C_1^2 \Delta t / 16 \cot\varphi$

Where C_1 is the speed of sound in water

D is the internal diameter of pipe

K_h is the profile correction factor

$K_h = V_{pipe} / V_{path}$

Where V_{pipe} is the actual average velocity in the pipe

V_{path} is the average velocity measured along the path

Transducer Configuration:

Fig. 52 show the transducer configuration used in industry.

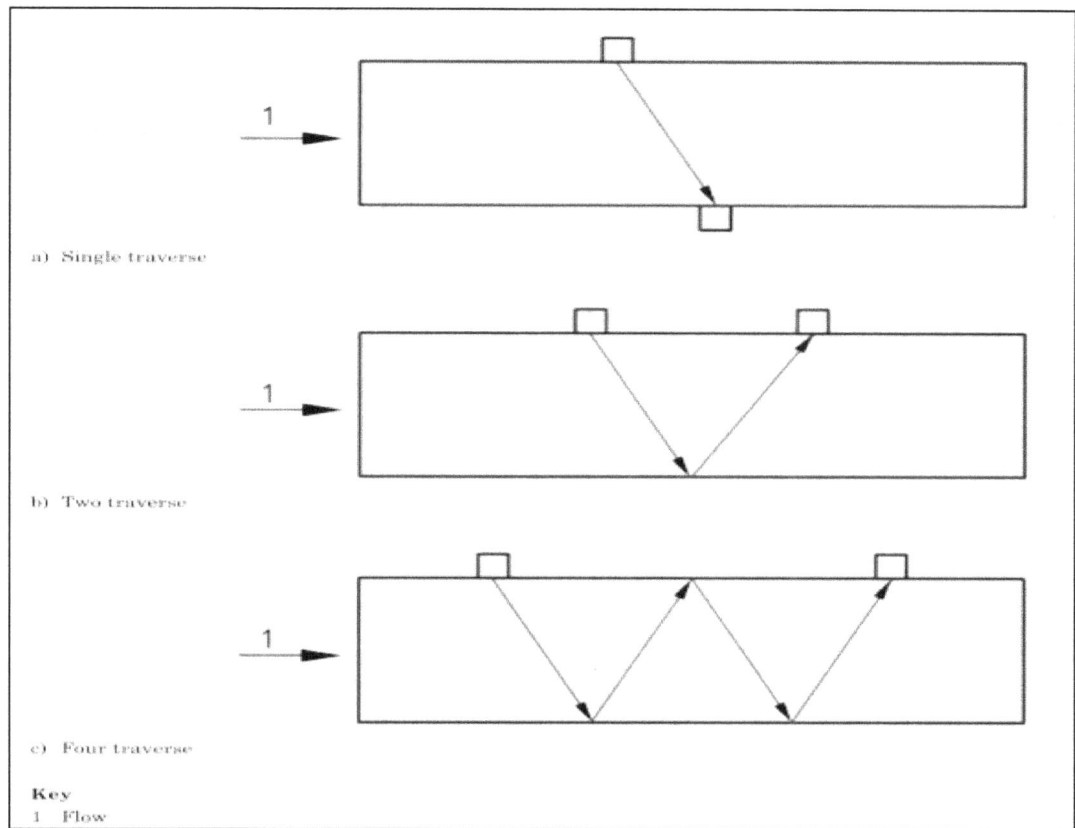

Fig.52 Acoustic Path Configuration

Use of Couplant during installation of Clamp on Ultrasonic Flowmeter:

The purpose of couplant is to provide a reliable transmission of ultrasound between the transducer and pipe wall. Different types of couplant used for long term and short term.

Long term couplants are silicon rubber pads, epoxy resin without fillers that will diffuse the sound.

Short term couplants are silicon grease, axle grease etc. With short term couplant, it is necessary to check that it should not dry out.

Pipe Size and Transducer Frequency:

a) 1 MHz transducers appear to be standard. Normally covers pipe size from 50mm to 2000mm.

b) 2 MHz transducers generally used for smaller diameter pipes from 10mm to 100mm.

c) 0.5 Mhz transducers used for large diameter pipes from 500mm to 5000mm. Generally used when fluid contains bubbles or particles.

Velocity Profile:

Ultrasonic Flowmeter are affected by variation in flow profile because uncertainty in the velocity path causes an error in Vm. This error may affect both the linearity and flow rate. Velocity profile variation can be caused by change in flow rate, wall roughness, temperature, viscosity etc. These errors can be reduced by increasing the number of acoustic paths.

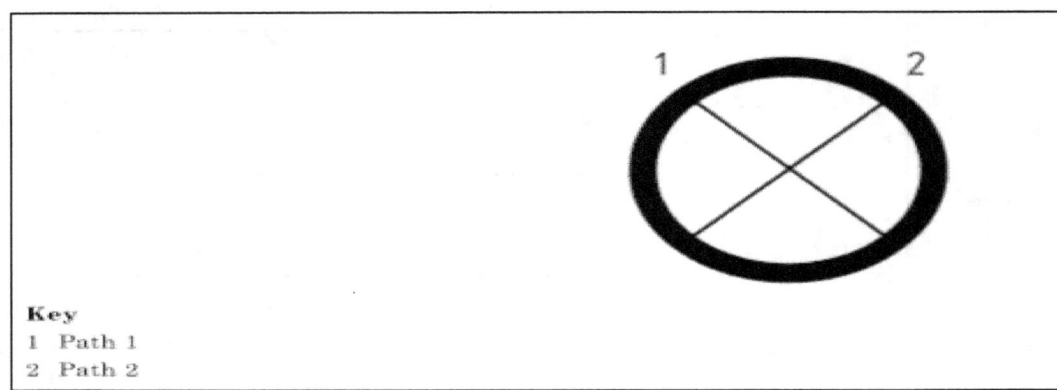

Fig.53 Two Path Measurement

Fig.53 shows for improving the measurement under distributed flow condition, take average of the two measurements.

3.40 Data Sheet of Ultrasonic Flowmeter

	RESPONSIBLE ORGANIZATION	ULTRASONIC FLOWMETER w/wo SWITCHES Device Specification		SPECIFICATION IDENTIFICATIONS	
1	RESPONSIBLE ORGANIZATION	ULTRASONIC FLOWMETER	6	SPECIFICATION IDENTIFICATIONS	
2	(ISA)	w/wo SWITCHES	7	Document no	
3		Device Specification	8	Latest revision	Date
4			9	Issue status	
5			10		
11	PRIMARY HEAD OR TUBE		56	TRANSMITTER OR TOTALIZER w/wo SWITCHES Continued	
12	Primary head/Tube type		57	Cert/Approval type	
13	End conn nominal size	Rating	58	Mounting location/type	
14	End conn termn type	Style	59	Failure/Diagnostic action	
15	Transducer conn type		60	Calibration mode	
16	Head/Tube material		61	Data logger points	
17	Flange material		62	Measurement compensation	
18	Gasket/O ring material		63	Enclosure material	
19			64		
20			65		
21	TRANSDUCER		66	PERFORMANCE CHARACTERISTICS	
22	Measurement type		67	Max press at design temp	At
23	Beam/Sound track style		68	Min working temperature	Max
24	Nominal range size		69	Flow accuracy rating	Ref
25	Temperature LRL	URL	70	Min flow/Velocity URL	Max
26	Mounting hardware		71	Minimum gas content	Max
27	Housing material		72	Minimum solids content	Max
28			73	Min ambient working temp	Max
29			74	Contacts ac rating	At max
30	CONNECTION HEAD		75	Contacts dc rating	At max
31	Housing type		76	Max sensor to receiver lg	
32	Enclosure type no/class		77		
33	Signal conn termn style		78		
34	Cert/Approval type		79	ACCESSORIES	
35	Enclosure material		80	Acoustic coupling matl	
36			81	Sound velocity sensor	
37	LEAD WIRE AND EXTENSION		82	Wall thickness sensor	
38	Extension type		83	Mounting kit	
39	Cable length		84	Tools	
40	Min cable operating temp	Max	85		
41	Signal termination type		86	SPECIAL REQUIREMENTS	
42	Cable jacket material		87	Custom tag	
43			88	Reference specification	
44	TRANSMITTER OR TOTALIZER w/wo SWITCHES		89	Compliance standard	
45	Configuration type		90	Certificates	
46	Aux input signal type		91	Software configuration	
47	Output signal type		92		
48	Enclosure type no/class		93	PHYSICAL DATA	
49	Local operator interface		94	Estimated weight	
50	Digital communication std		95	Face-to-face dimension	
51	Signal power source		96	Overall height	
52	Measurement type		97	Removal clearance	
53	Contacts arrangement	Quantity	98	Signal conn nominal size	Style
54	Integral indicator style		99	Mfr reference dwg	
55	Signal termination type		100		

	CALIBRATIONS AND TEST		INPUT OR TEST			OUTPUT OR SCALE	
110	CALIBRATIONS AND TEST		INPUT OR TEST			OUTPUT OR SCALE	
111	TAG NO/FUNCTIONAL	MEAS/SIGNAL/SCAL	LRV	URV	ACTION	LRV	URV
112		Meas-Analog output 1					
113		Meas-Analog output 2					
114		Meas-Analog output 3					
115		Meas-Analog output 4					
116		Meas-Scale 1					
117		Meas-Scale 2					
118		Meas setpoint 1-					
119		Meas setpoint 2-					
120		Meas setpoint 3-					
121		Meas setpoint 4-					
122		Failure signal-Output					
123							

	COMPONENT IDENTIFICATIONS		
124	COMPONENT IDENTIFICATIONS		
125	COMPONENT TYPE	MANUFACTURER	MODEL NUMBER
126			
127			
128			
129			

Rev	Date	Revision Description	By	Appv1	Appv2	Appv3	REMARKS

Form: 20F2341 Rev 0 © 2004 ISA

6. Data Sheet of Ultrasonic Flowmeter as per ISA

3.41 References

A) Hand book of Fluid Flow Metering by Ing C.J.Benard

B) Industrial flow measurement by Crabtree, Michael Anthony

C) Measurement of conductive liquid flow in closed conduits- Flanged Electromagnetic Flowmeter ISO 13359:1998

D) Measurement of Fluid Flow by means of Coriolis Mass Flowmeter ASME MFC – 11 M -2003.

E) Moore Process Automation Solution

F) Measurement of fluid flow by means of pressure differential devices inserted in circular cross-section conduits running full – Part 4 Venturi Tube ISO 5167-2003

G) Measurement of fluid flow by means of pressure differential devices inserted in circular cross-section conduits running full – Part 1 General principle and requirement ISO 5167-2003

H) Measurement of fluid flow by means of pressure differential devices inserted in circular cross-section conduits running full – Part 2 Orifice Plates ISO 5167-2003

I) Specification Forms for Process Measurement and Control Instruments- ISA TR 20.00.01-2001.

J) Measurement of Fluid Flow using Variable Area Meters ASME – MFC- 18M- 2001

K) Measurement of liquid flow in closed conduits- Method by collection of the liquid in a volumetric tank- ISO 8316

L) Measurement of conductive liquid flow in closed conduits – Method using electromagnetic flowmeter ISO 6817- 1997

M) Measurement of fluid flow in closed conduits- Guidance for the use of electromagnetic flowmeters for conductive liquid ISO 20456- 2017.

N) Measurement of fluid flow in closed conduits — Velocity area method using Pitot static tubes ISO 3996:2008

O) The Engineer's Guide to DP Flow Measurement 2020 by Emerson.

P) Use of clamp-on transit time metering techniques for liquid application BS 8452:2005.

Q) Measurement of Liquid Flow in closed conduit using Transit Time Ultrasonic Flowmeters ASME MFC-5M (2011).

R) Flow Measurement Handbook Second Edition by Roger C. Baker

S) Instrument's Engineer Handbook Volume 1 by Bela G. Liptak

4

Level Measurement

After completing this chapter, you should be able to:

Know about Level Measurement through Non-Contact type technology Radar and Ultrasonic; Contact type technology Bubbler type and Pressure or Differential Pressure Based Level Measurement

4.1 Radar

Radar provides a non-contact sensor that is virtually unaffected by changes in process temperature, pressure or the gas and vapour composition within a vessel.

In addition to above, the measurement accuracy is unaffected by changes in density, conductivity of the product being measured. These benefits have become more significant to the process industry.

The term "radar" is generally understood to mean a method by means of which short electromagnetic waves are used to detect distant objects and determine their location and movement. The term RADAR is an acronym from

RAdio **D**etection **A**nd **R**anging

Frequency & Wavelength:

To characterize electromagnetic waves, the relevant factors in addition to intensity are their frequency f and wavelength λ.

C= f * λ
 Where C is the velocity of light.

Electromagnetic frequency spectrum:

Microwaves are generally understood to be electromagnetic waves with frequencies above 2 GHz and wavelengths of less than 15 cm (6"). For technical purposes, microwave frequencies are used up to approx. 120 GHz – a limit that will extend upwards as technology advances.

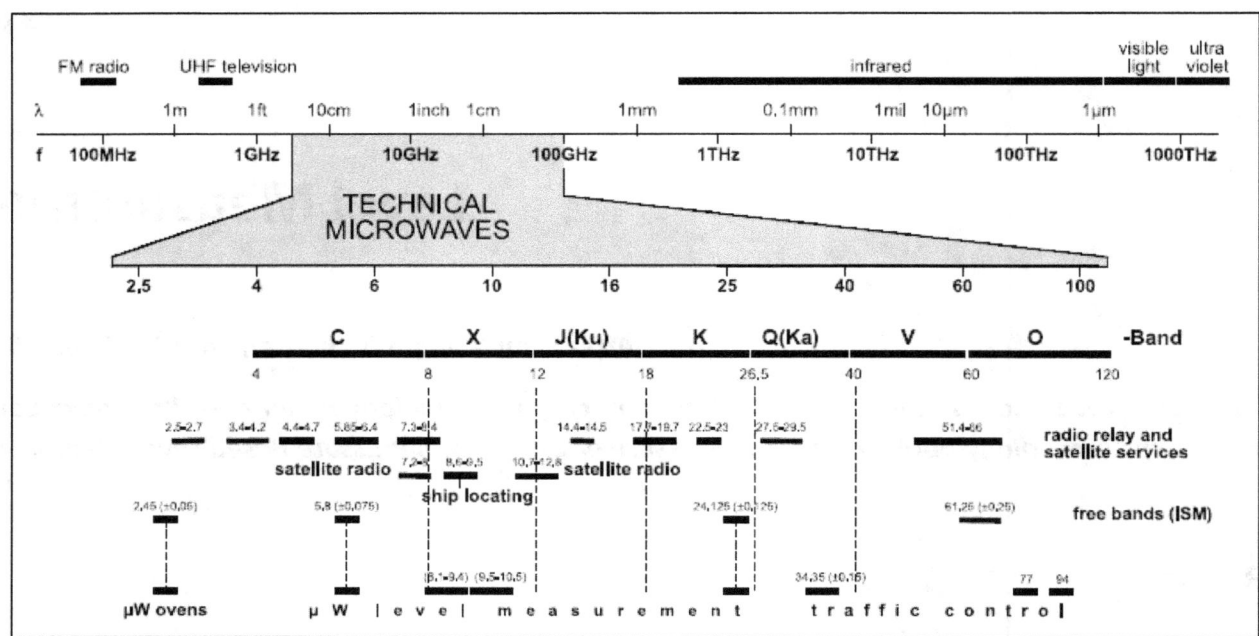

Fig.1 Electromagnetic frequency spectrum with typical applications in the microwave range

Types of Radar:

a) Pulse Radar
b) Frequency Modulated Continuous Wave Radar

Physics of Radar:

The velocity of light in free space is 3*10^8m/s. All forms of electromagnetic radiation travels with the speed of light in free space. This applies equally to long radio waves transmission, microwaves, infrared, ultraviolet, visible, X rays and Gamma Rays.

According to Maxwell, velocity of light in free space is expressed as:

C0=1/sq.rt($\mu*\varepsilon$)
Where C0- velocity of electromagnetic wave in free space
μ- permeability of free space ($4\pi*10^{-7}$ henry/meter)
ε- permittivity of free space ($8.854*10^{-12}$ farad/meter)

The velocity of an electromagnetic wave is the product of frequency and wavelength

$C = f * \lambda$

The frequency of pulse radar level transmitter may be 26GHz, the wavelength is 1.15centimeters.

The frequency remains uninfluenced by the changes in the propagation medium. However, velocity and wavelength can change depending on the propagation medium change in which electromagnetic waves are travel. The speed of electromagnetic waves in medium as:

$C = C0 / sq.rt(\mu_r * \varepsilon_r)$

Where: C is the velocity of electromagnetic waves in the medium
C0 velocity of electromagnetic waves in free space
μ_r relative permeability
ε_r relative permittivity

Reflection of Electromagnetic Waves:

If a liquid or solid is non-conductive the value of the dielectric constant or relative permittivity becomes more important. The amount of reflection from the dielectric layer can be calculated as:

$\Pi = 1 - 4 * Sq.rt (\varepsilon_r)/(1 + sq.rt \ \varepsilon_r)^2$

$\Pi = W1/W2$

Where W1- Transmitter Power
W2- Reflected Power

4.2 Frequency Modulated Continuous Wave Radar

The FMCW radar system transmits a series of continuous frequency modulated waves through the antenna and receives the signal from the target. The frequency of the transmitted wave changes according to the modulation voltage in the time domain. Commonly used modulated signals are sine wave, sawtooth wave and triangular wave.

When triangular or sawtooth waves are used as FM waves, it is called as LFMCW (linear frequency modulated continuous wave) Radar.
Working principle of LFMCW Radar:

When a target object is relatively stationary, the transmitted signal is reflected back after hitting the object and an echo signal is generated. The echo signal has the same shape as that of transmitted signal but it is delayed in time by t.

T= 2*D/c

Where D- distance of target from Radar; c- velocity of light

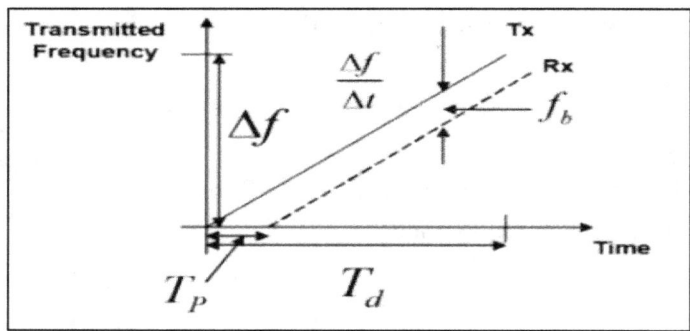

Fig.2 Frequency Modulated Continuous Wave Radar Stationary Target Echo Signal

Fig.2 shows the difference between transmitted frequency signal f_t and the reflected frequency signal f_r is fb which is called beat frequency

$F_b = f_t - f_r$

$T_d = 2 D/c$

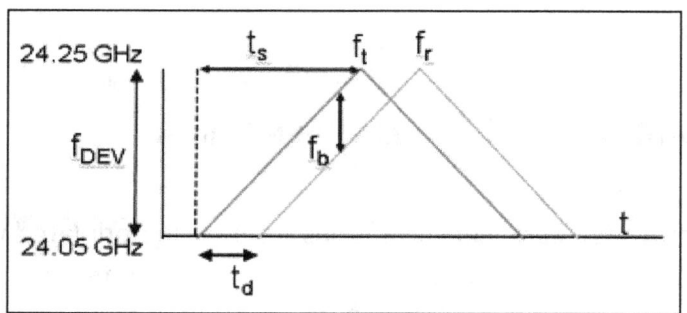

Fig.3 Sweep Frequency, Beat Frequency and Time Delay Relationship of FMCW Radar

Where f_{dev} is sweep frequency, t_s sweep time
$F_{dev}/t_s = f_b/t_d$

$R_{res} = c/2\ f_{dev}$
Where R_{res} is Range Resolution

Range is a function of fb.
$R = f_b * t_s * c/2f_{dev}$

Where R is the target range

For 24 GHZ ISM band, sweep frequency is 24 GHZ – 24.25 GHZ
Bandwidth 250 MHz
Range resolution= 0.6 mtrs.

For 24 GHZ UWB Band 24.25 GHz to 26.65 GHz
Bandwidth 2.4 GHz
Range resolution = 0.06 mtrs.

Accuracy of Radar:

The accuracy of the distance measurement depends essentially on the noise or better: the size of the noise in relation to the impulse. This quantity is described by the signal-to-noise ratio (SNR). The size of the noise itself also depends on the bandwidth. The slope of the pulse edge also depends on the bandwidth. For a signal-to-noise ratio of considerably higher than 1, the following relationship exists between these variables.
$\delta R = C/\ (2B*Sq.rt(2SNR))$
where - δR – Measuring Error
C- Velocity of Light
SNR – Signal Nosie Ratio
B- Bandwidth

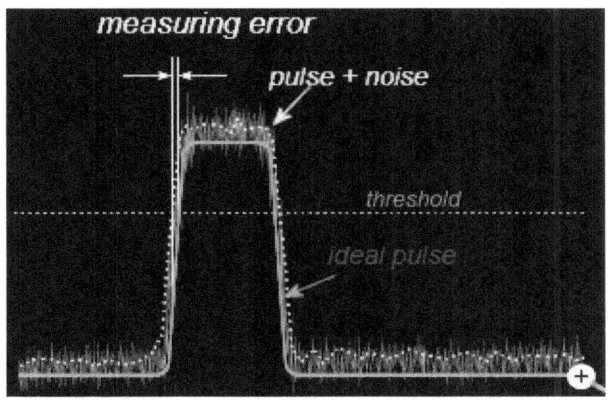

Fig.4 Falsification of the pulse edge due to
Superimposition of noise

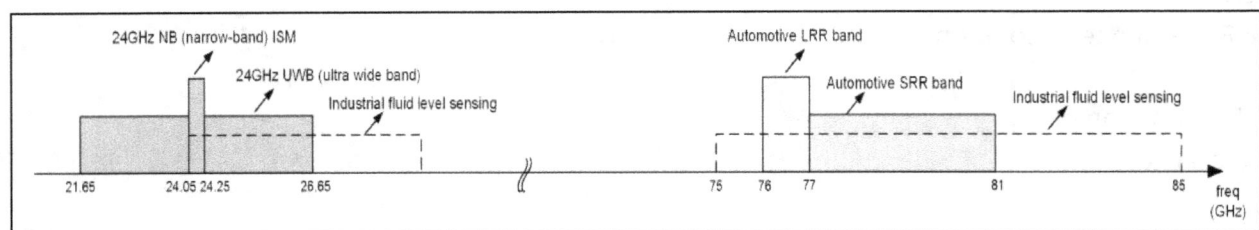

Fig.5 24 and 77 GHz frequency band usage in automotive radar

Frequency range	Band	Frequency (f)	Wavelength (λ.)
Low	C and X-band	6–11 GHz	50–30 mm (≈2-1.2-in.)
Mid	K-band	24–29 GHz	~10 mm (≈0.4-in.)
High	W-band	75–85 GHz	4–3.5 mm (≈0.15-in.)

Fig.6 Radar Frequency Band

When object is moving, then according to doppler effect the frequency difference is shown in fig.7.

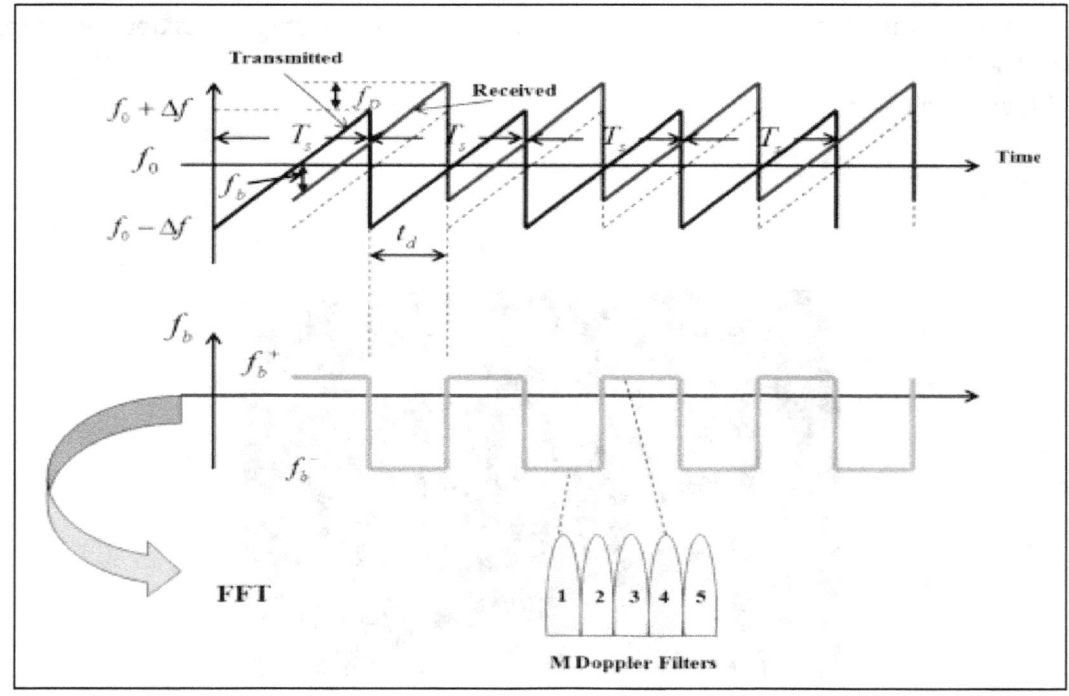

Fig.7 FMCW Radar Moving Target Echo Signal

$F_{b+} = f_b - f_d$

$F_{b-} = f_b + f_d$

$F_d = 2fv/c$

Where f is the nominal radar frequency
v is the target velocity relative to the radar.
F_d is doppler frequency

$F_{b+} = 2*f_{dev}*R/c*t_s - 2fv/c$

$F_{b-} = 2*f_{dev}*R/c*t_s + 2fv/c$

$f_b = f_{dev}*t_d/t_s + 2fv/c$

$f_b = 2Rf_{dev}/t_s c + 2fv/c$

$R = (f_b - 2fv/c) t_s c / 2 f_{dev}$

$R = (f_b - f_d) t_s c / 2 f_{dev}$

$v = \lambda/4 (f_{b-} - {}_{fb+})$

Assuming object is moving with a speed of 180km/hr, Radar of 24 GHz used. Then doppler frequency f_d is 8kHz. Means the speed of vehicle is directly proportional to the doppler frequency, lower the speed, lower will be the doppler frequency.

Simple storage applications usually have a large surface area with very little agitation, no significant false echoes from the internal structure of the tank and relatively slow product movement. These are the ideal conditions for which FM - CW radar was originally developed. However, in process vessels there is more going on and the problems become more challenging.

In an active process vessel, the various echoes are received as frequency differences compared with the frequency of the transmitting signal. These frequency difference signals are received by the antenna at the same time. The amplitude of the real echo signals are small compared with the transmitted signal. A false echo from the end of the antenna may have a significantly higher amplitude than the real level echo. The system needs to separate and identify these simultaneous signals before processing the echoes and making an echo decision.

The separation of the various received echo frequencies is achieved using Fast Fourier Transform (FFT) analysis. This is a mathematical procedure which converts the jumbled array of difference frequencies in the time domain into a frequency spectrum in the frequency domain.

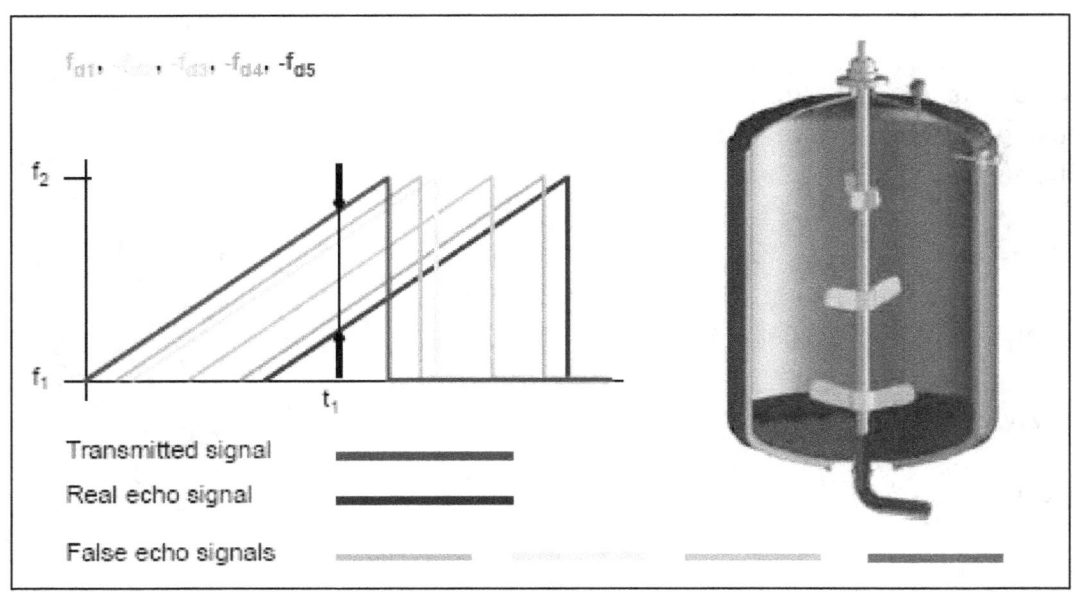

Fig.8 FMCW Radar level transmitter in an active process vessel

4.3 Pulse Wave Radar

Pulse wave radar has been widely used for distance measurement. The basic principle of pulse wave radar is on time of flight. Short pulses typically of milliseconds or nanoseconds are transmitted and the transit time to and from the target is measured.

The number of waves and length of the pulses depends on the pulse duration and the carrier frequency that is used. These regularly repeating pulses have a relatively long-time delay between them, to allow the return echo to be received before the next pulse to be transmitted.

Low amplitude signals and false echoes are common in chemical reactors where there is agitation and low dielectric liquids. Solids applications can be troublesome because of the internal structure of the silos and undulating product surfaces which creates multiple echoes.

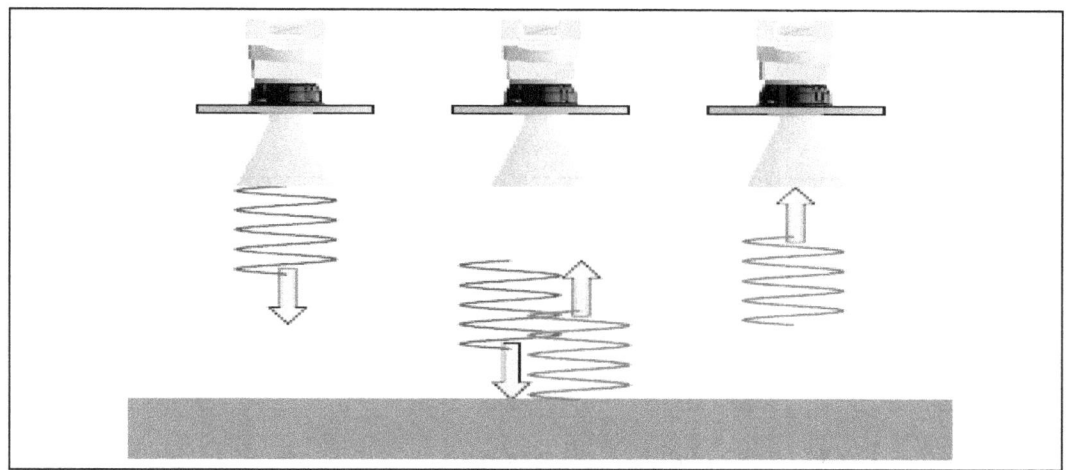

Fig.9 Pulse Radar operates purely within the time domain. Millions of pulses are transmitted every second and a special sampling technique is used to produce a 'time expanded' output signal

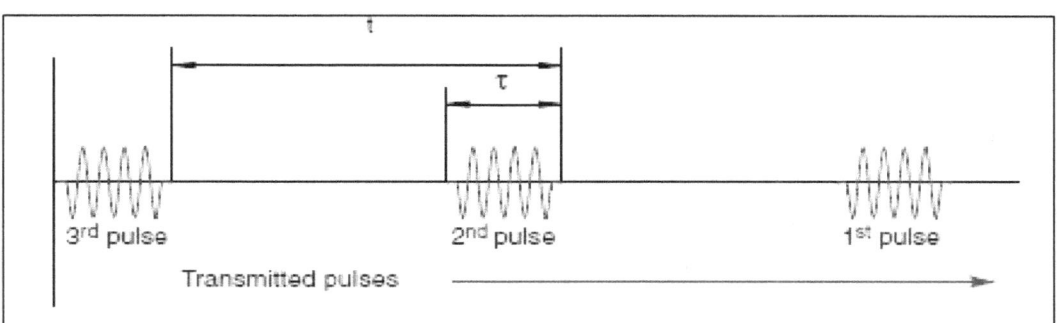

Fig.10 Basic Pulse Radar

Fig.10 shows pulse radar principle where t is inter pulse period and Ʊ pulse width.

Distance from target:

R= T*c/2

Where T is the time taken from transmission to received pulse.

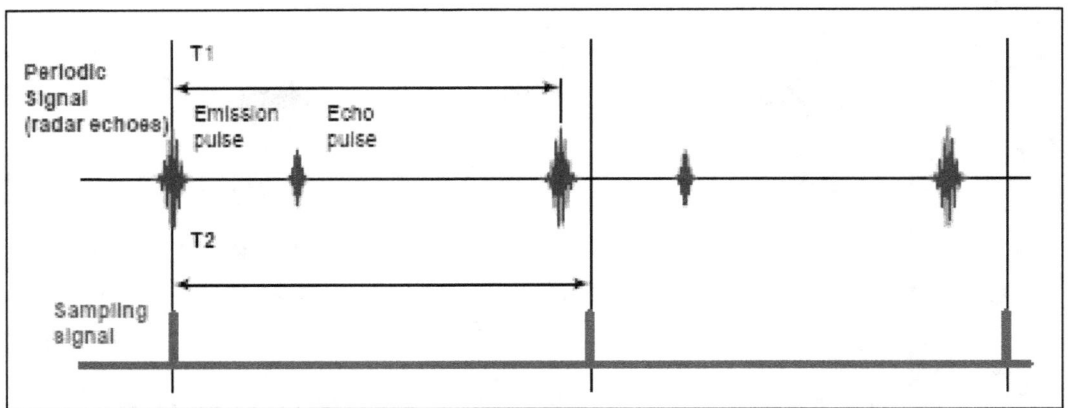

Fig.11 Sequential Sampling of a pulse radar echo curve

Fig.11 Sequential Sampling of pulse radar echo curve. Millions of pulses per second produce a periodically repeating signal. A sampling signal with a slightly longer periodic time produces a time expanded image of the entire echo curve.

4.4 Radar Antenna Size and Beam Angle

The higher the frequency of Radar better will be the focus or smaller beam angle for the same size of antenna. For eg. 1.5-inch horn antenna radar of 26 GHz frequency has beam angle equal to 6-inch horn antenna radar of 5.8 GHz.

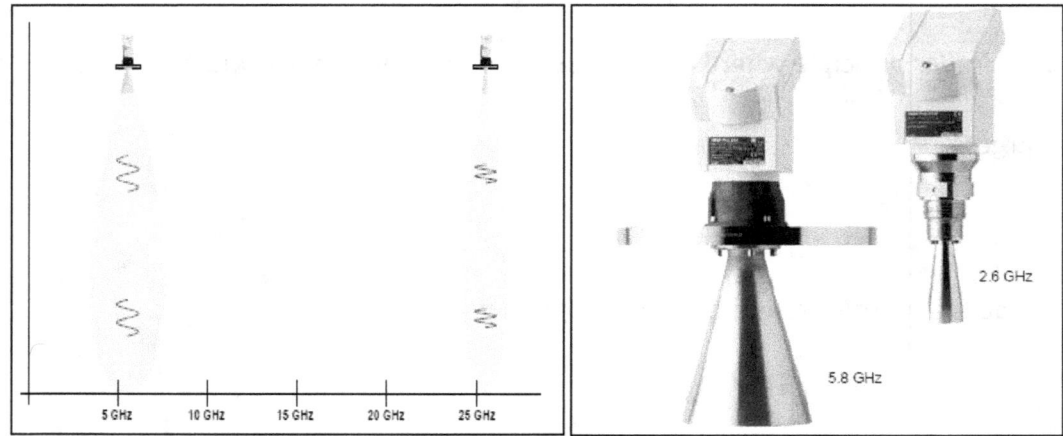

Fig.12 For a same size antenna, a higher frequency gives a more focused beam

The beam angle and beam width is determined by the antenna design in combination with the microwave frequency. High frequency signals can achieve small beam angles with small antennas. Equally, small beam angles can be achieved with low frequency radars, but this requires larger antennas. The benefit of a small beam angle in level measurement is that it can make it easier to avoid hitting installations in the tank. However, a narrow beam width can also be a disadvantage. For example, if there is an obstruction directly below the radar a narrow beam will be completely blocked, but a radar with a wider beam will be only partially blocked and still able to measure the product level.

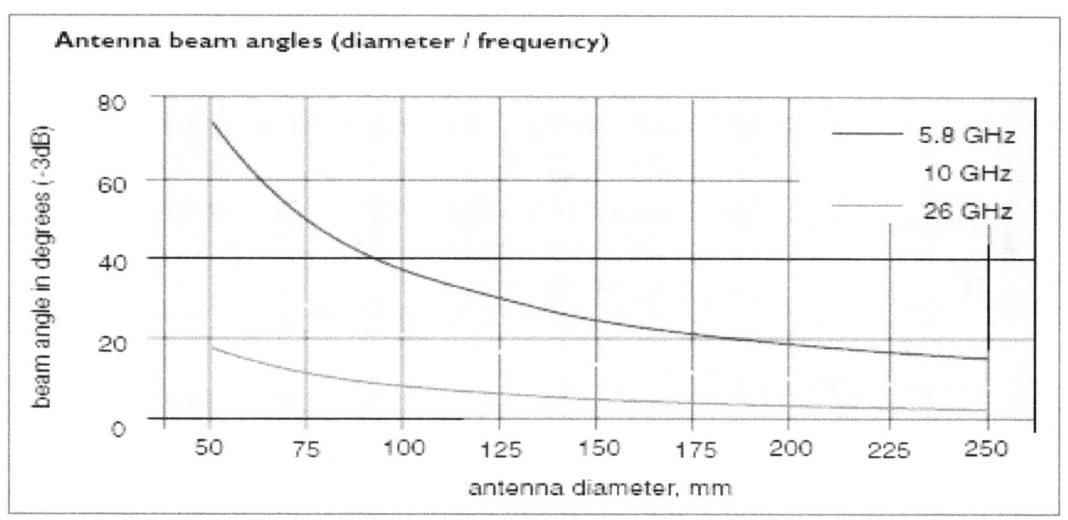

Fig.13 Graph showing relation between horn antenna diameter and beam angle for 5.8 GHz, 10Ghz and 26GHz radar

4.5 Antenna focusing and False Echoes

The wavelength of 26 GHz radar is only 1.15 cms compared to radar of frequency 5.8 GHz which have wavelength of 5.2cms. The short wavelength of the 26 GHz radar means that it will off many small objects that may be effectively ignored by 5.8 GHz radar.

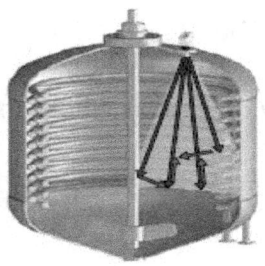

a) fig.14 shows low frequency radar which has wider bear angle and therefore if the installation is not proper, it will see more false echoes.

b) fig.14 shows high frequency radar which has more focus beam angle for a given antenna size. The focus beam is necessary because the short wavelength 1.5 cms for 26 GHz radar will reflect more with internal structures of the vessel.

c) fig.14 shows high frequency radar transmitter which are susceptible to scatter of signals when hits with unwanted or small objects inside the vessel. It is importance that radar software can handle these false echoes. By comparing with 5.8 GHz radar, it is not adversely affected by the agitated liquid surfaces. Lower frequency radar is better suited for solid level measurement applications.

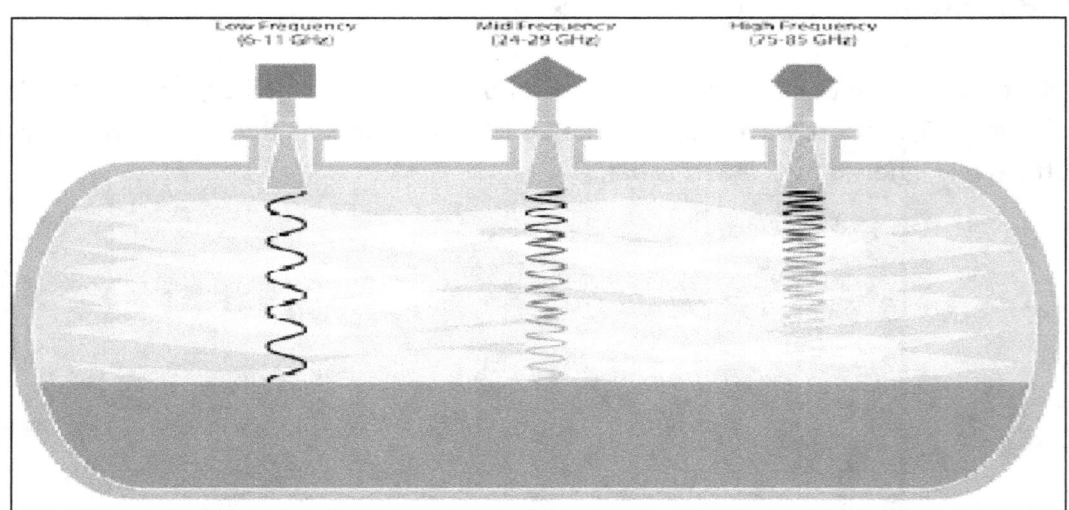

Fig.15 Loss of Signal Strength Due to Attenuation

Fig.15 shows when radar signals propagating through a media, some signals are absorbed and signal strength decrease. High frequency radar signals suffer more attenuation than a low frequency radar.

4.6 Radar Antenna

The function of the antenna is to direct the maximum amount of microwaves energy towards the target and to capture maximum amount of echoes for analysis within the electronics.

A measure of how well antenna is directing microwave energy towards target is called antenna gain.

Beam Angle $\emptyset = 70° * \lambda / D$

Where D: horn antenna diameter; λ – microwave wavelength

Beam width decreases with increasing center frequency in the case that the diameter of the antenna is kept constant. Furthermore, in the case of keeping the frequency constant, the beam width also decreases with increasing diameter of the antenna.

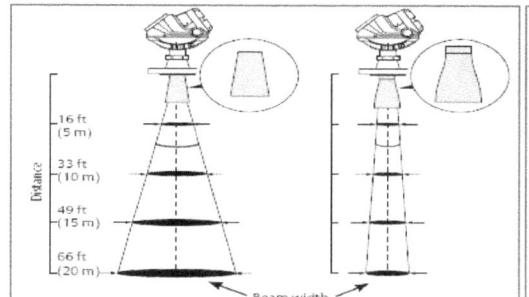

Distance	Antenna		
	4 in. (DN 100) cone /rod	6 in. (DN 150) cone	8 in. (DN 200) cone
	Beam width ft (m)		
16 ft (5 m)	11.5 (3.5)	6.6 (2.0)	4.9 (1.5)
33 ft (10 m)	23.0 (7.0)	13.1 (4.0)	9.8 (3.0)
49 ft (15 m)	32.8 (10)	19.7 (6.0)	14.8 (4.5)
66 ft (20 m)	42.7 (13.0)	26.2 (8.0)	19.7 (6.0)

Fig.16 Beam Width at Various Distances from the Flange

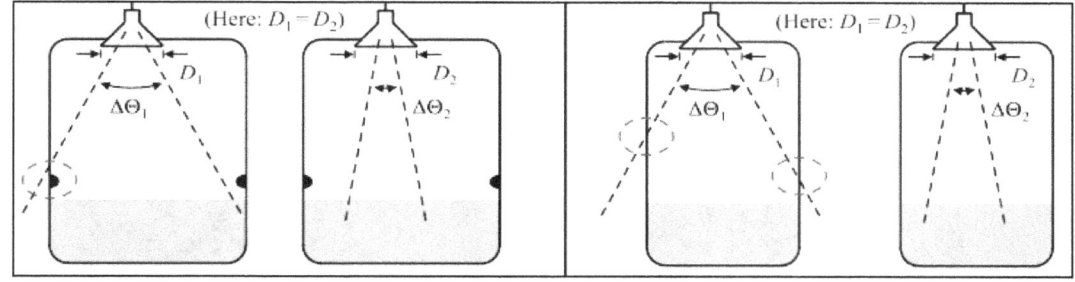

Fig.17 Measurement in Narrow Tanks

Horn Antenna:

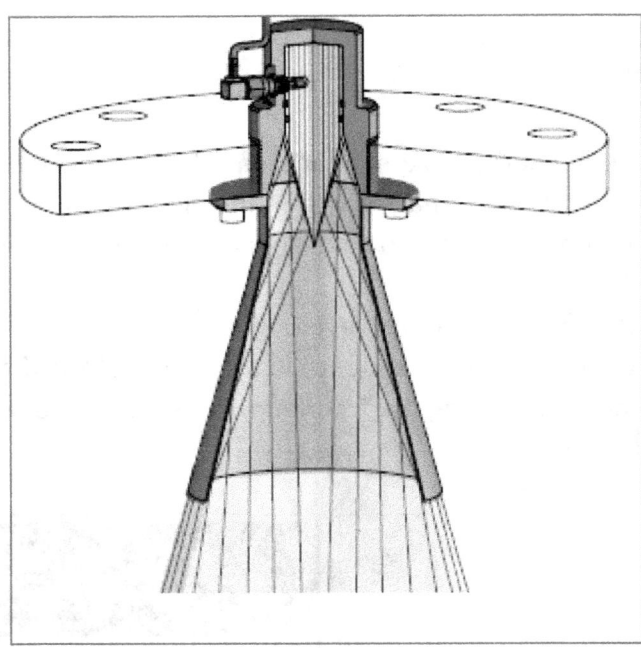

Fig.18 The transition of microwaves from the low dielectric waveguide into the metallic horn where they are focused towards the product being measured

Fig.18 The transition of microwaves from the low dielectric waveguide into the metallic horn where they are focused towards the product being measured.

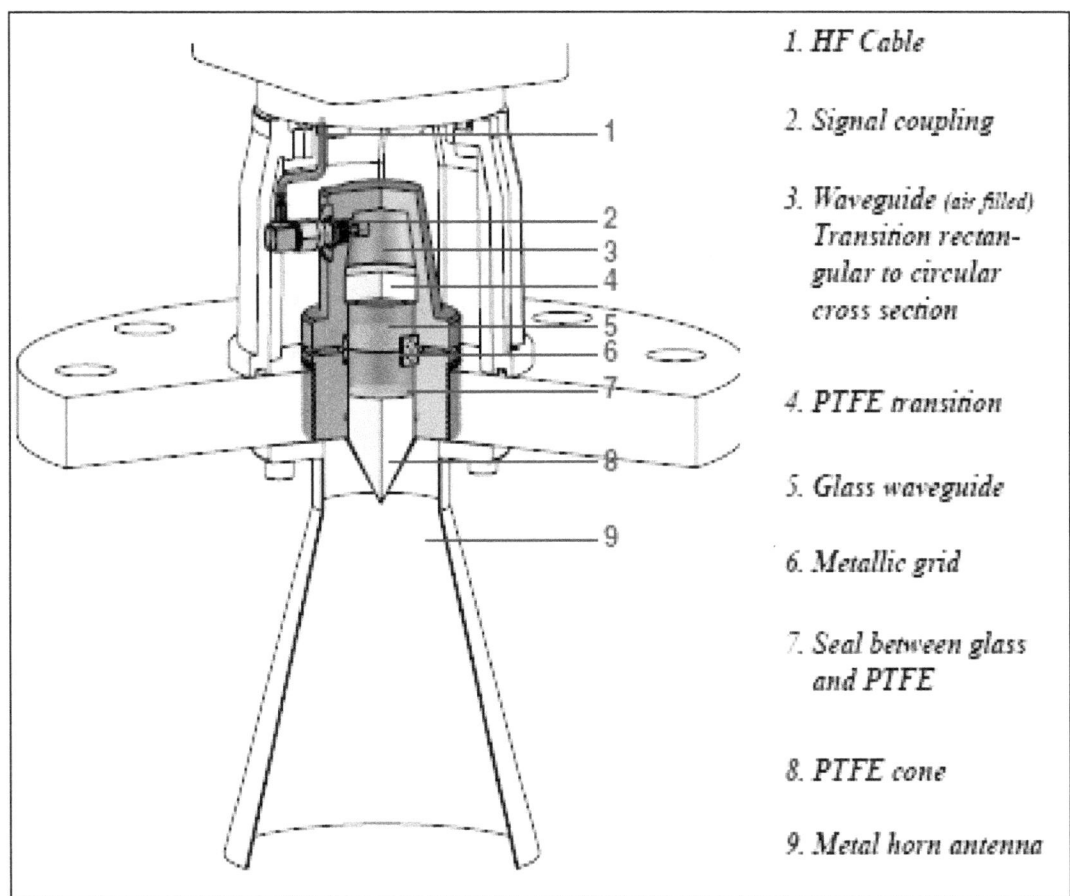

1. HF Cable

2. Signal coupling

3. Waveguide *(air filled)* Transition rectangular to circular cross section

4. PTFE transition

5. Glass waveguide

6. Metallic grid

7. Seal between glass and PTFE

8. PTFE cone

9. Metal horn antenna

Fig.19 Horn Antenna Design

Fig.19 show the microwaves that are generated within the microwave module are transmitted down a high frequency cable for encoupling into a waveguide. The metal waveguide then directs the microwaves towards the horn of the antenna. A low dielectric material such as PTFE, ceramic or glass is often used within the waveguide.

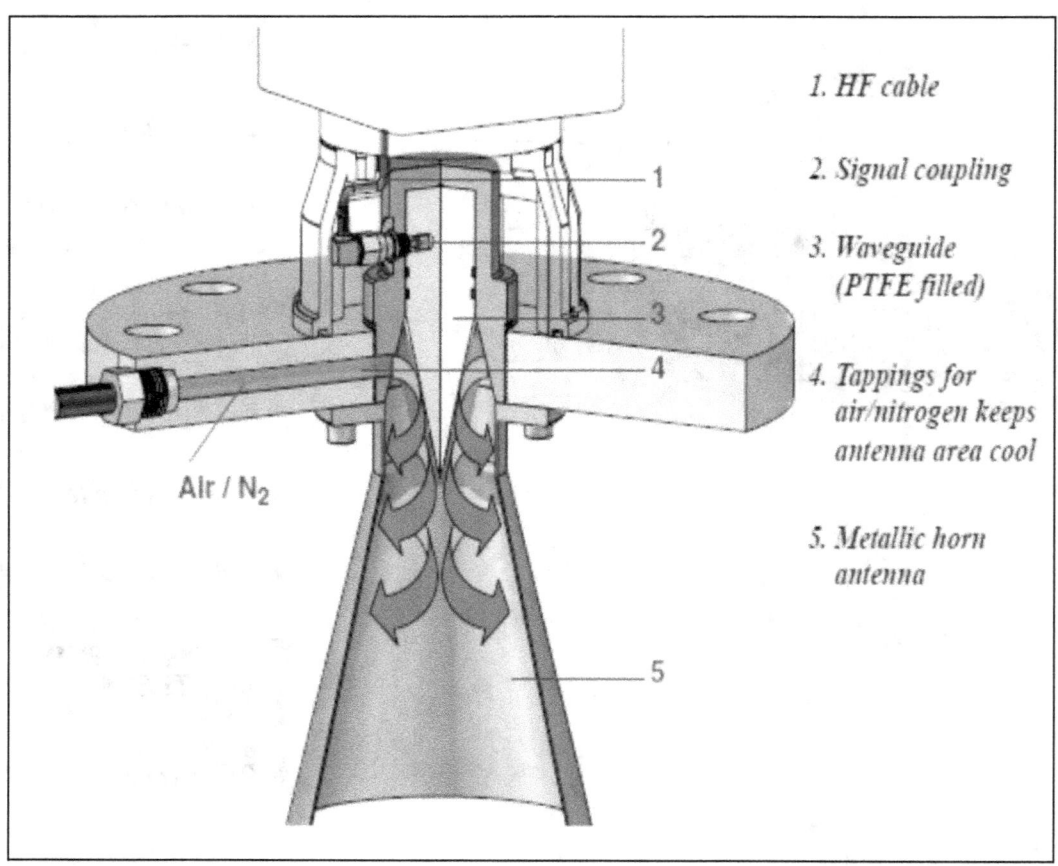

1. *HF cable*

2. *Signal coupling*

3. *Waveguide (PTFE filled)*

4. *Tappings for air/nitrogen keeps antenna area cool*

5. *Metallic horn antenna*

Air / N$_2$

Fig.20 Horn Antenna Design-2

Fig.20 shows Horn Antenna design-2 which is used for very high temperature, ambient pressure application with nitrogen or air cooling through flange. The flow of air or nitrogen prevents hot gases from affecting the PTFE and the viton seal and it effectively cools the entire flange and horn area. This technique has been used successfully with very high temperatures, including 1500° C + in the steel industry with applications such as blast furnace burden level and molten iron ladle levels. The microwaves are unaffected by the air movement within the horn area.

Fig. 21 shows 26 GHz High Frequency Radar Antenna Design

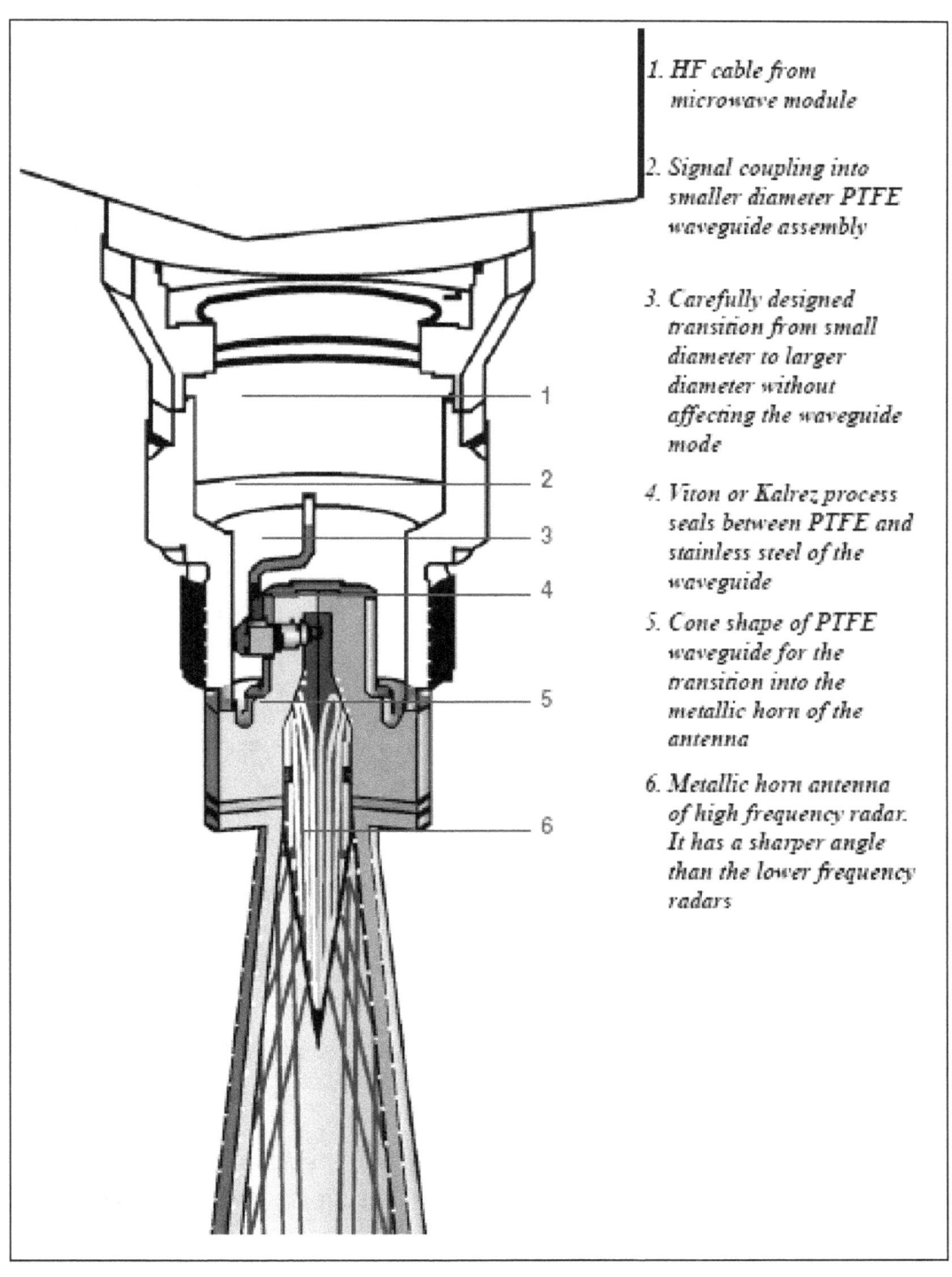

1. HF cable from microwave module

2. Signal coupling into smaller diameter PTFE waveguide assembly

3. Carefully designed transition from small diameter to larger diameter without affecting the waveguide mode

4. Viton or Kalrez process seals between PTFE and stainless steel of the waveguide

5. Cone shape of PTFE waveguide for the transition into the metallic horn of the antenna

6. Metallic horn antenna of high frequency radar. It has a sharper angle than the lower frequency radars

Fig.21 High frequency (26GHz) horn antenna design

4.6 Measuring Tube Antennas

There are applications within the process industry where the installation of an antenna directly within a vessel is not suitable for reasons of vessel design or radar functionality. In these cases a measuring tube (bypass tube or a stand pipe within the vessel) may be an alternative. Normally 40 mm and 50 mm tubes do not require horn, for 80 mm and above appropriate horn antenna is designed and fits into measuring tube. Fig.22 shows installation of horn antenna radar into stand pipes.

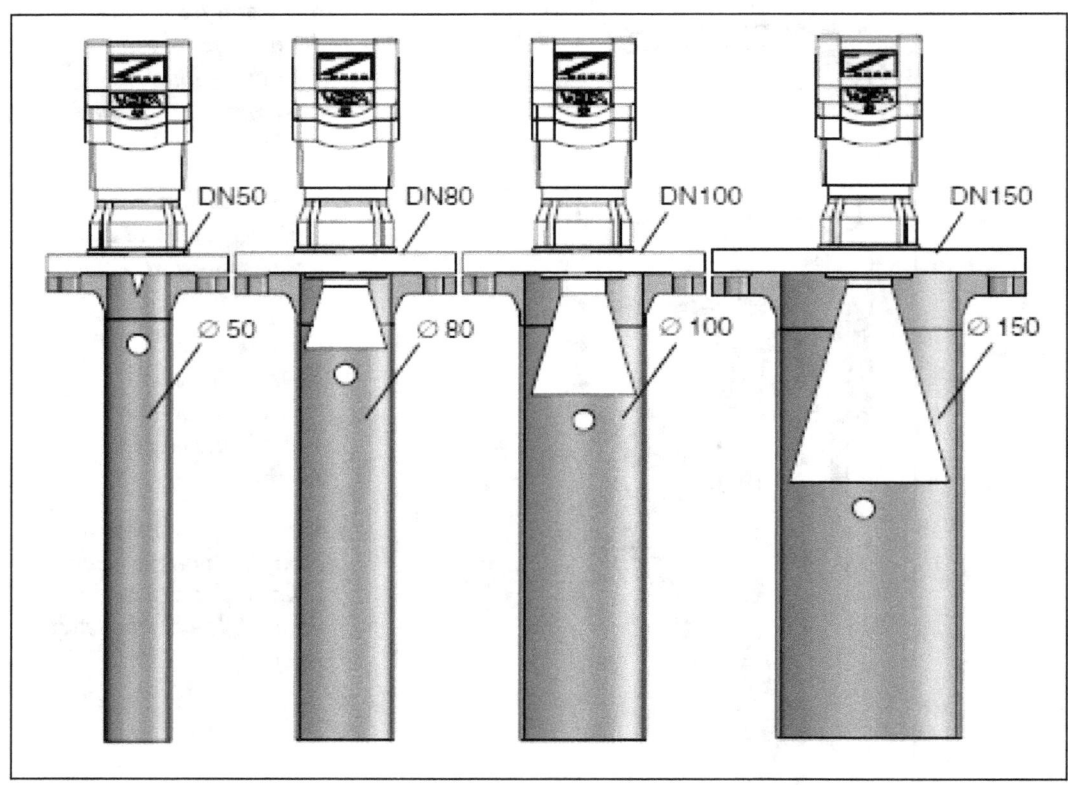

Fig.22 Installation of horn antenna radars into stand pipes or bypass tube

Fig.23 shows the microwaves bounces off from the side of walls of the tube. That is why speed of microwaves is slower in measuring tubes as compared to free space. Hence it is important to know the diameter of measuring tube to allow microwave propagation without disturbance. A waveguide extension should use when a radar device with horn antenna in a long one nozzle is installed see fig. 25 (6).

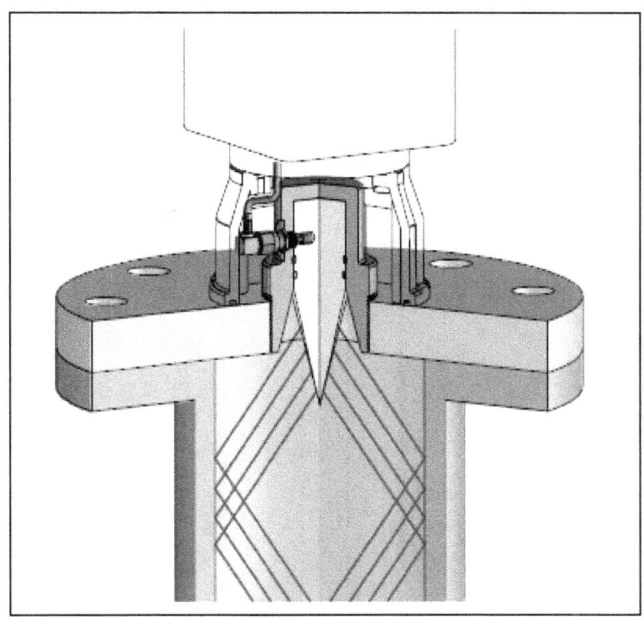

Fig.23 The transit time of microwaves is slower within a stilling tube. This effect must be compensated within the software of the radar level transmitter

The value of the critical diameter of measuring tube depends on the wavelength of microwaves. The higher the frequency of microwaves, the smaller measuring tube can be used.

For eg. With a frequency of 26GHz microwave signal radar with a wavelength of 11.5mm, requires measuring tube diameter of 6.75mm while radar having frequency of 5.8 GHz with wavelength of 52mm requires measuring tube diameter of 31mm.

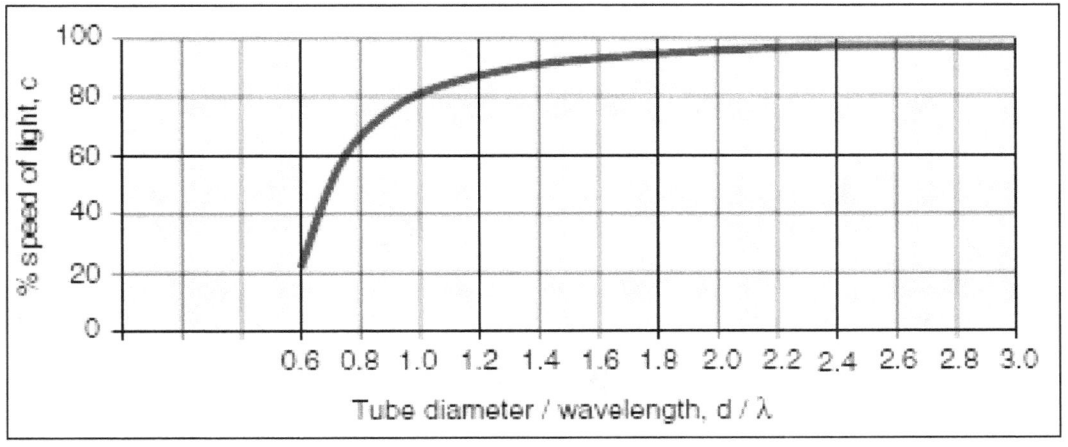

Fig.24 Effect of Measuring tube diameter on microwaves speed

4.7 Installation of Radar

Fig.25 shows correct installation methods of Radar in process application.

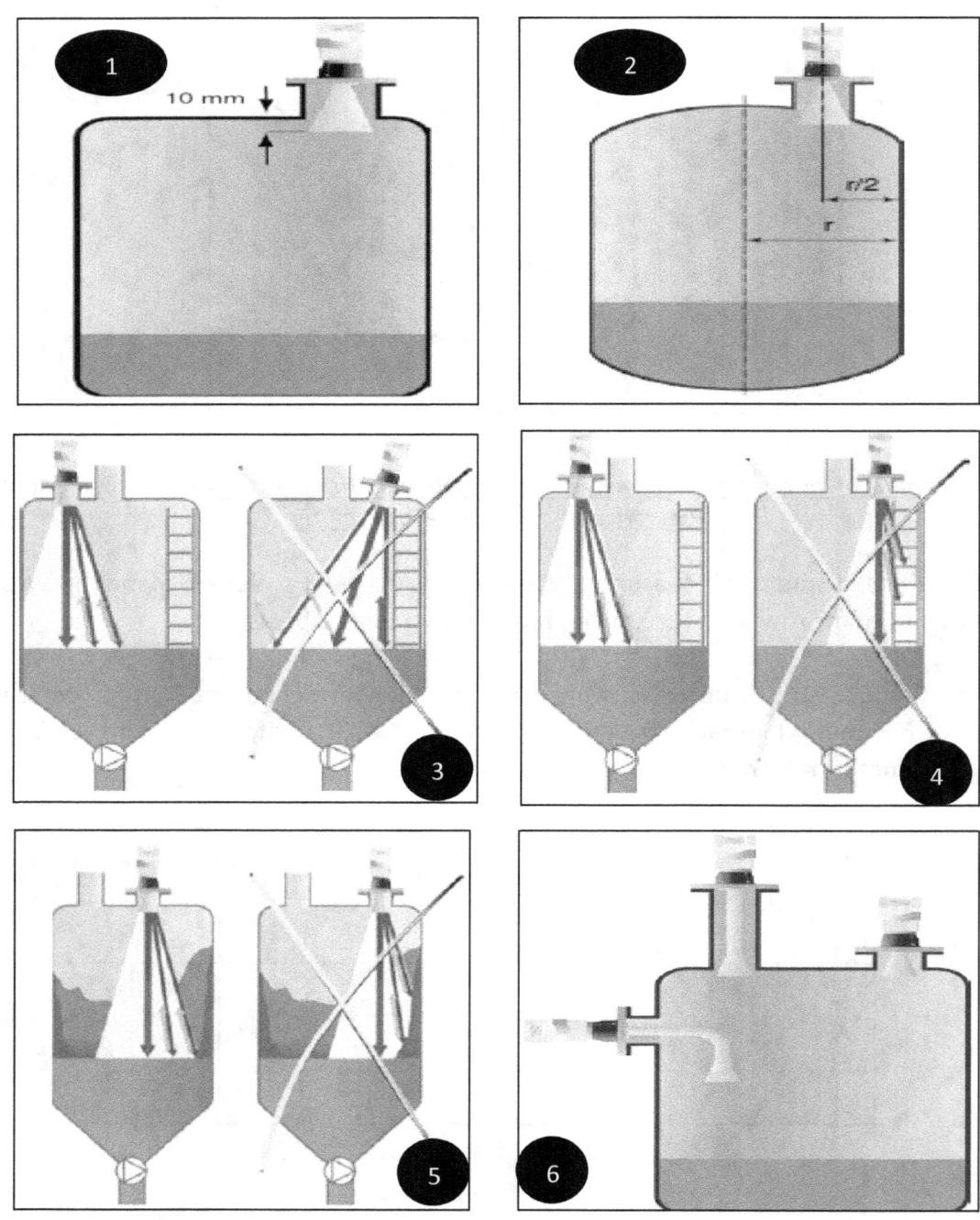

Fig.25 Installation Guideline or Practices for Unguided Wave Radar

4.8 Ultrasonic Level Transmitter

Ultrasonic measuring technology operates on the simple principle of measuring the time it takes sound to travel a distance.

The sound signals are caused by the mechanical vibration of the object. The vibration is transferred to the gas modules in the surrounding medium within which it is contained. If there is no gas, as in a perfect vacuum, then there will be no propagation of sound.

Measurement Principle of Ultrasonic Level Transmitter:

A piezoelectric crystal inside the transducer converts and electrical signal into sound energy, firing a burst of sound waves (typically 50KHZ) into the air where it travels to the target, after which it is reflected back to the transducer. The transducer then acts as a receiving device and converts the sonic energy back into an electrical signal. An electronic signal processor then analyses the return echo and calculate the distance between target and transducer.

$D = t*c/2$

Where D- Distance

t- time taken for transmission and receiving of ultrasonic waves

c- velocity of sound

for eg. Speed of sound in air 344 m/sec with ambient temperature condition. Therefore, if it takes 58.2 ms for the echo to be detected, then D is 10 mtr.

Sound Velocity and Temperature:

Temperature changes affect the velocity of sound in air, and the variation in temperature requires compensation to achieve accurate measurement. Fig.26 shows temperature vs velocity of sound in air curve.

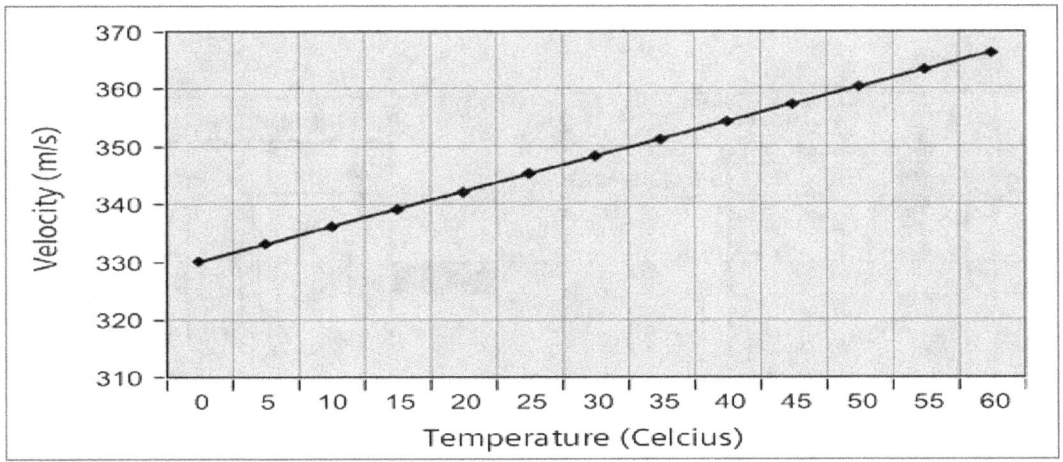

Fig.26 Speed of Sound vs. Air Temperature

Beam Width:

For ultrasonic level measurement, wide dispersion is undesirable. The narrower the beam width, the less likely vessel obstructions will be detected.

For short and wide vessels, a 12-degree beam width is ideal to simplify aiming. For tall, narrow vessel 5-6-degree beam width is good.

$Sin\alpha/2 = 1.2\lambda/D$

Where α divergence of beam; λ- wavelength; D- Diameter

Blanking Distance:

Ultrasonic sensors are a little farsighted. By nature of the technology, most do not measure surfaces that are within a few inches of the transducer face. This is known as a *blanking distance*.

Ultrasonic sensors produce and detect vibrations. They pulse to produce sound – much like a speaker – and they listen for the returning sound. When the sound wave does return, it produces a small vibration on the face of the sensor. It is this second vibration that the sensor associates with the target surface. The time the sensor waits for the return vibration is used to calculate distance. The sensor can then determine tank levels. The sensor uses the same transducer to both produce and listen for the sound wave, it can become confused without a little logic. When the transducer first pulses, it naturally causes residual vibration for a split second. The sensor must be programmed to ignore signals for as long as it takes the residual vibration to stop. This programmed time equals a certain distance. Therefore, the waiting period is called a blanking distance. Without a programmed blanking distance, the ultrasonic sensor would immediately signal a false surface right up against the transducer face. The blanking distance is part of the logic of the sensor, and is how it distinguishes between pulse vibrations and return vibrations.

The length of the blanking distance is tied directly to the frequency of the transducer and the power of the pulse. A small, low-range transducer will produce a smaller pulse that will not vibrate as long. Therefore, smaller ultrasonic sensors enjoy shorter blanking distances, as little as 4 inches. Large ultrasonic can have blanking distances as long as 1½ feet or more.

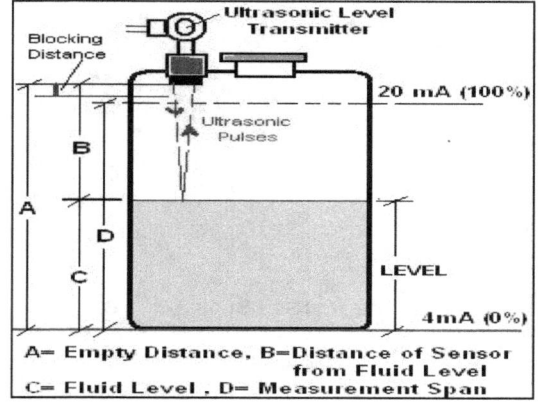

Fig. 27 (a) Blanking Distance of Ultrasonic Level Transmitter

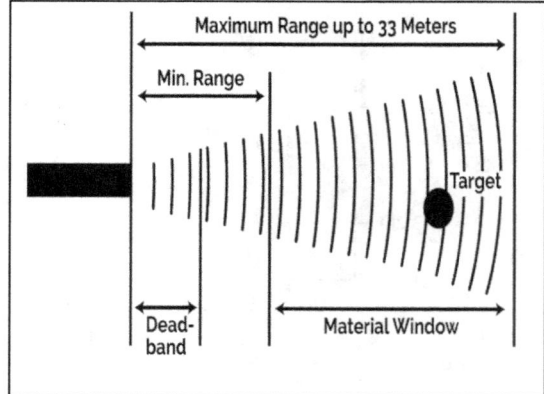

Fig. 27 (b) Blanking Distance of Ultrasonic Level Transmitter

4.9 Data Sheet for Ultrasonic Level Transmitter

	RESPONSIBLE ORGANIZATION	NON-CONTACT ULTRASONIC LEVEL TRANSMITTER w/wo SWITCHES Device Specification		SPECIFICATION IDENTIFICATIONS
1			6	
2	(ISA)		7	Document no
3			8	Latest revision Date
4			9	Issue status
5			10	

	PROCESS CONNECTION			PERFORMANCE CHARACTERISTICS
11			60	
12	Body/Fitting type		61	Max press at design temp At
13	Process conn nominal size Rating		62	Min working temperature Max
14	Process conn termn type Style		63	Accuracy rating
15	Wetted material		64	Level lower range limit URL
16	Flange material		65	Min measurement span Max
17	Seal/O ring material		66	Beam angle
18			67	Min ambient working temp Max
19			68	Contacts ac rating At max
20			69	Contacts dc rating At max
21	SENSING ELEMENT		70	Max sensor to receiver ig
22	Detector type		71	
23	Detector style		72	
24	Insertion length		73	
25	Integral cable length		74	
26			75	
27			76	
28			77	
29	CONNECTION HEAD		78	
30	Type		79	
31	Enclosure type no/class		80	
32	Cert/Approval type		81	
33	Mounting location/type		82	
34	Enclosure material		83	
35			84	ACCESSORIES
36			85	Connecting cables length
37			86	Enclosure heater
38	TRANSMITTER w/wo SWITCHES		87	Mounting hardware
39	Housing type		88	
40	Measurement compensation		89	
41	Output signal type		90	
42	Enclosure type no/class		91	
43	Control sequence		92	SPECIAL REQUIREMENTS
44	Characteristic curve		93	Custom tag
45	Digital communication std		94	Reference specification
46	Signal power source		95	Compliance standard
47	Contact arrangement Quantity		96	Calibration report
48	Failsafe style		97	
49	Integral indicator style		98	
50	Cert/Approval type		99	
51	Mounting location/type		100	PHYSICAL DATA
52	Failure/Diagnostic action		101	Estimated weight
53	Signal processing		102	Overall height
54	Enclosure material		103	Removal clearance
55			104	Signal conn nominal size Style
56			105	Mfr reference dwg
57			106	
58			107	
59			108	

	CALIBRATIONS AND TEST		INPUT OR SETPOINT			OUTPUT OR SCALE	
110							
111	TAG NO/FUNCTIONAL IDENT	MEAS/SIGNAL/TEST	LRV	URV	ACTION	LRV	URV
112		Level–Analog output					
113		Level-Scale					
114		Level setpoint 1-Output					
115		Level setpoint 2-Output					
116		Level setpoint 3-Output					
117							

	COMPONENT IDENTIFICATIONS		
118			
119	COMPONENT TYPE	MANUFACTURER	MODEL NUMBER
120			
121			
122			
123			
124			
125			

Rev	Date	Revision Description	By	Appv1	Appv2	Appv3	REMARKS

1. Data Sheet of Ultrasonic Level Transmitter as per ISA

4.10 Level Measurement through Pressure or Differential Pressure

Differential pressure (DP) installations are a straight forward level measurement technique that is easily verified and calibrated. For open or vented tanks, the measurement is done using a gage or differential pressure transmitter installed with an impulse line or a single remote seal. For closed or pressurized tanks, the measurement is done using a differential pressure transmitter with two impulse lines, two seals, or one of each (impulse line and remote seal). The measurement can also be made using two gage pressure transmitters that are linked together digitally and DP is calculated in one of the two devices.

DP level applications may be categorized into four sections:

• Direct mount level transmitters where a flange, seal, and fill fluid assembly is mounted to an open or vented tank.

• Wet systems where impulse piping connects the vessel to the transmitter and is filled with process fluid or condensed vapours.

• Balanced seal system transmitters where seals are attached to both tank connections and an equal length of capillary is used to connect the seals to the transmitter.

• Electronic Remote Sensors, where two-gauge (or absolute) pressure sensors are used to calculate DP electronically. One sensor is direct-mounted at the bottom of the vessel. The other is direct mounted at the top of the vessel, and the two sensors are connected with an electrical wire.

More detail covered in chapter 1 "Pressure Measurement".

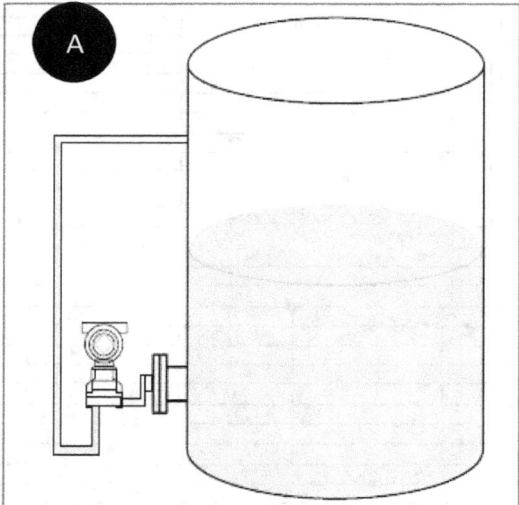
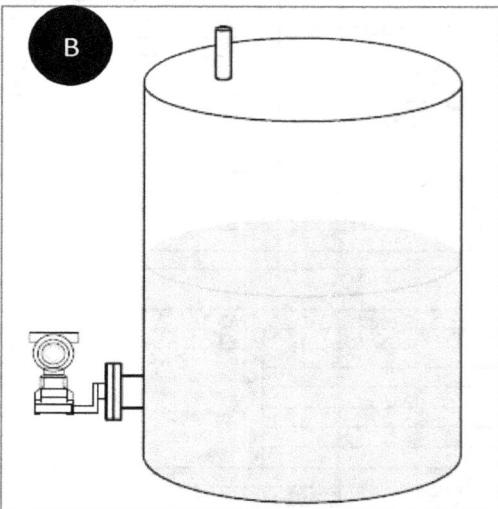

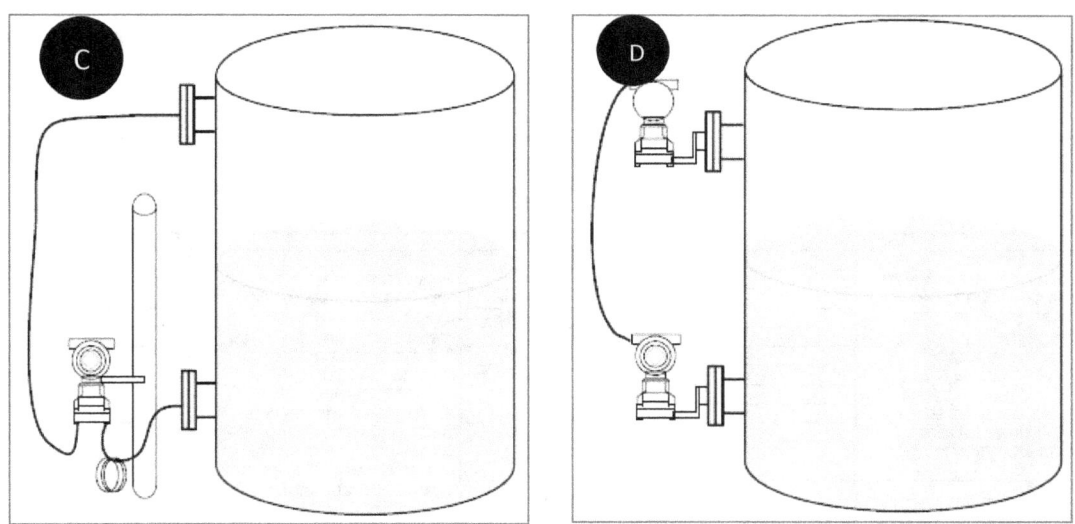

Fig.28 a) Wet System b) Direct Mounted c) Balanced Seal d) Electronic Remote Sensors

4.11 Data Sheet for Level Measurement through Differential Pressure

#	RESPONSIBLE ORGANIZATION	DIFFERENTIAL PRESSURE LEVEL TRANSMITTER - FLANGE MOUNTED Device Specification	#	SPECIFICATION IDENTIFICATIONS
1	(ISA)		6	
2			7	Document no
3			8	Latest revision — Date
4			9	Issue status
5			10	

#	TRANSMITTER BODY		#	REMOTE SEAL	
11	TRANSMITTER BODY		60	REMOTE SEAL	
12	Body/Flange type		61	Seal type	
13	Process conn nominal size	Rating	62	Process conn nominal size	Rating
14	Process conn termn type	Style	63	Process conn termn type	Style
15	Vent/Drain location		64	Diaphragm extension lg	
16	Mounting type		65	Instrument conn nom size	
17	Body/Flange material		66	Flushing conn quantity	
18	Vent/Drain material		67	Capillary/Fitting dia	Length
19	Bolting material		68	Diaphragm material	
20	Flange adapter material		69	Capillary-armor material	
21	Gasket/O ring material		70	Bolting material	
22	Mounting kit material		71	Upper housing material	
23			72	Lower housing/Flange matl	
24			73	Gasket/Oring material	
25			74	Fill fluid material	
26	SENSING ELEMENT		75		
27	Detector type		76	PERFORMANCE CHARACTERISTICS	
28	Min diff pressure span	Max	77	Max press at design temp	At
29	Diaphragm/Wetted material		78	Min working temp	Max
30	Fill fluid material		79	Accuracy rating	
31			80	Diff pressure LRL	URL
32			81	Fill fluid sp gr	At temp
33			82	Ambient temperature error	
34	TRANSMITTER		83	Min ambient working temp	Max
35	Output signal type		84		
36	Enclosure type no/class		85		
37	Characteristic curve		86		
38	Digital communication std		87	ACCESSORIES	
39	Signal power source		88	Air set filter style	
40	Transient protection		89	Air set gauges	
41	Integral indicator style		90	Heating kit style	
42	Signal termination type		91	Remote indicator style	
43	Cert/Approval type		92		
44	Span-Zero adjust loc		93		
45	Failure/Diagnostic action		94	SPECIAL REQUIREMENTS	
46	Enclosure material		95	Custom tag	
47			96	Reference specification	
48			97	Special preparation	
49			98	Compliance standard	
50	INTEGRAL SEAL		99	Software configuration	
51	Seal type		100		
52	Process conn nominal size	Rating	101	PHYSICAL DATA	
53	Process conn termn type	Style	102	Estimated weight	
54	Diaphragm extension lg		103	Overall height	
55	Diaphragm material		104	Removal clearance	
56	Upper housing/Flange matl		105	Signal conn nominal size	Style
57			106	Mfr reference dwg	
58			107		
59			108		

#	CALIBRATIONS AND TEST		INPUT OR TEST			OUTPUT OR SCALE	
110	CALIBRATIONS AND TEST		INPUT OR TEST			OUTPUT OR SCALE	
111	TAG NO/FUNCTIONAL IDENT	MEAS/SIGNAL/TEST	LRV	URV	ACTION	LRV	URV
112		Diff press-Analog output					
113		Diff press-Scale					
114		Diff press-Digital output					
115		Press-Digital output					
116		Temp-Digital output					
117							

#	COMPONENT IDENTIFICATIONS		
118	COMPONENT IDENTIFICATIONS		
119	COMPONENT TYPE	MANUFACTURER	MODEL NUMBER
120			
121			
122			
123			
124			
125			

Rev	Date	Revision Description	By	Appv1	Appv2	Appv3	REMARKS

2. Data Sheet of Level Measurement by Differential Pressure as per ISA

4.12 Bubbler Level Measurement

Liquid level is determined by measuring the pressure required to force a gas into the liquid at a point beneath the surface. In this way, the level may be obtained without the liquid entering the piping or instrument. A source of clean air or gas is connected through a restriction to a bubble tube immersed a fixed depth in the tank. The restriction reduces the air flow to a minute amount, which builds up pressure in the bubble tube until it just balances the fluid pressure at the end of the bubble tube. Thereafter, pressure is kept at this value by air bubbles escaping through the liquid. Changes in the measured level cause the air pressure in the bubble tube to build up or drop. A pressure instrument connected at this point can be made to register the level or volume of liquid.

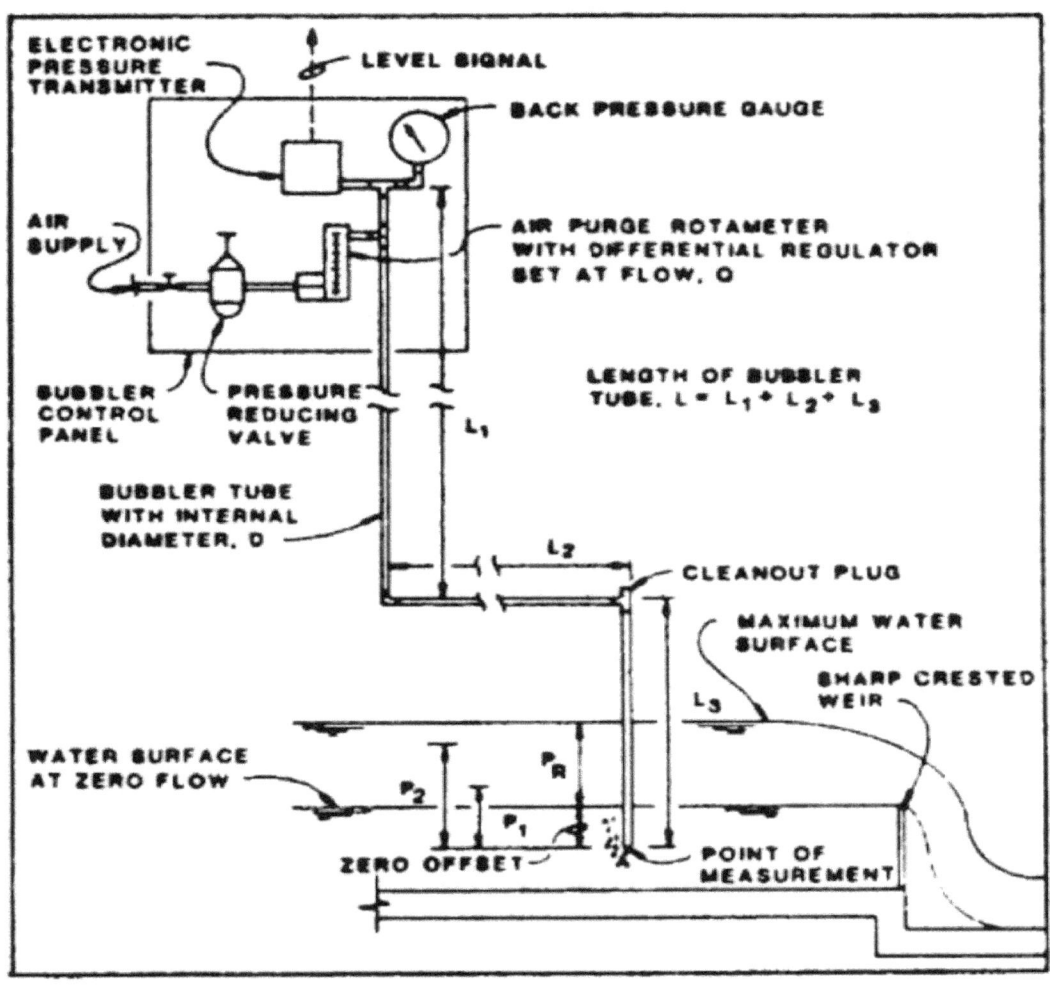

Fig.28 Bubbler Level Measurement

Fig.28 shows when the water depth over the bottom of the bubbler tube increases from P1 to P2, the back pressure in the bubbler tube also increases from P1 to P2. The time the back pressure increases from P1 to P2 in response to the depth changes from P1 to P2 is equal to the time the mass of air changes in the bubbler tube from M1 to M2.

P1/M1 = P2/M2

The rate of changes of the mass of air in the bubbler tube is equal to:

Dm/Dt = DP/dt (V/RT)

Where V is the volume of air in the bubbler tube, R is the gas constant and T is the air temperature.

To calculate level:

P= ρ * g * h

Where ρ is density of water; g is gravity 9.8m/sec2; h is the height of liquid in a tank; P is the back pressure of compressed air/instrument air/nitrogen.

4.13 References

- Radar Level Measurement by Peter Devine.
- FMCW Radar Design by M. Jankiraman
- Rosemount 5400 Manual
- White Paper on FMCW vs. Pulse Radar by Hawk Measure
- Understanding Ultrasonic Level Measurement by Stephen Milligan, Henry Vandelinde and Michael Cavanagh

5

Process Gas Analyzer

After completing this chapter, you should be able to:

Know about Sample Handling System, Different Components used in Sample Handling System, Principle of NDIR, Paramagnetic and Thermal Conductivity Based Analyzer

Maintenance and Troubleshooting Methods

Know about International Standards for Process Gas Analyzer Design and Selection

5.1 Process Gas Analyzer

Process monitors that measure and transmit information about chemical composition, physical properties are known as process analyzers. A process monitors usually requires sample conditioning system, a process analyzer and one or more output data devices.

Process Analyzer Selection Design Requirements:

1. Economics: Analyzer system can improve product quality, increase the yields of products with higher economic value and reduce energy cost. Process analyzer system should be considered for product quality control when frequent and rapid measurement are required because of fast variation in process stream.

2. Environment and Safety: Today environment norms are very stringent. The use of analyzers for safety or environmental monitoring should be considered for analytical application techniques that comply with the government regulations. Process analyzers are used to detect hazardous plant conditions.

3. Application Requirement: Primary factors are considering while selecting process analyzer;

a) Measured Variable
b) Measurement Range
c) Measurement Purpose
d) Repeatability and Accuracy

e) On Stream Factor
f) Overall Response Time
g) Sample Conditioning
h) Installation
i) Maintenance

5.2 Sample Conditioning System

Sample conditioning system are comprised of all components necessary to extract a sample and to condition the sample for measurement by the analyzer. The main function of sample conditioning system is to condition the sample by adjusting the flow, pressure, temperature, filtering etc.

Design Factors:

• **Sample Point Location**: The following factors should be considered in determining optimum sample point location. Locate the analyzer as nearer to the sample point to minimise the transportation time. Location should be considered where maintenance can be easily done. Fig.1 shows preferred sample point location in vertical and horizontal pipes or ducts.

When deciding upon the location of gas analyzer, there are two possibilities extractive and in-situ. Some analyzers are mounted remotely from the sample point; these are known as extractive type gas analyzers. Extractive type gas analyzer draws a sample from remote location through a sample line to the analyzer. Extractive type gas analyzer requires sample conditioning system to make sample clean before it get enters into gas analyzer for analysis. Analyzers that are mounted in line are referred to as in-situ analyzers. With in-situ types, the instrument analysing the sample is at the process and does not have to extract the sample. This eliminates the problem of maintaining sample condition system.

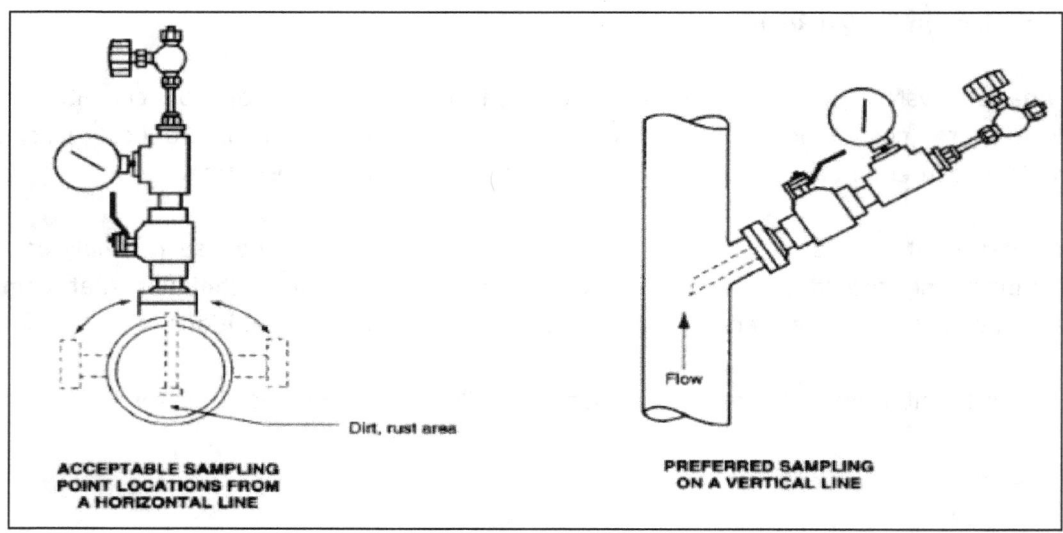

Fig.1 Acceptable Sampling Locations

- **Sample Probes:** The sample probe is a component of the process analyzer conditioning system. Sample probe acts as a first stage of filtration in sample conditioning system.

a) Open probes: These probes are basically short length made of stainless steel pipe or tube, used for temperature up to 540 Degree Celsius.

b) Multi port averaging sample probe: Multiport probe normally used in flue gases and in stack for large boilers and heaters.

c) Filter probes: These probes are generally used in gas stream such as combustible applications when the stream contains significant quantities of particulate material. The filter material used are primarily sintered or stainless steel or various ceramics based on the application. These probes located in the process duct in such a manner that their exposure to particulates minimises otherwise require frequent cleaning. A filter probe must be always operated at a temperature well above the dew point temperature of the stream in which probes is inserted. Filter probe can be clean through scavenging system. The medium used for cleaning of filter and probe must be inert or depends on the application. This probe head can be heated by a multi-wattage ring-heater that wraps around the probe head. Fig.2 shows sampling probe with filter.

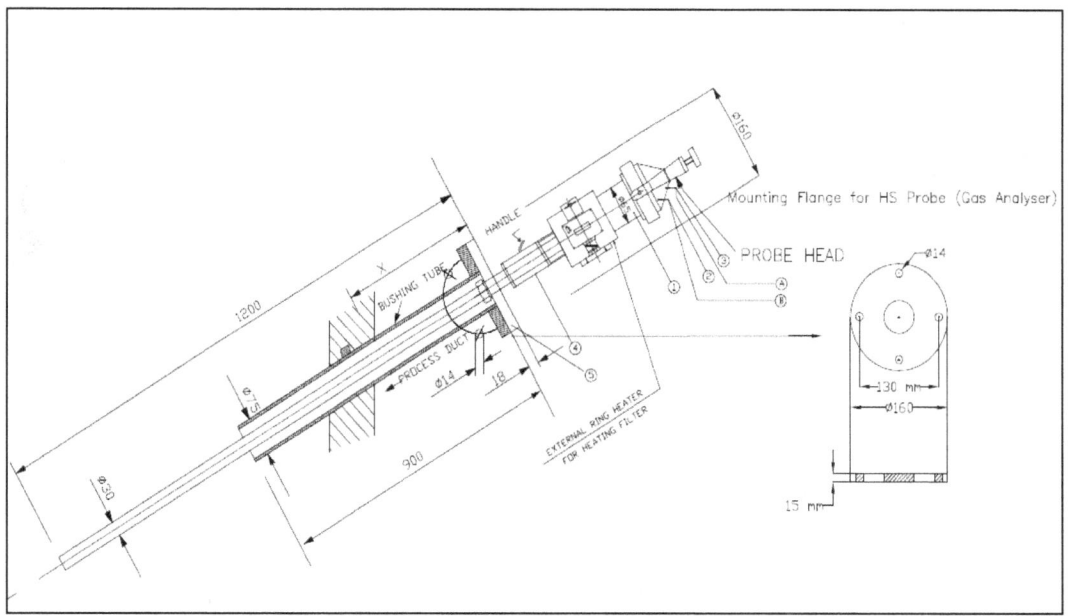

Fig.2(a) Guideline for Sample Probe Installation in Process Pipe

In most applications for online analysis, the probe is cut at a 45° angle and installed with the entry port facing downstream. As shown in Figure 2 (b), this orientation reduces the amount of solids or liquids in the extracted sample, extending the service life of sample conditioning devices.

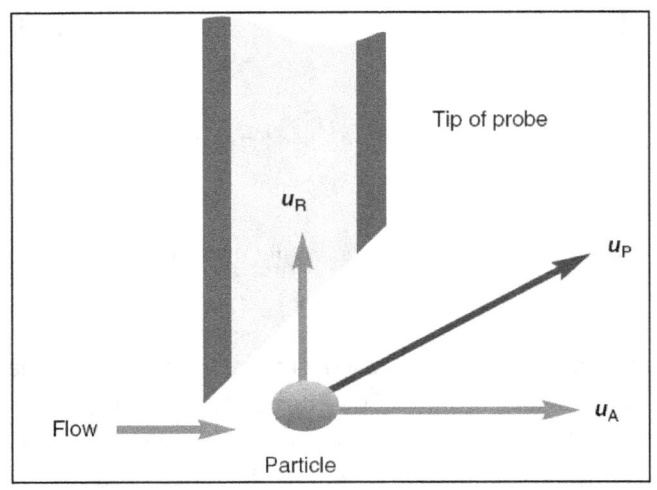

Fig.2(b) Particle Trajectory

Fig. 3 (a), (b) and (c) illustrates different components used in sampling probe.

Fig.3(a) Sampling Tube

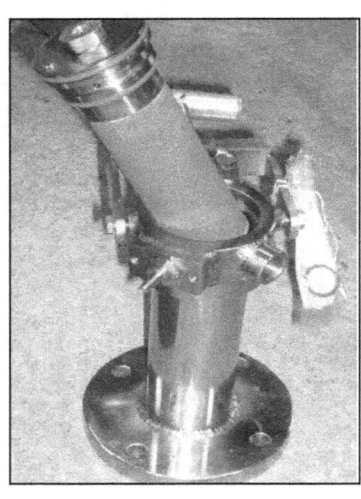

Fig.3(b) Filter Unit

Fig.3(c) Ring Heater

Gas	Ambient to 750⁰ F	Temperature at Sample Plane 750⁰ F to 1050 ⁰ F
Oxygen	Carbon Steel	Stainless Steel
CO	Carbon Steel	Cooled Stainless Steel 350-900⁰ F
CO2	Carbon Steel	Stainless Steel 350-1000⁰ F
NO	Stainless Steel	Stainless Steel 350-1050⁰ F
NO2	Stainless Steel	Stainless Steel 350⁰ F

Table:1 Minimum Probe Material to Avoid Sample Reaction

- **Scavenging System:** It is also called back blow arrangement. This system is used for cleaning of sample probe and probe filter on regular interval either through manually or automatically. It consists of solenoid valves which are controlled through PLC or local controller.

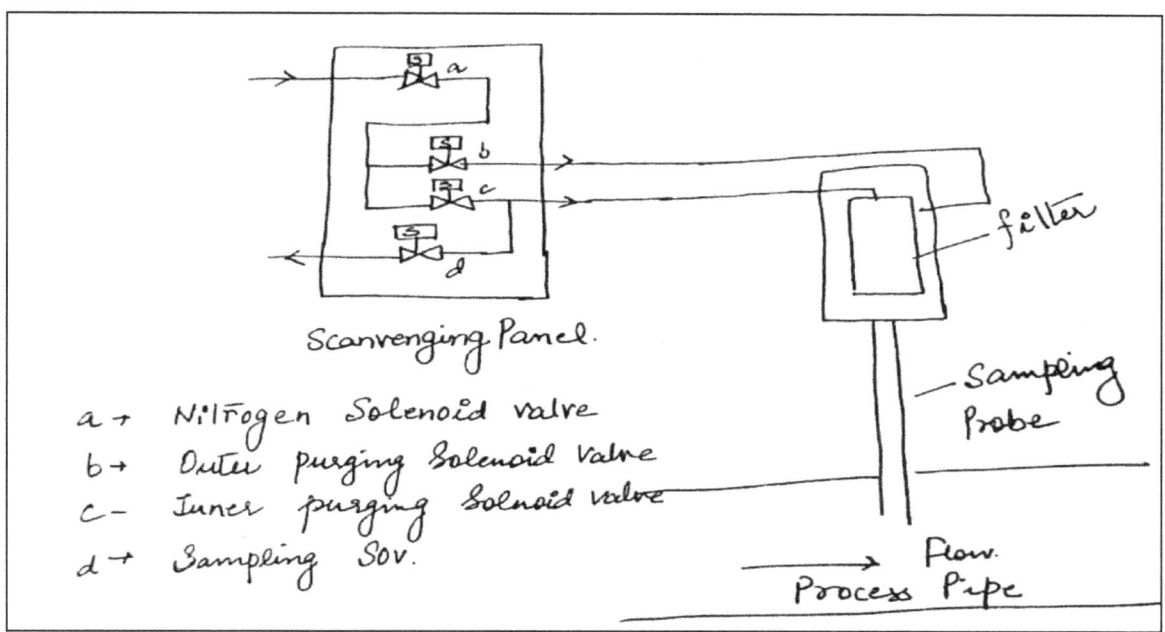

Fig.4 Scavenging System

- **Sample Transportation Time:** In installations where the analyzer and sample system must be located at some distance from the sample point, the sample transport time must be considered. The total measurement response time is defined as the sample transport time plus the analyzer response time. The analyzer response time for continuous analyzer is often expressed as T90 value, designating the time required for an analyzer output signal to achieve 90% of the final measured value for a step change in input. A convenient method to reduce the sample transport time is the use of fast vent loop.

Simplified Formula for Sample Transport Time is:

$T = V/F$

Where T is sample transport time in secs.
V- Volume of Sample
F- Pump Capacity

$V - \pi d^2 L/4$

Where d is diameter of sample tube
L is the length of sample tube

Detailed Formula for Transport Time Calculation is:

T (liquid) = V*L/ Flow Rate

T (vapour) = V*L (P1+14.7)*520/F_s*P2*460+T

Where T is the sample transport time in mins.

P1- is the pressure in sample volume whose transport time is to be calculated (psig)

P2- is the pressure at point of flow measurement in psig

V is the tube volume cc/ft

L is the tube length in ft.

F_s is the flow rate in cc/min at standard conditions

T is the temperature in Degree F.

- **Sample Tube:** These are electrically heated tubes, used in gas analysis technique for transportation of the sample gas from the sample point to the gas conditioning system. They protect the system against freeze. The most common types of tracing systems employed in chemical and refinery industry use steam or electric. There are three type of heat trace tube used in industry.

a) Parallel resistance self-regulating

b) Parallel resistance constant wattage

c) Parallel resistance power- limiting

The most popular from the above is self-regulating.

Parallel resistance self-regulating: Parallel resistance self-regulating heating cables utilize a continuous extrusion of a temperature – sensitive semiconductor polymer matrix between two current carrying wires and have no preset "zone" length. Therefore, self-regulating cables can be cut to any length without regard of the heating zone length. Most self-regulating cables are damaged due to elevated temperature and tend to deteriorate after extended period of extreme thermal cycling.

There are two types of self-regulating cables used in industry. The energised minimum operating temperature are 150-degree F /65-degree C and 250-degree F with corresponding deenergised maximum temperature of 185-degree F/ 85-degree C and 375-degree F. The 150-degree F rated cable is used for low heat process application and freeze protection while 250-degree F used for high heat process application.

Another problem with self-regulating is cold start power requirement. Their resistance is designed to decrease with decrease in temperature and increase with increase in temperature, hence maximum current draws during start up. Fig.5 illustrates self-regulating heat trace tube construction.

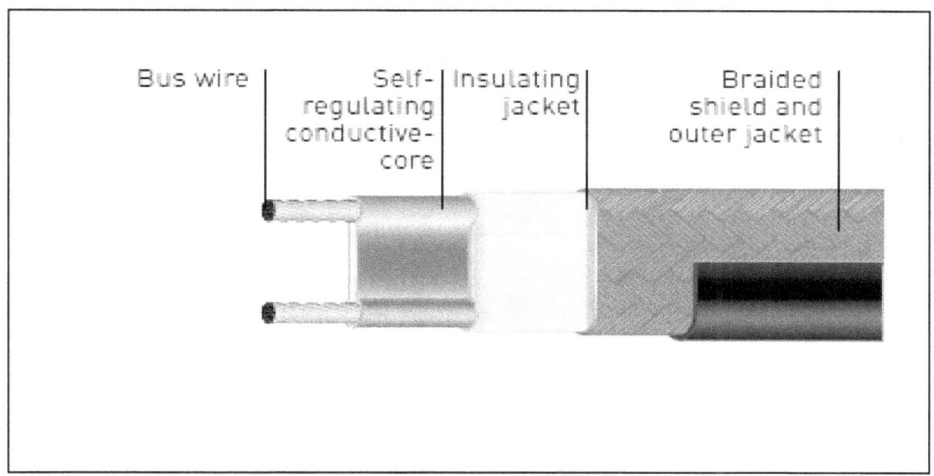

Fig.5 Self-Regulating Heat Trace Tube

- **Particulate Filters:** Process Analyzer sample streams all require filtration prior to being analysed by a process analyzer. Gaseous sample require filtration to remove particulate matter. Prefilter are used together with sample probes for continuous gas sampling where dust loads are high. These prefilter are selected on the basis of application which includes temperature, pressure, gas velocity, dust loading and grain size. Other kinds of filters are also used in sample conditioning system just before the sample entering into analyzer, to make the sample free from any kind of dust or moisture as shown in fig.6.

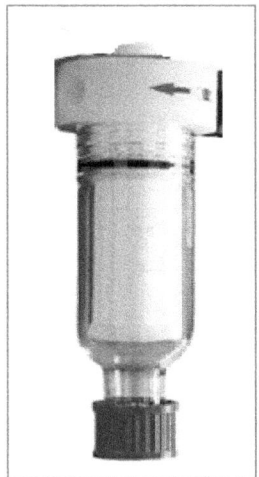

Fig.6 Particulate Filter

- **Sample Coolers:** Sample coolers can be air powered or electric powered. The function of these sample coolers is to remove the moisture from the sample which is passing through them. Typically, the sample cooler cannot operate below 0 Degree Celsius as moisture in sample will freeze and block the passage.

a) **Air Powered Sample Cooler**: Vortex Cooling tubes provides small quantities of cool air at the expense of using huge quantities of ambient temperature compressed air. Compressed air that rotates around an axis (like a tornado) is called a vortex. A vortex tube creates cold air and hot air by forcing compressed air through the generation chamber and centrifugally along the inner walls of the tube at a high rate towards the control valve. A percentage of high speed air is permitted to exit at the control valve but the remainder of the air stream is forced to counter flow through the center of the high-speed air stream giving up heat, finally exiting through the opposite end as extremely cold air. Fig.7 a and b shows vortex cooling tubes.

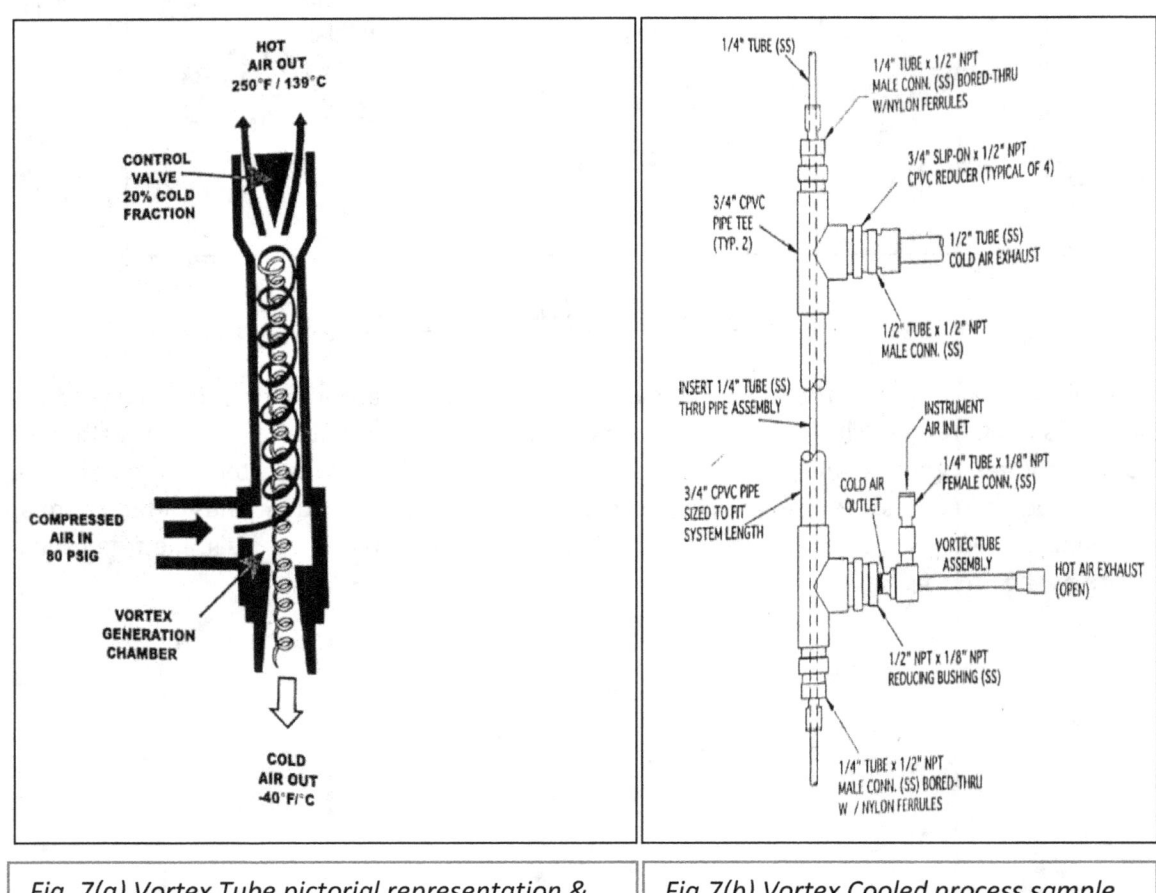

Fig. 7(a) Vortex Tube pictorial representation & operation description	*Fig.7(b) Vortex Cooled process sample stream Cooler for a gaseous process*

b) **Electric Powered Sample Cooler**: These sample coolers can either be mechanical or solid state (Peltier Element) refrigeration and conventional refrigeration (motor driven compressor). The cooling of both these devices is dependent on six factors:

1) Process Sample Gas Dew Point
2) Process Sample Gas Temperature
3) Process Sample Gas Flow Rate
4) Process Sample Gas Pressure
5) Ambient Temperature
6) Heat Exchanger Efficiency

Peltier gas cooler can reduce the dew point of a sample gas by a maximum of 17 degree Celsius. Compressor driven gas cooler can be used to achieve a constant dew point between 2 Degree Celsius to 6 Degree Celsius at an ambient temperature condition up-to 50 Degree Celsius.

Conventional Refrigeration gas coolers operate on the same principle as a home refrigerator. The refrigerator unit consists of an electric motor driven compressor, an air-cooled condenser coil and a heat exchanger. The temperature of a cooling chamber is controlled by a thermal switch that turn on the refrigeration compressor when switch senses a positive temperature deviation in the chamber. The compressor pulls the warm refrigerant from the evaporator and pumps it into the condenser, compressing it in the process. The compression causes the temperature of the refrigerant to rise sharply, along with the pressure. The warm refrigerant then passes through condenser coil, where heat is transferred from the refrigerant to an air medium resulting in high pressure saturated liquid. A pressure controlling device is used to reduce high pressure liquid to low pressure liquid. As the refrigerant is metered from capillary tube into the evaporator, it enters the low-pressure side of the cycle. The pressure drop on the evaporator side allows the liquid refrigerant to begin expanding or evaporating into a vapour and then return to compressor. During the process of vaporising, the refrigerant absorbs heat.

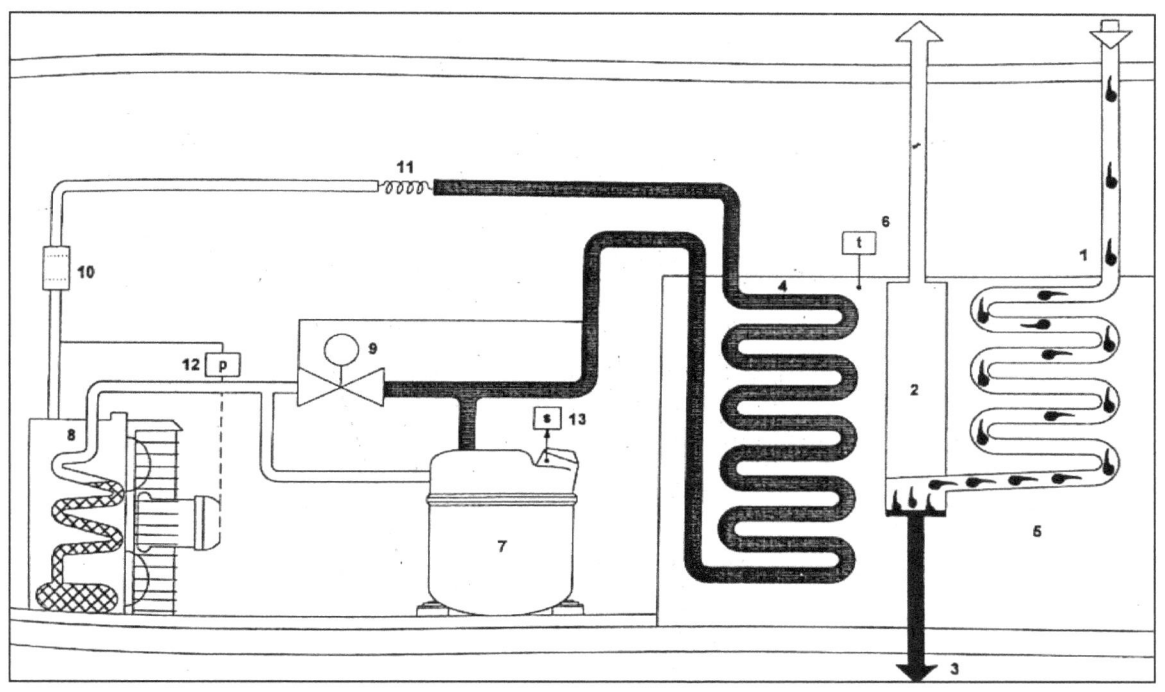

Fig.8 Schematic of a refrigerated process sample gas cooler illustrating refrigeration cycle and condensate removal from a gaseous sample.

Representation of figure 8.
1- Sample inlet
2- Condensate separator
3- Condensate removal port
4- Evaporator
5- Heat Exchanger

6- Microprocessor controlled temperature device

7- Compressor Unit

8- Condenser Coil

9- Pressure regulating unit

10- Dryer

11- Capillary Tube

12- Pressure Switch

A Study of Heat Exchanger Pressure on the exit dew point of a chilled gas sample:

The pressure within the heat exchanger is affected by the placement of the sample pump before and after the chiller unit. The sample line length or restrictions will also affect the pressure within the chiller. If the sample pump before the chiller unit, the chiller can be operated under pressure. If the sample pump is pumping a hot and wet sample, diaphragm pump are damaged by the presence of droplets in the head area. One way the sample pump can be protected from liquid droplets within the head area is to place the sample pump between two heat exchangers using a dual stage chiller. The first chiller condenses a major portion of the water vapour present in the sample, the second heat exchanger can be operated under pressure with the resulting improvement in exit dew point of the sample.

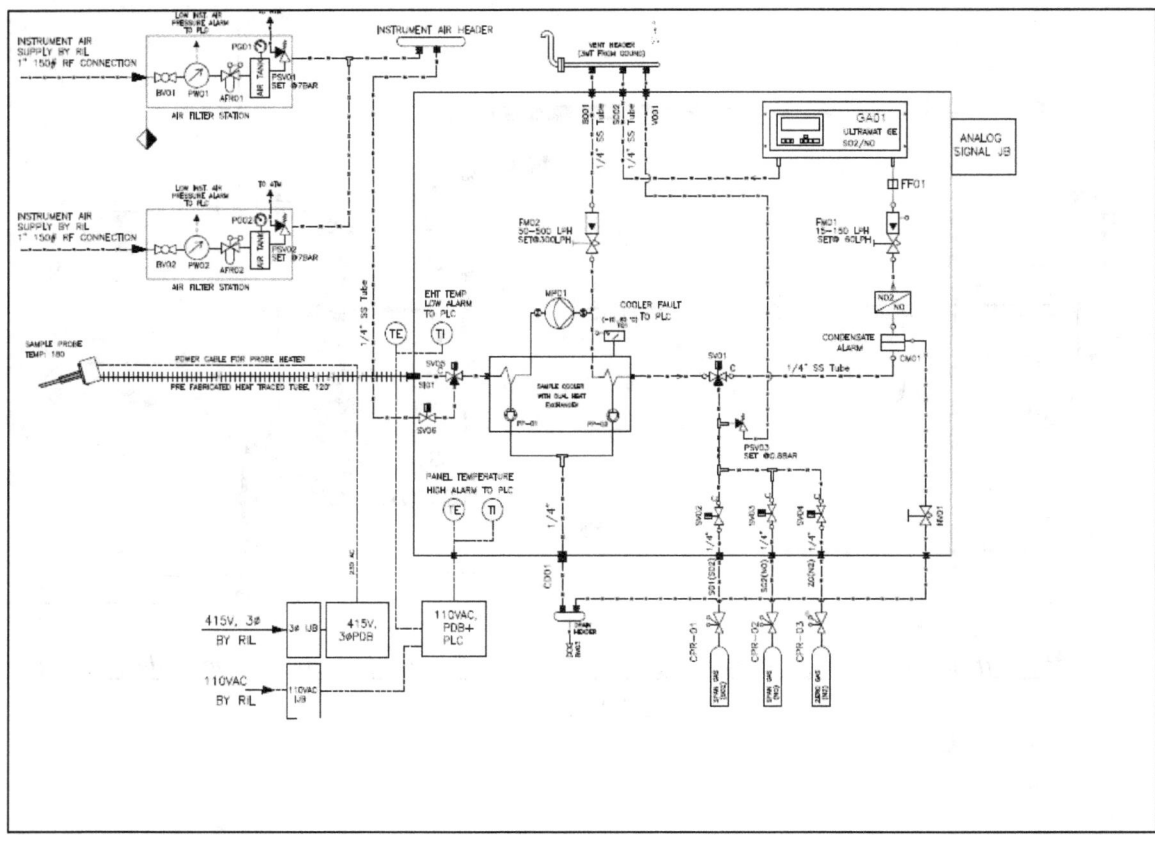

Fig.9 Schematic of Dual Channel Gas Cooler for Single Probe System

- **Nafion Dryer:**

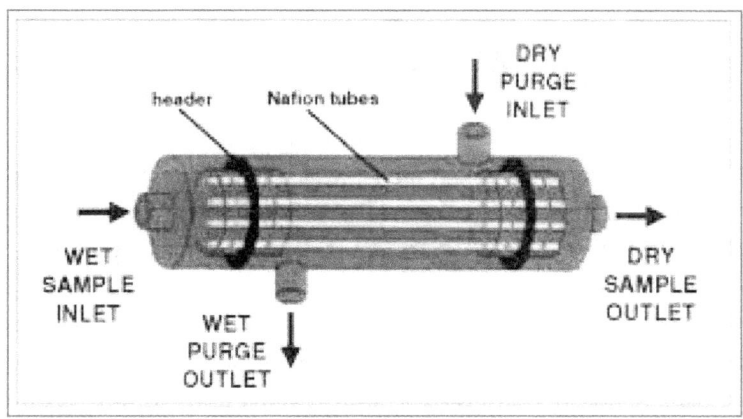

Fig.10 Nafion Dryer

Gas dryers are shell and multi tube moisture exchangers that transfer water vapour between two counter current flowing gas streams. The dryers consist of a bundle of Nafion polymer tubes surrounded by an outer tube. Dry gas flowing over the exterior surface of Nafion tube continuously extract water vapour from the gas sample. The driving force is the difference in water concentration on opposite side of the tubing wall.

Following rules applying while installing Nafion Dryer:

a) Sample gas pressure must be equal or greater than the purge gas pressure. Sample gas pressure do not exceed 80psig while purge gas pressure does not exceed 10psig.
b) Purge air must be having a dew point of -40 degree Celsius with flow rate 2-3 times more than that of sample flow.

Disadvantage with Nafion Dryer:
Nafion tubes have very minute Internal diameters, so chances of choking of tubes due to dust particles is higher.
Nafion removes only 90% of water vapour in sample gas. 10% is still retained.
It is mandatory to use filters before the nafion dryer. If not drained properly, acid mists can damage the nafion dryer.

- **Condensate Removal**: Condensate removal is initially by gravity separation of the condensed liquid from the flowing gas sample stream at a temperature below the dew point. When the sample pressure is negative or process is in suction you must use a condensate reservoir or vessel and thereby it should be removed with the use of peristaltic pump.

- **Sample Pump:** The sample pump is used in sample conditioning system to take the sample from the process media in draft conditions. Pressurised system do not required sample pump.

Following types of Pumps are used in Sample Conditioning System
a) Bellows
b) Diaphragm
c) Peristaltic Pump

d) Piston

e) Syringe

The most commonly used pump in sample conditioning system are Diaphragm Pump and Peristaltic Pump.

Diaphragm Pump: A diaphragm pump is one example of reciprocating pump. The diaphragm can be used for pumping gaseous, its output is discontinuous and highly pulsating. An enclosed volume (the pump chamber) is formed by the rigid part on one side and a flexible diaphragm on the other. The volume contained within the pump chamber is variable and controlled by the piston of the diaphragm. A rotating eccentric transfers a reciprocating action to the diaphragm through a connecting rod. Fig.11 shows diaphragm pump.

As diaphragm moves away from the head, an increase in volume and a reduction in pressure occurs inside the pump chamber, causing the outlet check valve to close and inlet check valve to open. This induces the flow of fluid from inlet port to pump chamber. Once the diaphragm is fully extended, a maximum volume is enclosed in the pump chamber and flow into the pump stops. As

The diaphragm moves towards the head, the volume enclosed is reduced and an over pressure is developed. In response to diaphragm movement the inlet check valve closes and outlet check valve opens. The sample contained in the pump chamber now flows from the pump outlet port. If resistance is presented after outlet port, back pressure is developed which can damaged the diaphragm and seize the motor bearings. If resistance is on inlet side, a vacuum condition is developed.

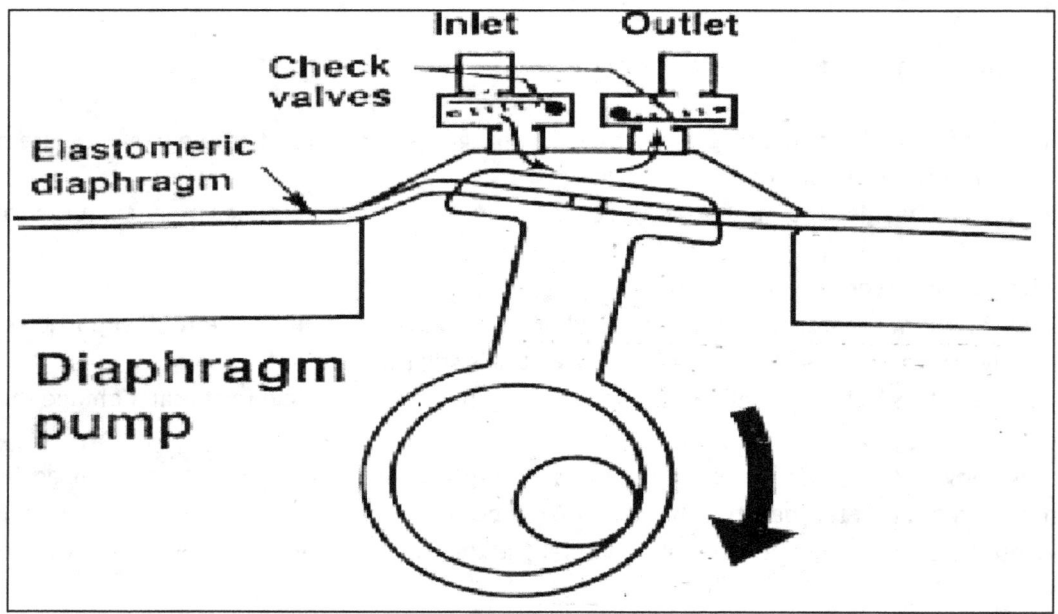

Fig.11 Diaphragm Pump

Peristaltic Pump: These pumps uses a flexible tube that is compressed by a series of rollers to induce liquid flow within a system. As rollers travel along the tube, fluid is forced through the tubing. Contained fluid can only leak if tube punctures. These pumps are used to remove the condensate from condensate vessels. The advantage of peristaltic pump is the low cost for replacement of tubes and the relative low cost of the drive motor replacement. The significant disadvantage with peristaltic pump is the need of routine maintenance. Fig.12 shows peristaltic pump.

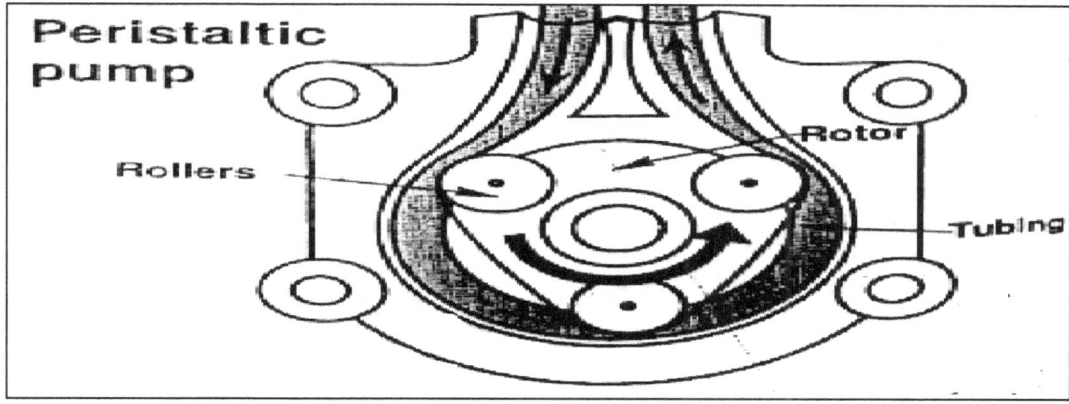

Fig.12(a) Peristaltic Pump

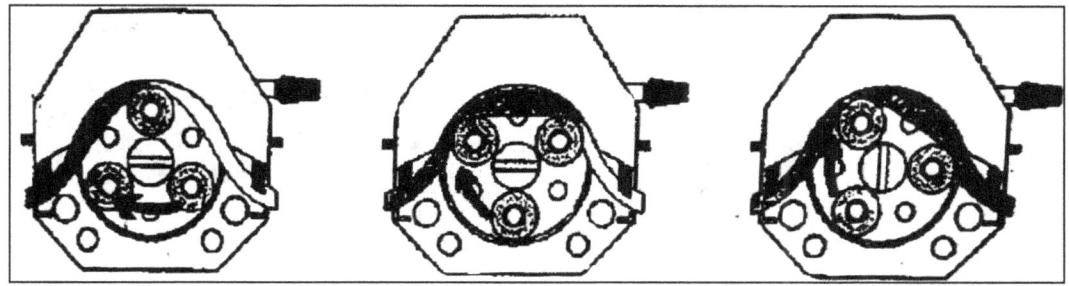

Fig.12(b) Pump head consists of only two parts: the rotor and housing. The tubing is placed in the tubing bed between the rotor and the housing. The rollers on the rotor moves across the tubing. The tubing behind the rollers recovers its shape, creates vacuum and draws fluid in behind it. A pillow of fluid is formed between the rollers.

Ejector (Jet) Pumps: An ejector pump also knowns as aspirator or jet pump. Is only used for pulling gaseous sample from the process stream. The principle involved here is the Venturi, a transfer of momentum from the inlet compressed air or nitrogen to provide motion to the process fluid.

Ejector Pump are very reliable because they are cast or machined from a solid material. The only problem is plugging of the sample inlet or the pump outlet with particulate matter present in the process sample. Ejector Pumps always need to be install after analyzer this means suction of process sample will be always through process gas analyzer. Fig.13 a and b shows ejector pump cutaway and operational illustration.

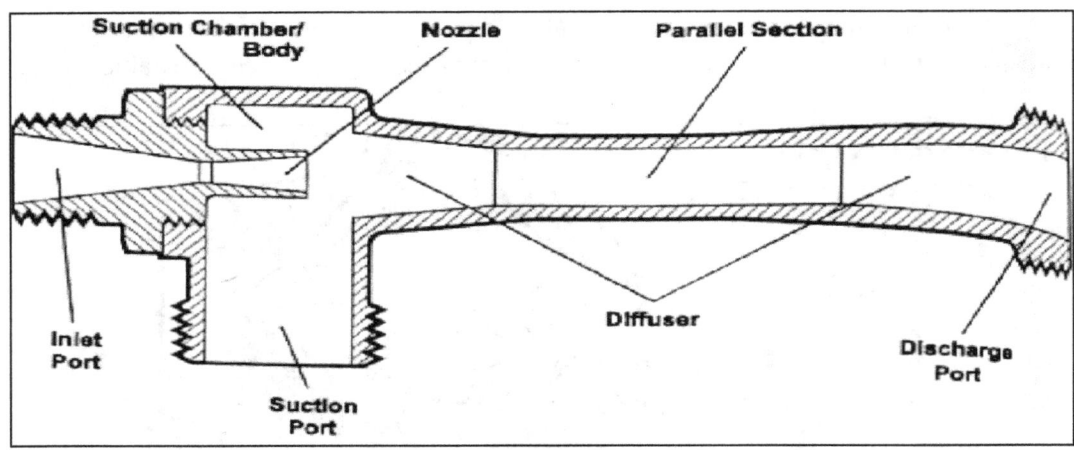

Fig.13(a) Ejector Pump cutaway with internal components parts

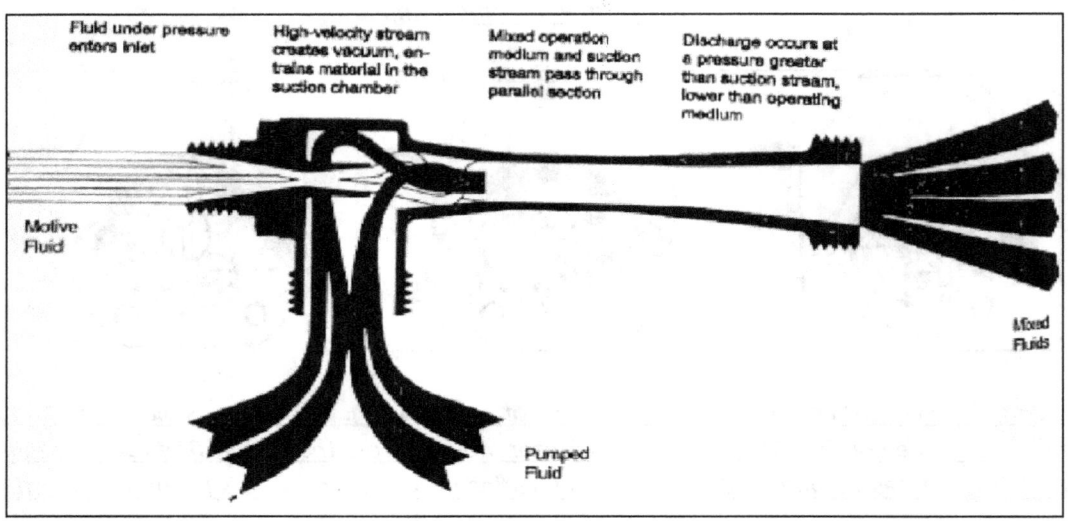

Fig.13(b) An ejector pump operational illustration showing fluid moving in and pumped out.

Sample Flow Rate Measurement and Control: Valves are one of the most important component in sample conditioning system. An isolation valve is either full open or close, its function is to isolate the process gaseous sample from the analyzer stream. The sample conditioning system component isolating valves are ball, plug or needle valves. A regulating valve is designed to adjust the rate of flow passing through. These valves are normally referred as needle valves.

The most commonly used flowmeter in process gas analyzers are rotameter. Rotameters are specialized form of rudimentary flowmeter used to measure volumetric flow rate. They allow the visual measurement of volumetric flow rate by noting the float position on a scale. The internal float may be spherical, obloid or cylindrical with a cone bottom and made of materials

That vary in density to allow measurement of gaseous. The flow range or capacity of a tube can be varied by changing the float material. Material with lower density such as Pyrex glass or sapphire give a lower flow capacity than material with higher density such as stainless steel or tantalum.

Fig.14 Rotameter used for flow measurement

- **Sample Pressure Measurement and Control**: Accurate pressure control could be vital for process analyzer sample conditioning system based on measurement characteristics. The common device used for pressure regulation in sample conditioning system is diaphragm type pressure regulator.

Fig. shows pressure reducing single stage regulator and b shows internal components.

Fig.15 (a) Pressure Reducing Regulator

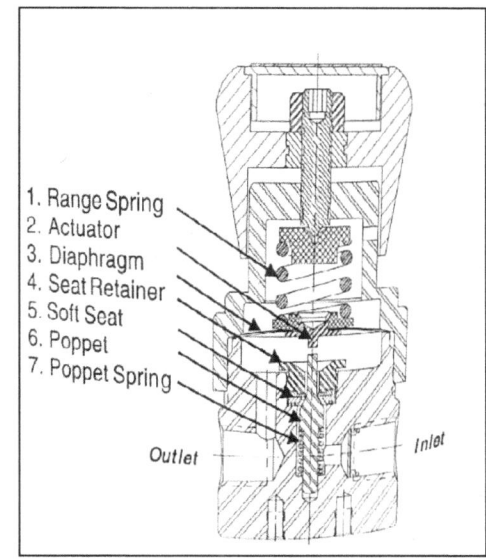

Fig.15(b) Pressure Reducing Regulator Internal Components

1. Range Spring
2. Actuator
3. Diaphragm
4. Seat Retainer
5. Soft Seat
6. Poppet
7. Poppet Spring

Outlet   Inlet

- **Condensate Monitor:** This device is used to detect the presence of moisture in the sample gas. The Condensate Monitor has a Glass Fiber Filter Paper of 2-micron particle size, this Filter Paper filters dust in sample gas up to 2 microns, such that the gas after the condensate monitor is free from dust and moisture.

The latter function is based on changes of the electrical conductivity between two electrodes. In the dry state, the filter diaphragm constitutes a high resistance path between the two electrodes. It takes on a low resistance when it becomes moist, i.e. the electrical conductivity increases and a small current flow between the electrodes. The voltage required for the Condensate Monitor is provided by the connecting Switching Unit.

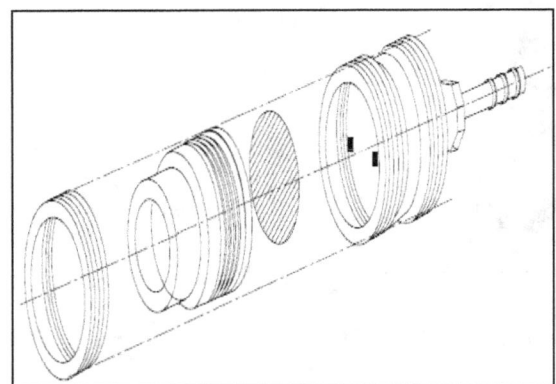 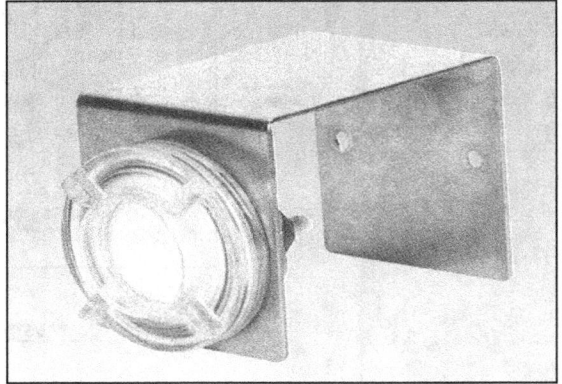

Fig.16 Construction of Condensate Monitor

- **Sample Disposal:** Once sample analyzed, it should be routed to an acceptable and secure location, preferably back to the process. All piping should be routed and designed to prevent condensation from accumulating in line pockets as well as to prevent rain and debris from entering the sample vents.

5.3 Gas Analyzer

Gas Analyzers are of several types used in process and chemical industries based on the application and complexity of the process. Following are the types of analyzer based on different principles:
a) Chromatographs
b) Spectrometers
c) Ultraviolet
d) Infrared
e) Electrochemical
f) Paramagnetic
g) Chemiluminescence
h) Thermal Conductivity
i) Tuneable Diode Laser Spectroscopy

This chapter covers only Infrared, Paramagnetic, Thermal Conductivity and Chemiluminescence based analyzer.

Non-Dispersive Infrared Gas Analyzer:
Infrared (IR) instruments were one of the first analyzers to be moved from the laboratory to the pipeline, and the technology is available for use with gas, liquid, or solid samples.
Following Components can be measure through this technique; CO, CO2, CH4, NO, SO2.

Absorbed energy is converted into oscillation or rotational energy inside the molecules. Absorption takes place only if the molecule has a dipole moment. Any component absorbs a specific wavelength only. The absorbed wavelength characterizes the gas component. Strength of absorption is a measure of concentration

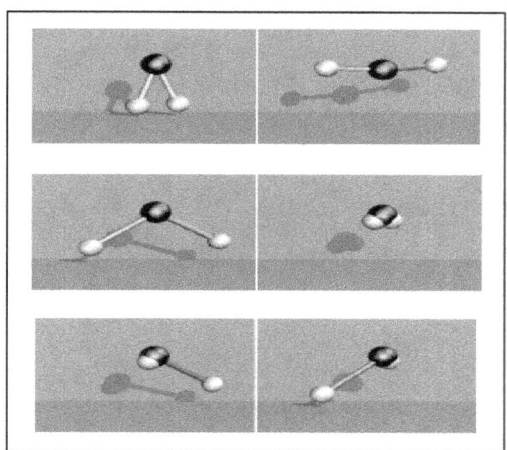

Fig.17 Molecular Vibration

NDIR analyzer works on the principle of Beer's Lambert Law which states that the amount of light absorbed to the sample's concentration and path length.

$$A = abc = \log_{10} \frac{I_0}{I}$$

Where A is absorbance
I IR power reaching to detector with sample in the beam path
I_0 IR power reaching to detector with no sample in the beam path
a is the absorption coefficient of pure component of interest at analytical wavelength
b is the sample path length
c is the concentration of sample component

The law states that concentration is directly proportional to absorbance at a given wavelength and path length at specified temperature and pressure.

Infrared covers the range of the electromagnetic spectrum between 0.78 µm to 1000 µm.

Wavenumber = 1/ wavelength in cms

It is useful to divide the infrared region into three sections.

Near IR, Mid IR and Far IR

Region	wavelength range (µm)	wavenumber range (cm-1)
Near	0.78 – 2,5	12800 - 4000
Middle	2,5 – 50	4000 - 200
Far	50 – 1000	200 – 10

The most useful IR region lies between 4000 – 670 cm-1 corresponding to 2.5 µm to 15 µm.

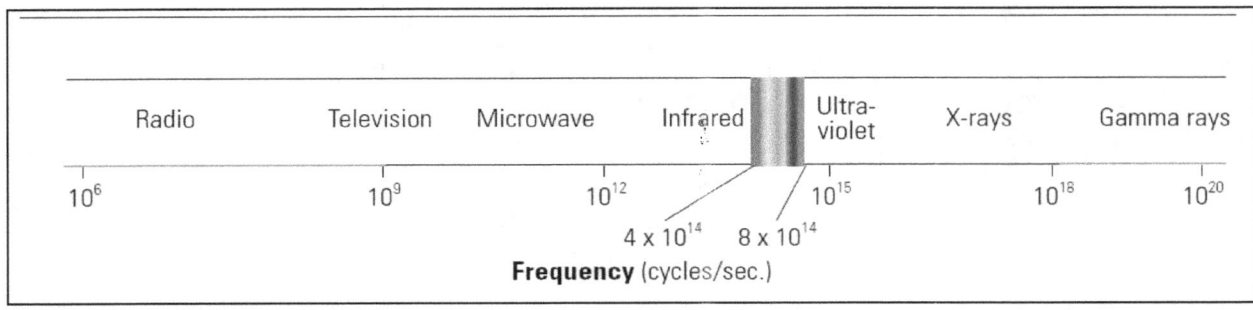

Fig.18 Location of Infrared Rays in Electromagnetic Spectrum

- **CO Gas Analyzer**: Carbon monoxide is also an indicator of incomplete combustion and, therefore, it is measured to optimize boilers and other combustion processes. For these applications, the most frequently used analyzer is the nondispersive infrared (NDIR) sensor. NDIR allows continuous analysis, because carbon monoxide absorbs the infrared radiation at a wavelength of 4.6 µm. Because infrared absorption is a nonlinear measurement, it is necessary for the analyzer to accurately linearize its output signal.

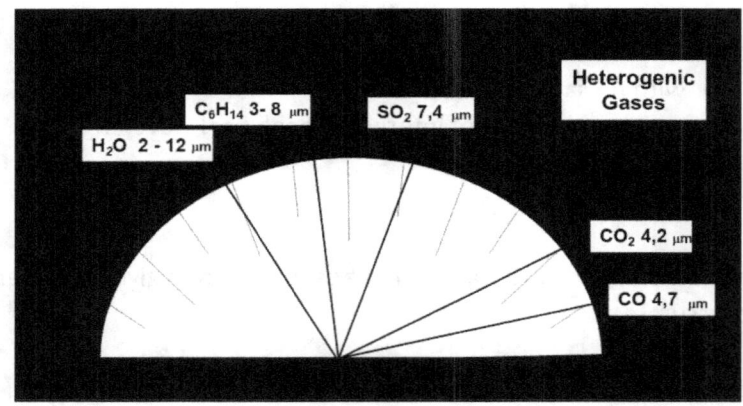

Fig. 19 Correlation between gases molecules and electromagnetic radiation

A schematic diagram of a typical NDIR analyzer is shown in Figure. Infrared radiation from a hot filament is chopped to pass alternately through sample and reference cells, to be absorbed in the detector cell divided by a pressure-sensitive diaphragm. If the sample contains carbon monoxide, it will absorb part of the radiation, causing that half of the detector to exert less pressure on the diaphragm, whose distortion is converted to an electrical signal for rectification and amplification.

In fig.20., the IR source (2 to 15-micron meter) is pointed at the beam splitter, which divides the beam into two parts. One part is sent to the moving mirror, the other part to the stationary mirror. The two reflected beams are recombined at the beam splitter. Because the beam from the stationary mirror travelled a constant distance, while the one reflected from the moving mirror travelled a variable distance, when they are recombined, an interference pattern is created, because some of the wavelengths are strengthened, while others are weakened by the recombination. The result is an interferogram, which is sent through the sample, where some wavelengths are absorbed, while others are transmitted and reach the detector. The detector signal is sent to a computer, where an algorithm called a Fourier Transform interprets it.

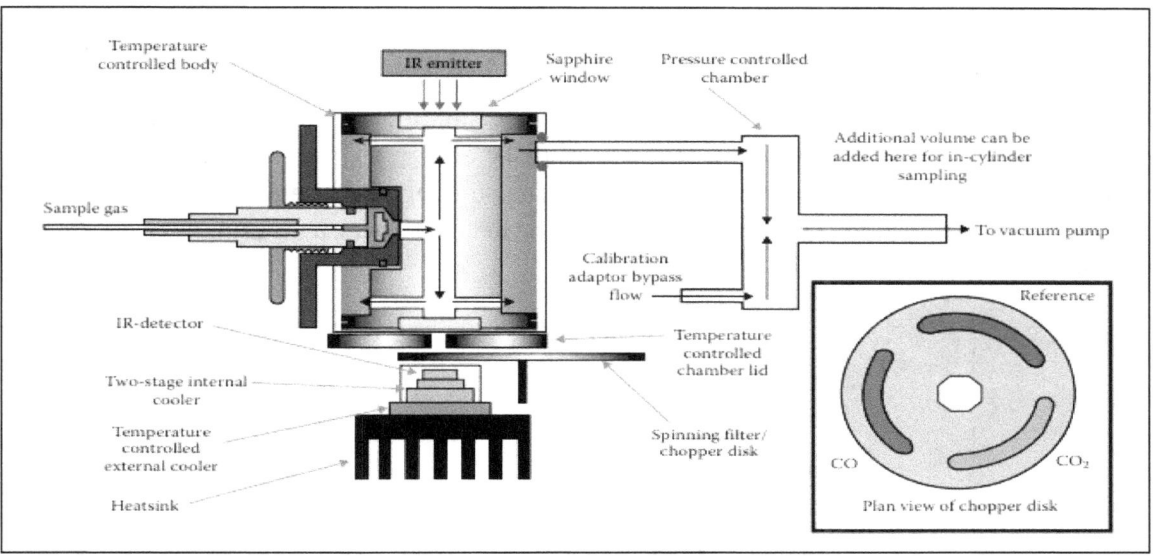

Fig.20 NDIR analyzers are often used to measure the carbon monoxide and carbon dioxide concentration

Analyzer Design Configuration:

a) Single Beam Configuration:

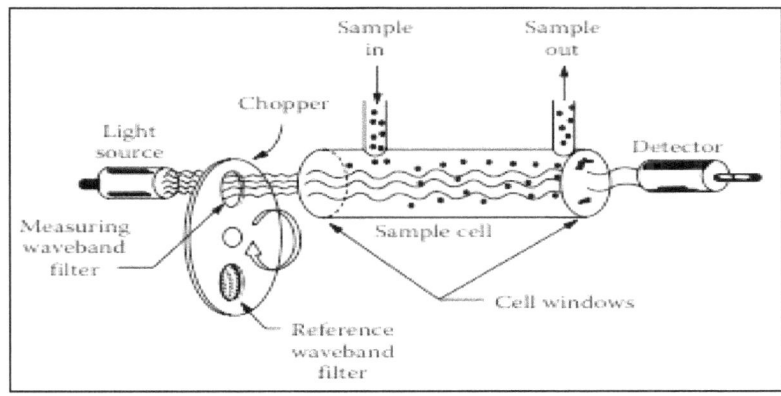

Fig.21 Single Beam Configuration

Single-beam analyzers are the main ones used in the process world. Filters are put on a chopper wheel in the light beam, and as the chopper alternatively spins one filter or the other into the optical path, the difference or ratio in the energies received at the detector will be a function of the concentration of the component of interest. Because filters normally change absorbance at a given wavelength with temperature, this approach must be temperature stabilized. Also note that dirt on the window will not affect the reading because it effects both the measuring and reference wavelength in the same way so that the ratio stays constant if the monitoring and reference wavelengths are close enough.

b) Dual Beam Configuration:

The detector and reference chambers are sealed with a target gas at a low pressure. IR transparent windows are fitted to seal the chambers and the same intensity of pulsed infrared radiation is received by both chambers when no target gas is present. When the sample containing target gas flows through the sample cell, a reduction in radiation energy is received by the detector chamber due to absorbance of IR energy by the gas molecules, which causes the temperature and pressure to drop in the detector chamber. The amount of temperature or pressure drop is in direct proportion to the gas concentration. a diaphragm separates the two chambers, a movement of the diaphragm causes a measurable change in capacitance.

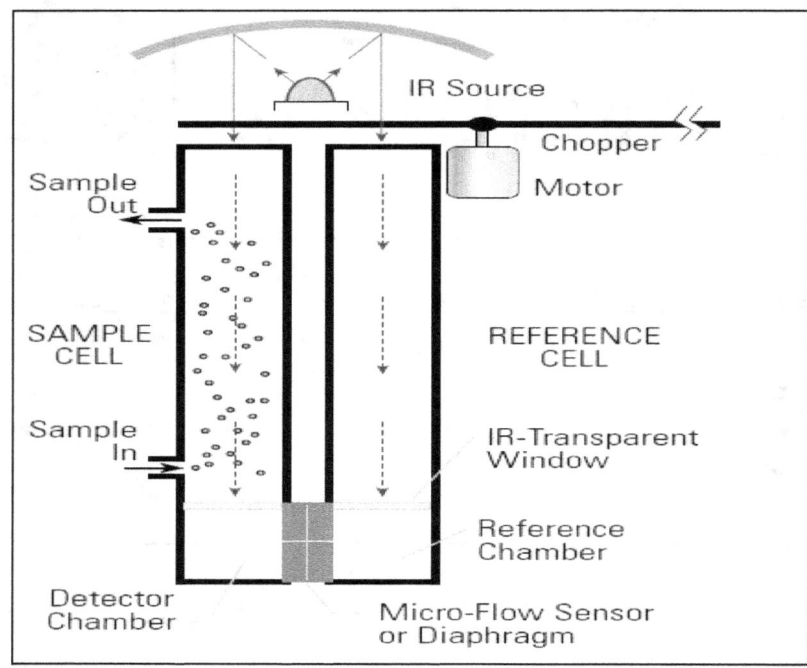

Fig.22 Dual Beam Configuration

Influence caused by pressure and temperature
• Increasing the temperature decreases the gas density and thereby the number of gas molecules present inside the measuring cell: The concentration of gas decreases.
– Measure cell temperature to compensate
• Higher gas pressure increases number of gas molecules inside the cell (1 mbar (0.014 psi)) changes the result by 0.1 %).

Cross Sensitivity: The overlapping of the absorption lines of the measured component and the interfering component is termed as Cross Sensitivity. Cross Sensitivity can be reduced or removed using optical filter.

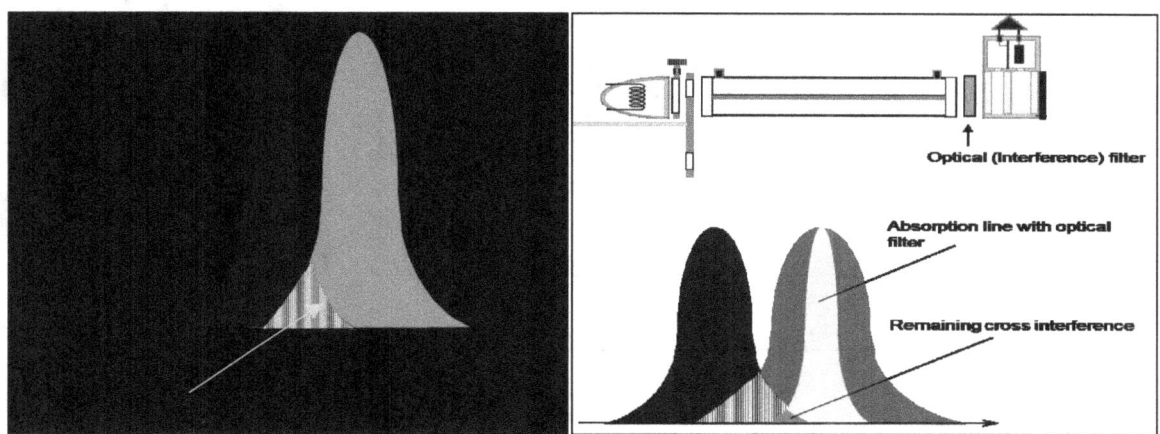

Fig.22 Cross Sensitivity due to Interference Component

Calibration of CO and CO2 Gas Analyzer:

Calibration is required at scheduled intervals to assure that the analytical signal is valid and that the instrument has not drifted due to environmental conditions, component aging or other factors. The calibration of an infrared analyzer is a two-point operation- zero and span. Analyzers are calibrated by adjusting the output reading of a fluid containing none of the components of interest so called zero calibration, generally done through nitrogen gas. Adjusting the output readings to match another fluid of interest so called span calibration, generally done on 80% of the full-scale value of analyzer.

Calibration of CO and CO2 Gas Analyzer through Calibration Cell:

Calibration of gas analyzer required test gases, this traditional method used test gases or standard gases which are filled up in heavy bottles which are made of carbon steel or aluminium based on the applications. Calibration without test gases can be possible with the use of calibration cell. Calibration cell is placed in the optical path between sample cell and detector as shown in fig. 23.A calibration cell consists of two chambers; one filled with complete nitrogen and other filled with nitrogen as well as target gas.

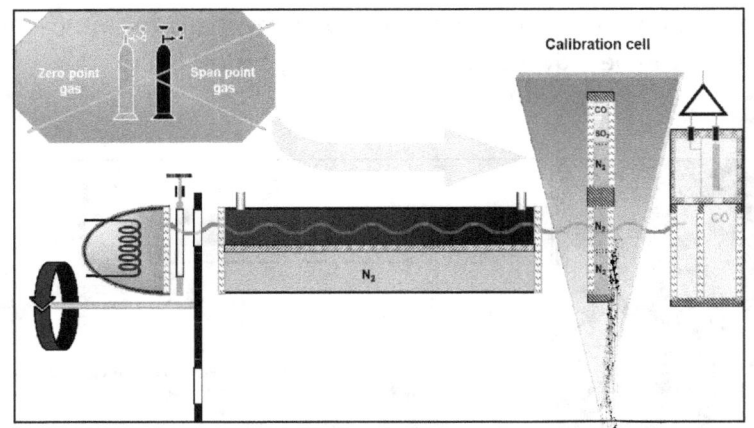
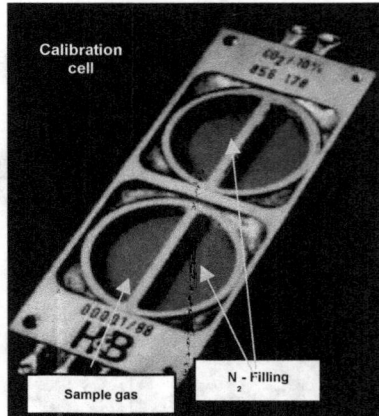

Fig.23 illustrates calibration of CO gas analyzer without using standard calibration gas cylinder

Paramagnetic Gas Analyzer:

The paramagnetic measurement technique is also known as "Magneto Dynamic" is dedicated to measuring oxygen. It is based on the principle that oxygen atom is paramagnetic, that is, are attracted into a magnetic field. There are two technologies available in market related with oxygen measurement on a dry basis; for the wet measurement of oxygen; zirconia cell should be used. The paramagnetic detector will experience measurement interference from NO, so it could be used for the measurement of NO in absence of oxygen. In addition, the detector is also susceptible to the presence of nitrogen dioxide, chlorine etc. due to paramagnetic properties, for this reason plant requires low measuring ranges (2% or less) should be aware about the background gases which may cause significant error.

- **Dual Gas Analyzer:**

The measurement of oxygen in a sample is measured by paramagnetic principle. O2 molecules, owing to their paramagnetic property are attracted in an inhomogeneous magnetic field in the direction of higher field strength.

Fig.24 shows, if two gases having different O2 concentration are brought together in a magnetic field, a pressure difference is generated between them. Once of the gases is the sample gas and the other is the reference gas (N2, air). Both gases (reference and sample) are introduced to the measuring chamber via two ducts. Since both ducts are connected, the pressure difference Δp is sensed by microflow sensor, which is proportional to the oxygen concentration in the sample gas. The microflow sensor consists of two nickel grids heated to approx. 120 ºC which form a Wheatstone bridge together with two supplementary resistors. The microflow sensor is located in the reference gas stream, the measurement is not influenced by the thermal conductivity, the specific heat or the internal friction of the sample gas. This also provides a high degree of corrosion resistance because the flow sensor is not exposed to the direct influence of the sample gas.

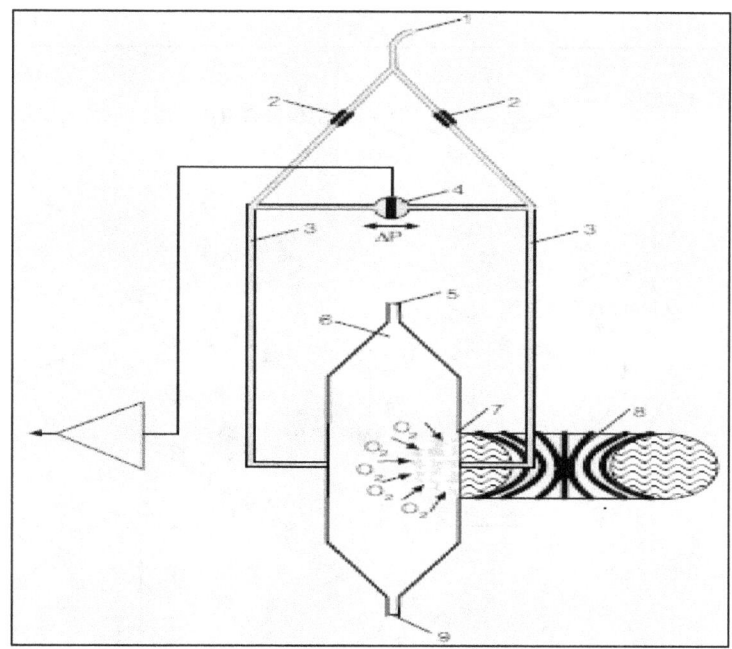

Fig.24 Dual Gas Oxygen Analyzer

- **Deflection Type Oxygen Analyzer:**

Fig.25 shows oxygen analyzer paramagnetic detector. Test body filled with N2. Test Body suspended in a non-uniform strong magnetic field.

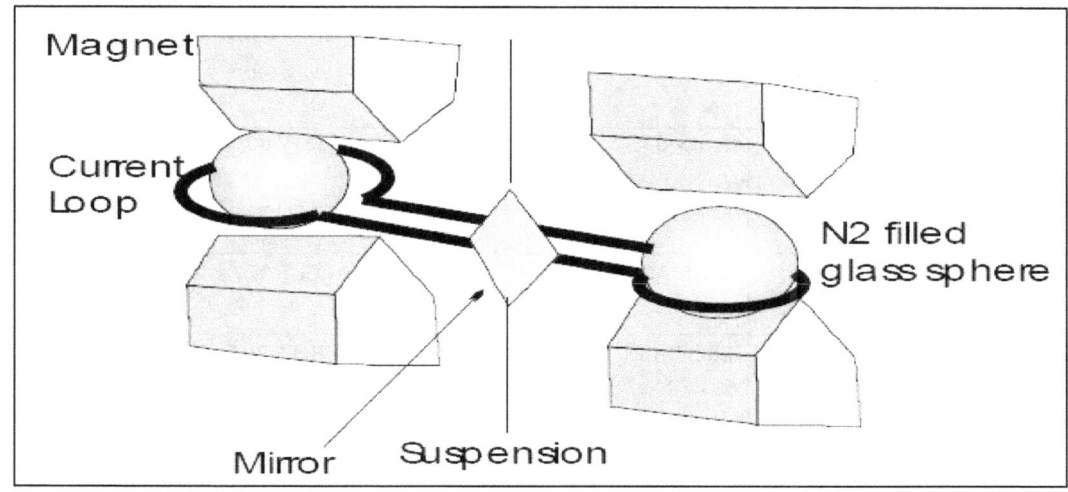

Fig.25 Deflection Type Oxygen Analyzer

Mirror is attached to the center of the test body. Light source/photocell combination reflects light off mirror and is used to measure the movement of the test body. Current loop used to pull the test body back to its "null" position.

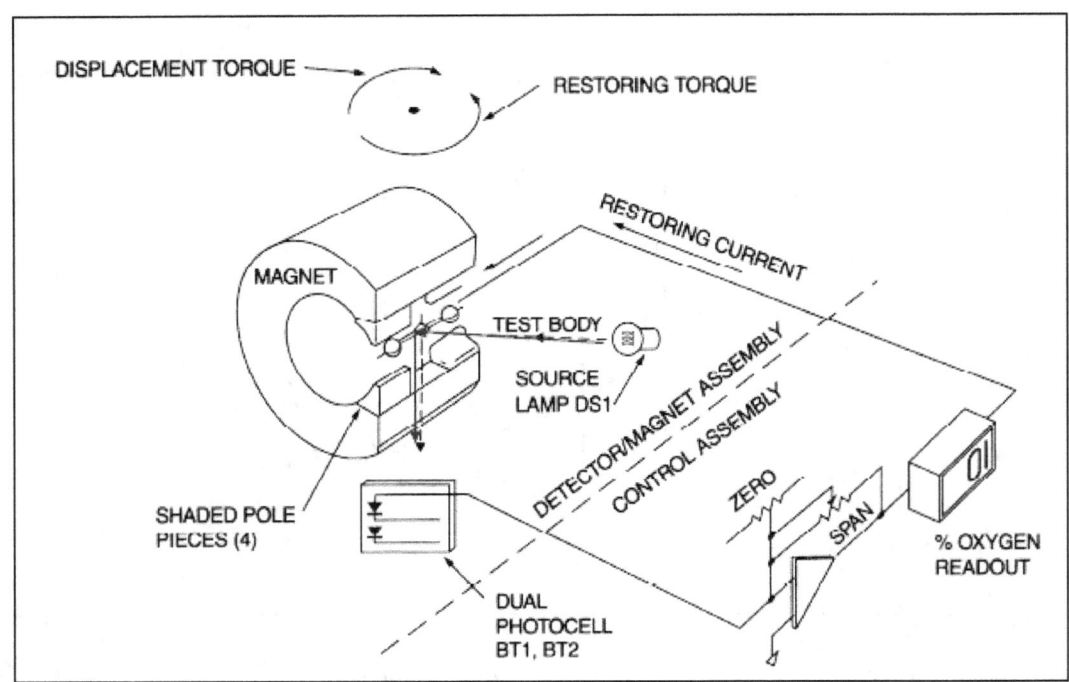

Fig.26 Deflection Type Oxygen Analyzer Operating Principle

Fig. 26 shows, when sample is introduced to the device, the oxygen present in the sample is drawn to the magnetic field. The buoyancy of the test body causes it to be displaced out of its zero position in the field. The light focuses a beam on the mirror of the test body and is reflected to the photocells. When the test body moves, the photocells register an imbalance and drive a restoring current to pull the test body back to its zero (null) position. The intensity of this restoring current is directly proportional to the concentration of oxygen present. By measuring this current, we can accurately derive the oxygen concentration of the sample.

ANALYZER WITH ALTERNATING PRESSURE METHOD	ANALYSER WITH DUMBEL TECHNOLOGY	ANALYSER ON RESISTANCE BALANCE
ACCURACY 0.1%	ACCURACY 0.5%	ACCURACY < 2%
ZERO DRIFT **0.5% PER MONTH**	ZERO DRIFT 3% PER WEEK	ZERO DRIFT 1% PER WEEK
SENSOR NOT IN CONTACT WITH SAMPLE GAS	SENSOR IS DUMBELL TYPE HANGING IN A THIN TUNGSTEN WIRE. HIGHLY SUSCEPTIBLE TO CORROSIVE GASES	SENSOR IS BASED ON RESISTANCE BALANCE. HOWEVER ALWAYS EXPOSED TO SAMPLE GAS

Table:2 Comparison Study between different paramagnetic gas analyzers

Thermal Conductivity Analyzer:

The thermal conductivity analyzer is based on the principle that different gases have different abilities to conduct heat. The detector measures the changed in thermal conductivity of the sample gas versus reference gas. The detector consists of heated temperature sensitive wires arranged in an elongated helix. Two wires exposed to sample gas and other two exposed to reference gas. An equilibrium is reached when the electrical power input creates heat that is equal to the thermal loss from the wire which is exposed to sample gas. The temperature rise of the wire is inversely proportional to the thermal conductivity of the sample gas. Another way of measuring, when the sample concentration changes, the temperature of the wire also changes, changing the resistance value of the wire. This resistance change is directly proportional to the gas concentration.

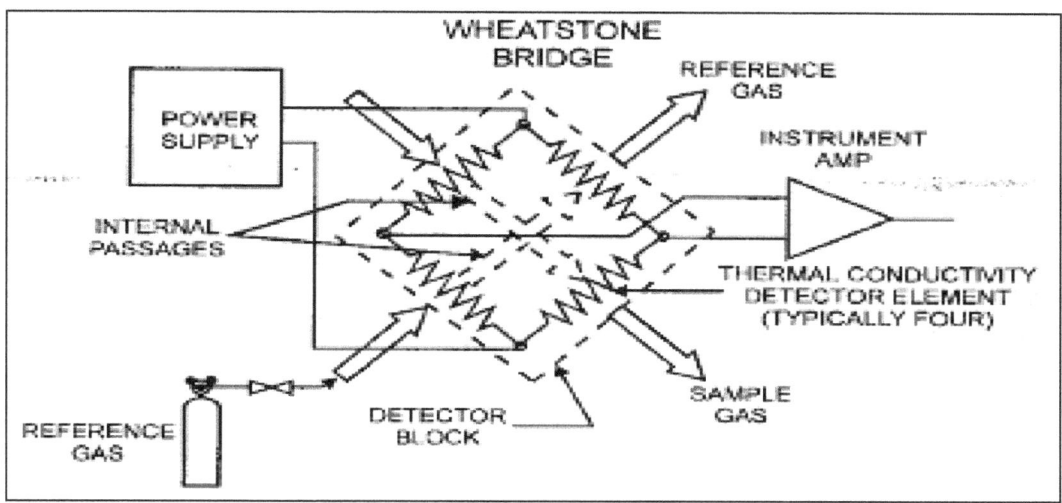

Fig.27 Thermal Conductivity Analyzer

Some analyzer manufactures made a sealed reference gas chamber which does not require flowing reference gas. Thermal Conductivity based analyzers have an accuracy from +/- 1% to +/- 5% of full scale and a response time of less than 30 seconds.

Heated wires or filaments are the heart of thermal conductivity analyzers. Two types of thermal conductivity detectors are used:

• Hot Wire Sensors: These are coaxial cylindrical coils of wire. Typical wire used may be tungsten, nickel-iron alloy, platinum. Depending on the application, the wire may be plated with gold and glass also. The wires can be operated with currents between 150 and 600mADC. Coil temperature may vary between 500 degrees Celsius to 900 degrees Celsius. Hot wire sensors have positive temperature coefficient of resistance.
• Thermistor Bead Sensors: These are small glass devices that enclose oxide materials such as manganese oxide, nickel oxide. These sensors are cold devices that are driven with electric currents; a few milliamperes to 15mADC. These sensors provide fast response as less mass than hot wire filaments. Because of fast response, these sensors are subject to more noise.

The Detector of the thermal conductivity analyzer are located in a metallic block. These blocks are made from non-reactive materials such as stainless steel. The temperature of the block is normally maintained at +/- 0.3 degree Celsius.

Typical applications where thermal conductivity gas analyzers used:

a) Steel Plant: Top Gas Analysis of hydrogen in Blast Furnace Process.

b) Petroleum refineries: Hydrogen in hydrocarbon gases.

Chemiluminescence Analyzer:

Chemiluminescence means the emission of light by the product of a chemical reaction. The analyzer is widely used for measuring stack gases and ambient air monitoring. Oxides of Nitrogen- nitric oxide and nitrogen dioxide are the most frequently measured components through chemiluminescence.
In stack, flue gases generally consist of 95-96% of NO, NO2 is very less so direct measurement is not possible. For that NO2 is again converted into NO.
$NO_2 + NO - NO_X$

By taking the difference between the direct NO measurement and indirect NO measurement gives the NO_2 concentration. NO can be directly measured though IR mythology and NO2 is converted in NO through convertor, which is further measured with IR.

The main advantage of chemiluminescence method includes:
a) Increased sensitivity
b) Rapid response time
c) Linearity over a wide dynamic range
d) Continuous Monitoring

Measurement Principle:
The gas phase reaction of nitric acids and ozone produces a characteristics luminescence via the following basic chemistry:
$NO + xO_3 - NO_2^+ + O_2$
$NO_2^+ - NO_2 + h\vartheta$

When NO reacts with O3 some electronical excited molecules NO2+ are generated. These molecules give off energy in the form of light ($h\vartheta$). When the emitted radiation is monitored, it becomes a measure of the concentration of the NO in the reacting sample. The light emission occurs between 600nm to 3000nm.

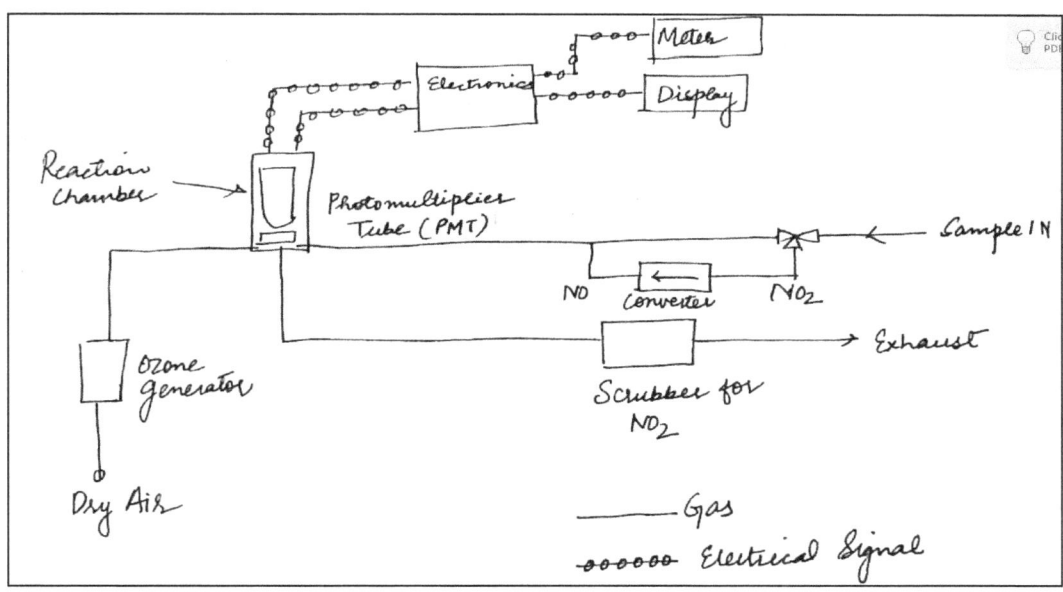

Fig.28 (a) Basic Chemiluminescence Instrumentation

Fig.28 (a) shows a simplified diagram of chemiluminescence analyzer. To measure NO concentration, the sample is mixed with ozone in a reaction chamber. The ozone required for the reaction is produced from the dry air within the analyzer.

The sample inlet generally has two flow paths. The first path is a direct path to the reaction chamber. Ideally only the NO is the sample reacts with ozone to produce NO2 in excited state, the excited state NO2 molecule emits a photon. In the second mode, NO2 is converted into NO through convertor. These photons are directed to PMT where large number of secondary electrons generated. The flow of electrons per second represents an electrical current output signal.

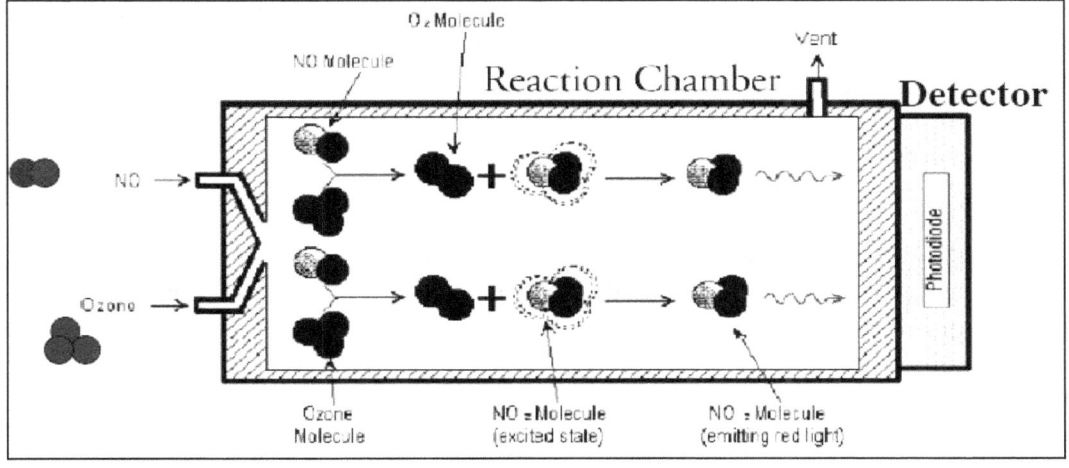

Fig.28 (b) Generation of Excited NO2 Molecule inside Reaction Chamber

Tuneable Diode Laser Spectroscopy Analyzer:

Gas Analyzers for industrial application must have high reliability and availability. Analyzer based on tuneable diode laser absorption spectroscopy meet the industrial requirement as they require less maintenance. These analyzers are in-situ type.
These analyzers can measure oxygen, carbon monoxide and carbon dioxide, hydrogen fluoride, hydrogen chloride and ammonia.

In-situ gas analyzers are of two types:
a) Point In-situ Type
b) Cross Stack Type

Point In-Situ Type:

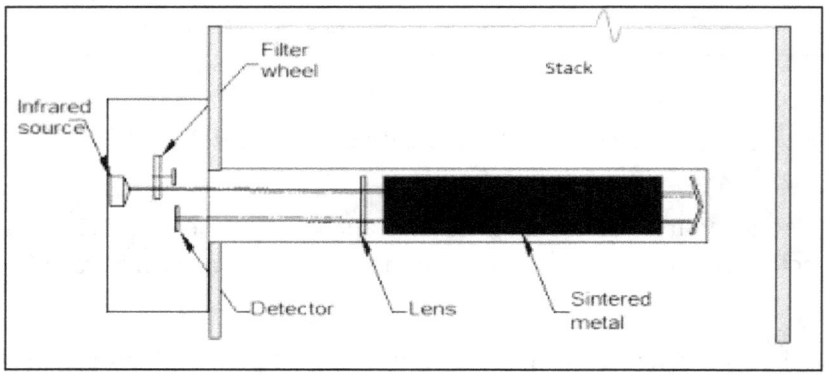

Fig.29 Point In-Situ Gas Analyzer

Fig.29 illustrates Point In-Situ Type TDLS Analyzer. Point in-situ systems perform measurements at a single point in the stack, as do extractive system probes. In certain CEMS, the measurement length is extended over the length of a probe (say 0.05mt to 1.00mt) to increase resolution and provide more coverage (representative measurement). In any case 1/3rd of the stack cross section ensures better representation.

Cross Stack Type:

Cross stack monitors measure over the entire stack or duct diameter. They are based on a beam of a certain wavelength that crosses the duct and is attenuated proportionately to the concentration of the target compound. There are two basic types of path systems: single pass and double pass where the beam is reflected back across the stack.
Fig. 30 (a) and (b) illustrates single pass and double pass.

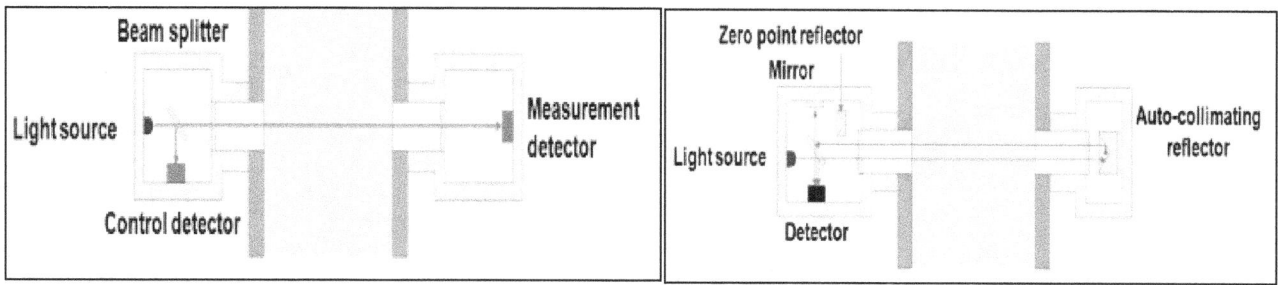

Fig.30(a) Single Pass Beam TDLS   *Fig.30(b) Double Pass Beam TDLS*

Single beam and double beam principle- Single-beam configuration is simplest where one light beam from source is passed to receiver. Dual or double-beam configurations internally split the light emitted from the source into two beams – one becomes measurement beam and another becomes reference beam. The measurement beam is projected through the optical medium of interest and is referenced to the second (reference) beam, which is totally contained within the instrument. There can be common or separate detectors for both the beam.

TDLS gas analyzer is based on Beer-Lambert Law:

$$T = \exp(-Sg(f)NL)$$

Where T is the transmittance
S is the absorption line strength
g(f) is the line shape function
N is the concentration of absorbing gas molecules
L is the optical path length

Concentration N can be calculated by measuring T and knowing S, g(f) and L. Unlike conventional ultra violet and infrared spectrographic instruments, laser based monitors employs the measurement principle known as "single line spectroscopy", where a single gas absorption line is chosen in the NIR spectral range and scanned by the single mode diode laser.

A typical TDLS system consist of the following units:
• Laser module with temperature controlled diode laser, driving circuit and lens with window to isolate from the process.
• Detector module, photo diode and focusing lens plus window to isolate from the process.

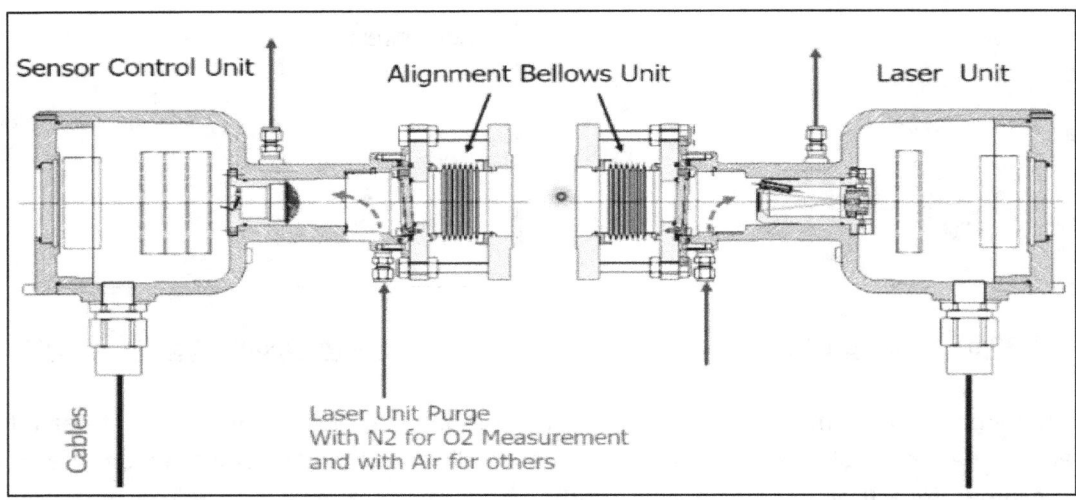

Fig:31 Layout of TDLS analyzer

Fig. 31 illustrates a laser transmitter and receiver (detector) unit are optically coupled optically with fiber at the tapping point with mounting flanges and intermediate purge and alignment units. A continuous flow of nitrogen is utilised to prevent the dust from fouling the optical windows, Fig.31 shows in-situ approach, where TDLS analyzer installed across pipes and stacks with typical path length ranging from 0.5-20 meters. This is the most preferred approach because it eliminates the need for sample conditioning system and therefore it is ideal for fast and reliable measurement. Laser source for CO measurement have wavelength 1200nm-1500nm, for O2 760nm and moisture 1500nm. Some manufacturers supply TDLS with bellows assembly and alignment studs in the large dust application and vibration prone area.

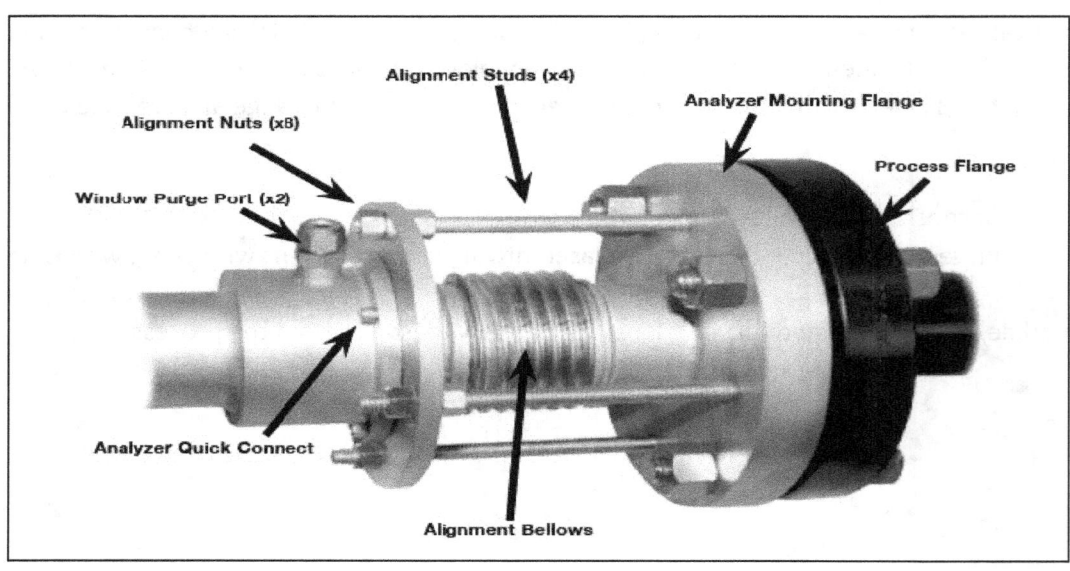

Fig.32 Avoid Line Vibration and Aligned TDLS Unit through Bellows

Some manufacturing companies provides an extractive set up with TDLS analyzer with use of ejector pump, flow control unit, and dynamic pressure and temperature compensation. This arrangement gives a flexibility for maintenance people to check and validate the analyzer on regular intervals, also alignment issues becomes zero.

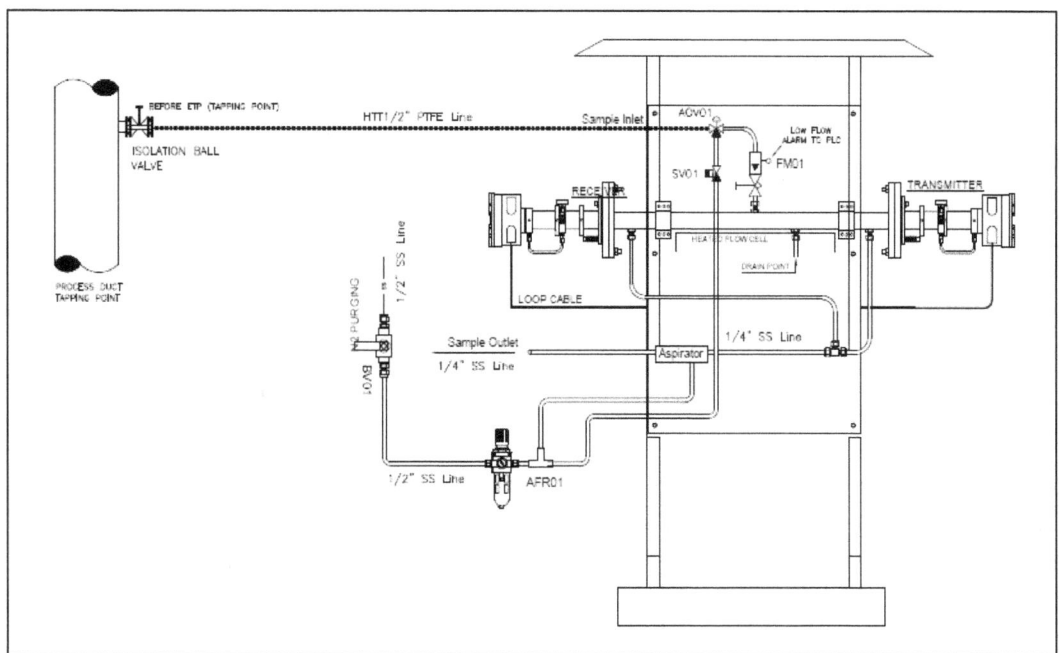

Fig.33 TDLS Gas Analyzer with use of Extractive System

The absorption signal peak shape, formed by amplitude and width, is influenced by a range of factors including analyte concentration, optical path length, pressure, temperature and variation in stream composition created by background gases. The amplitude is influenced by analyte concentration and optical path length while peak width is influenced by collision broadening, doppler broadening and natural line broadening.

Collision broadening of the absorption signal dominates industrial processes, depends on the collision frequency of the gas molecules. It is directly depending on gas pressure, temperature and composition. Accuracy and repeatability is always influenced by these parameters. Line broadening is also important in order to achieve measurement accuracy, for process measurement the effect of pressure and temperature is compensated by live temperature and pressure input to the TDLS analyzer.

Fig.34 shows initial factory calibration layout. Based on the pressure and temperature data submitted by process plant, a pressure dependence curve and a temperature dependence curve are generated in a controlled environment, using a standard calibration gas.

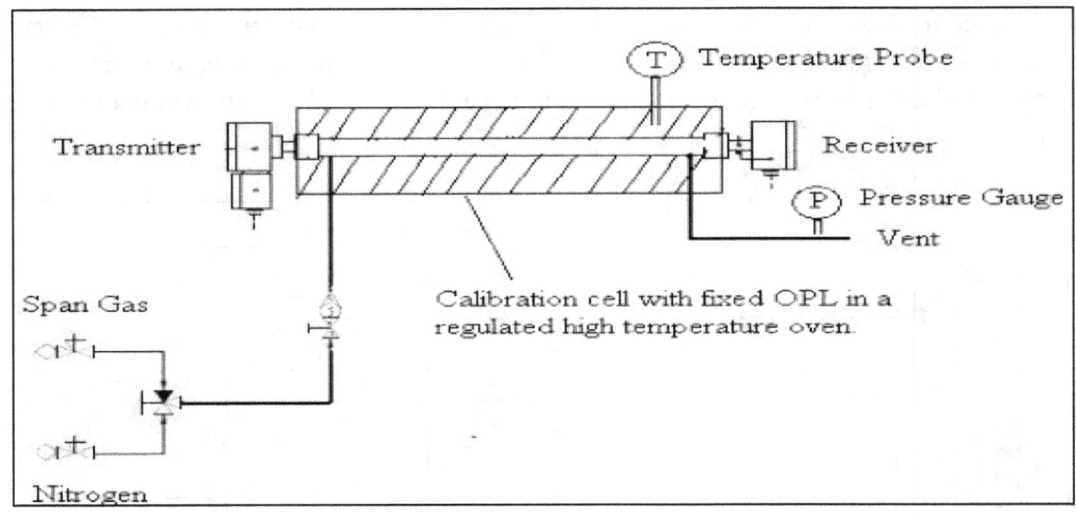

Fig.34 Calibration Setup of TDLS Gas Analyzer in Factory

Fig.35 illustrates the relationship between fluid pressure and detector signal. The signal increases at low pressure while the line broadening mechanism dominates at pressure above 1 bar where signal decreases. This profile is wavelength dependant; as different absorption lines will have different pressure dependence profile.

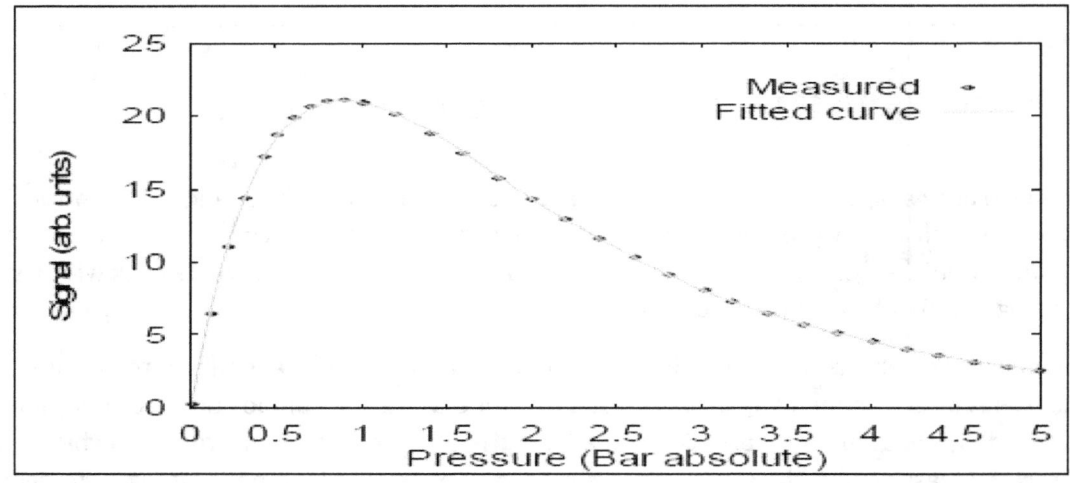

Fig.35 Pressure Dependence of the Absorption Signal

Fig.36 illustrates relationship between temperature and detector signal.

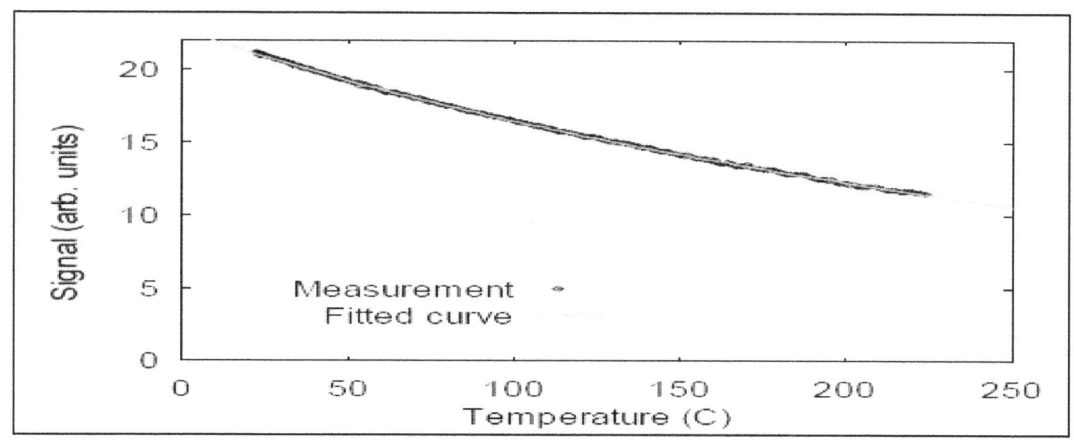

Fig.36 Temperature Dependence of the Absorption Signal

Point or Probe In-Situ Analyzer: As explained earlier about point in-situ analyzer, a detailed view and purge area and optical path length as shown in fig.37. The probe has the slots which is helpful for the free movement of the gas and to increase the strength of the probe. If sintered probe is used there is a possibility of chocking due to dust content in the gas.

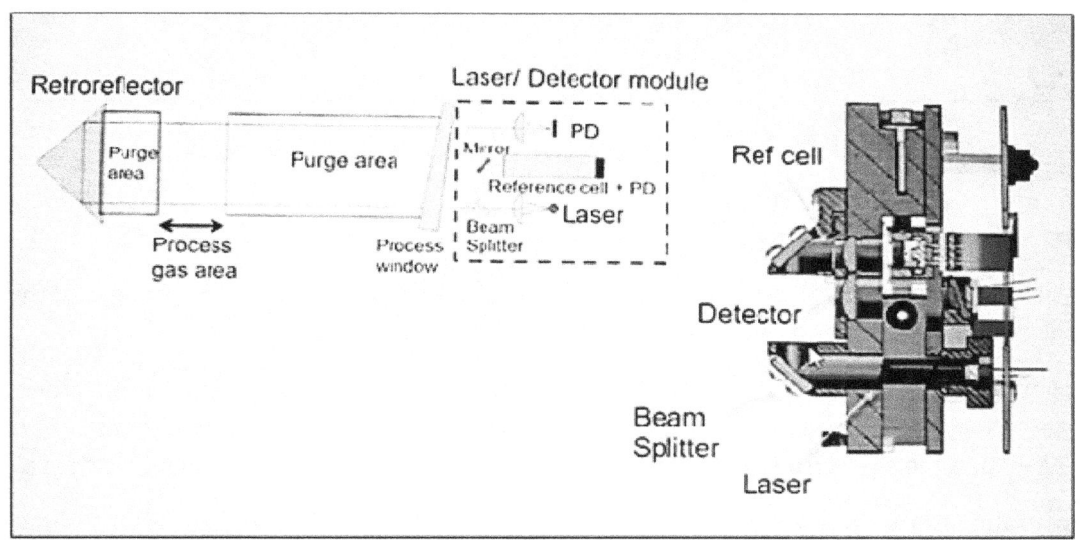

Fig.37 Detailed View of Point In-Situ Gas Analyzer

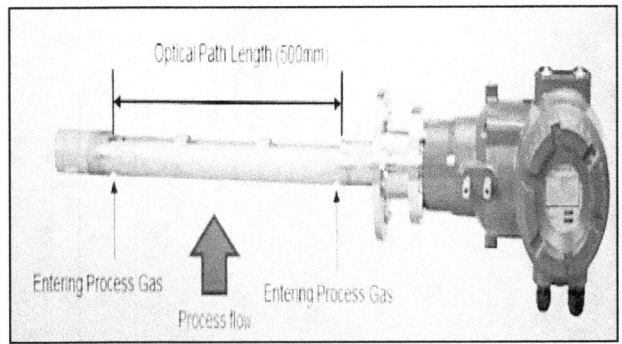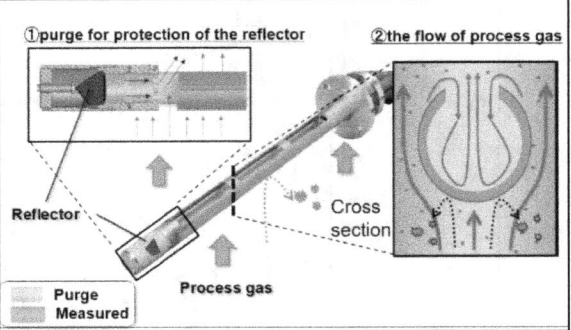

Fig.38 Flow of Process Gas and Purge Gas through the Probe

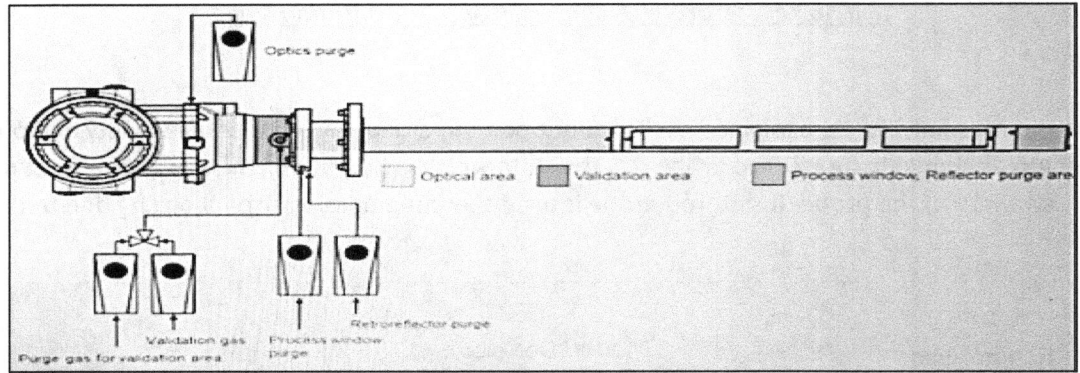

Fig.39 Purge Gas required for different components in Probe

`Calibration and Validation of TDLS Analyzer:

It is easy to validate TDLS analyzers installed in the extractive mode by introducing the spike gas at the sample inlet. For in-situ application a different approach is taken as shown in fig. 40 and 41.

For example, consider the following situation where oxygen measurement in the range of 0-6%, process optical path length is 1 meter, and chamber optical length is 0.01meter. The absorption signal from the gas cell chamber should be always stronger than the signal from the process gas in order to achieve a reliable validation check.

Effective Concentration of the spike gas is calculated using equation:

$CS = S * LC/LP$

Where CS Effective Concentration of the spike gas

S concentration of the spike gas

LC is the OPL of chamber

LS is the OPL of process duct

Therefore for 100% introduction of oxygen gas in 0.01-meter chamber corresponds to 1% oxygen for 1 meter process path length. However, if the process optical

length 0.2meter then for 100% oxygen on 0.01 meter opl chamber, corresponds 5% oxygen for O2. In this situation, it is viable to perform a reliable validation check using 100% O2 as the spike gas.

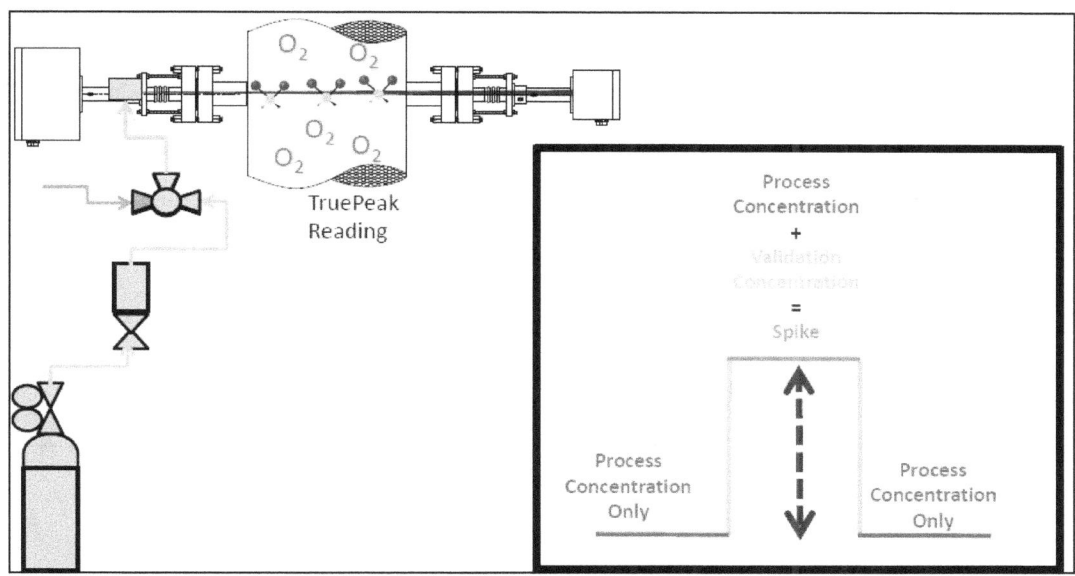

Fig.40 Calibration and Validation of TDLS Gas Analyzer In-Situ Type

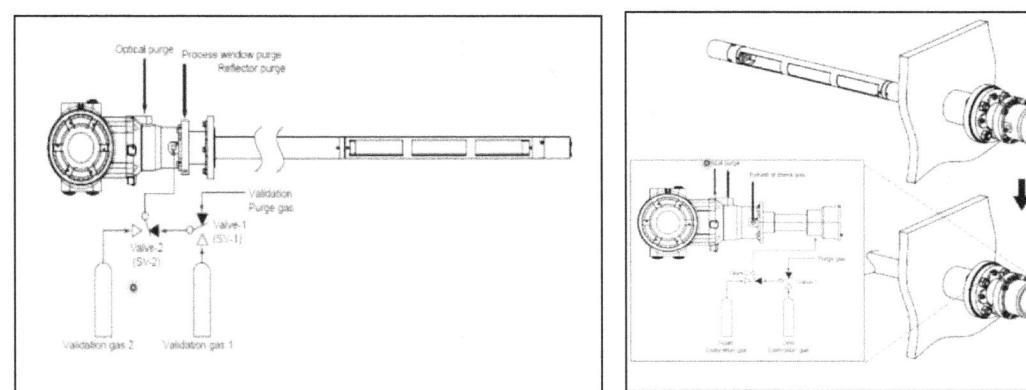

Fig.41 Point or Probe Type In-Situ TDLS Gas Analyzer Calibration or Validation Arrangement

Zirconia Gas Analyzer: The zirconia high temperature analyzer is designed to measure oxygen concentrations in the air and calculate the moisture content. The most common type of Zirconia Measurement is Potentiometric.

Fig.42 shows the basic potentiometric method which uses a solid electrolyte such as zirconia with platinum electrodes attached that when heated the device acts as an oxygen concentration cell. The platinum electrode has a special coating to protect it from deterioration caused by SOx, NOx contained in the process gas.

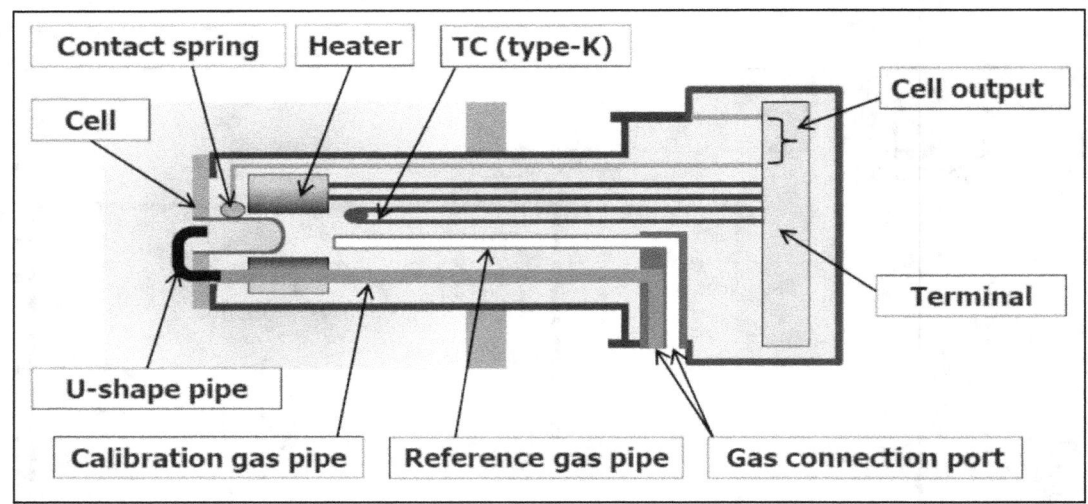

Fig.42 Zirconia Sensor Principle and Structure

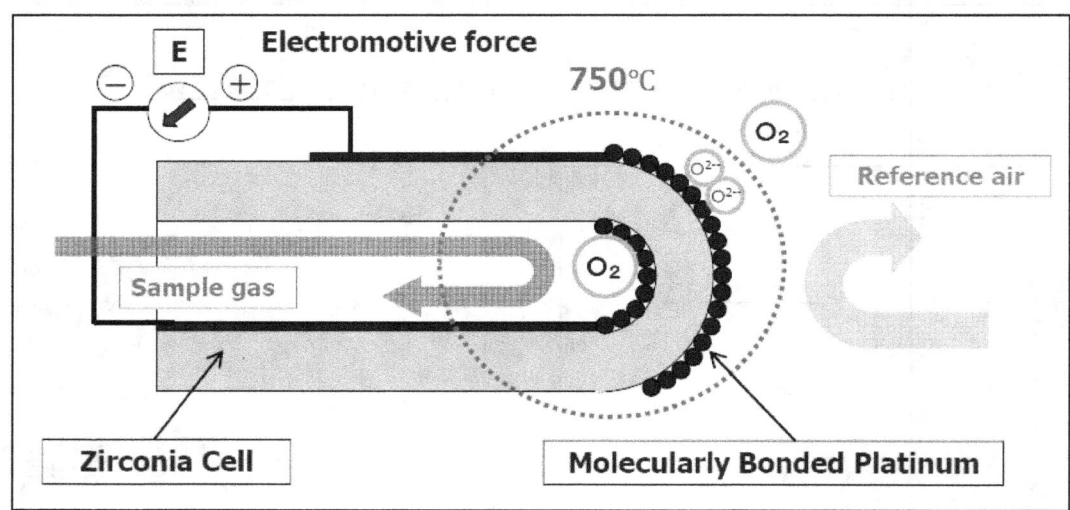

Fig.43 Zirconia Cell

Basically, a concentration gradient of oxygen ions is established within the zirconia lattice which produces a voltage potential between the platinum electrodes according to the Nernst equation:

$$E = - \frac{RT}{nf} \ln Px/Pa \qquad \text{--- Eq.-1}$$

Where R- Gas Constant

F- Faraday's constant

T- Absolute Temperature

Px- Oxygen Concentration in a gas (sample or process gas) in contact with the positive electrode

Pa- Reference Air in contact with the negative electrode

The Zirconia become conductive at high temperature above 600 degrees Celsius. A heater is used to control the temperature of Zirconia to 750 degrees Celsius. At that temperature, the eq-1 becomes E=-50.74log(O2/21) when instrument air is used as reference gas and pressure on both sides are equal. Fig.44(a) shows a graph between oxygen concentration and E(mv). However, if concentration is different on both sides then ions moves to the low concentration side because there is equilibrium between oxygen molecules and ions. The bigger the difference of oxygen concentration on both sides, the bigger voltage is generated. The analyzer measures this millivolt and the calculates the oxygen concentration.

Negative Electrode: $O_2+4e- - 2O^{2-}$

Positive Electrode: $2O^{2-} - O_2 +4e$

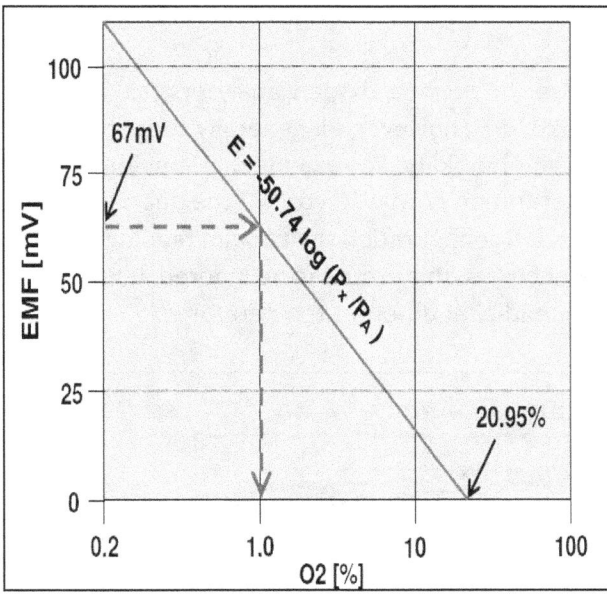

% O₂	mV	% O₂	mV
0.1	117.83	6	27.61
0.2	102.56	7	24.21
0.3	93.62	8	21.27
0.4	87.28	9	18.67
0.5	82.36	10	16.35
0.6	78.35	21	0
0.7	74.95	30	-7.86
0.8	72.01	40	-14.2
0.9	69.41	50	-19.2
1	67.09	60	-23.1
2	51.82	70	-26.5
3	42.88	80	-29.5
4	36.54	90	-32.1
5	31.62	100	-34.4

Fig. 44(a) Logarithmic Relationship between O2% and Cell EMF

Fig.44(b) Oxygen Concentration Vs. Cell Voltage

Since the sensor is heated upto 750 degrees celsius during measurement, if the process gas contains combustible components such as CO, Hydrogen, CH4 these gases burn at the detector and consume oxygen and show less reading than the actual. The ratio of consumption of combustibles to oxygen is 2:1; thereby if 1000 ppm combustibles is present in the flue gas, 500 ppm of O_2 will be consumed from the flue gas sample as they combust on the zirconium cell. The decrease of the oxygen reading due to the burning of combustibles is normally considered negligible.

If oxygen concentration at both sides of zirconia cell is same, then no voltage is generated.

Troubleshooting of Zirconia Gas Analyzer:

Following types of problem occurs in Zirconia Gas Analyzer:

a) Cell voltage goes outside limits
b) Heater goes below 730 degree celsius or above 780 degree celsius
c) Zero Calibration Failure
d) Span Calibration Failure

Cell Voltage Failure occurs when the cell voltage signal input to the convertor fall below -50 mv. Cell resistance should be less than 200 ohms.

Heater Temperature Failure: It generally due to thermocouple (K Type) connected to convertor with reverse polarity. The Heater resistance should be below 90 ohms (normally in the range of 57 ohms to 85 ohms) if healthy. Thermocouple resistance should be below 5 ohms in all case.

Nitrogen cannot be used as zero gas during zero calibration because Nitrogen gas grade available may contains traces of oxygen, the cell voltage becomes unstable and high.

Effect of Humidity on Reference Gas at different temperature: In zirconia oxygen analyzers, air is used as the reference gas. The oxygen concentration of dry air is constant at 20.95%; however, air generally contains water vapor, in that case the oxygen concentration varies with temperature and humidity. For example, at ambient temperature of 40°C and 100 % relative humidity the Oxygen reference concentration is only 19 vol%. If the analyzer is calibrated with instrument air and installed with ambient air natural convection configuration the Oxygen readings will read (21-19)/19 = 10 % too low. When instrument air is used as the reference gas, this error can be ignored, but if it cannot be used, care is required. Fig. 45 illustrates oxygen percentage vs humidity at different temperatures.

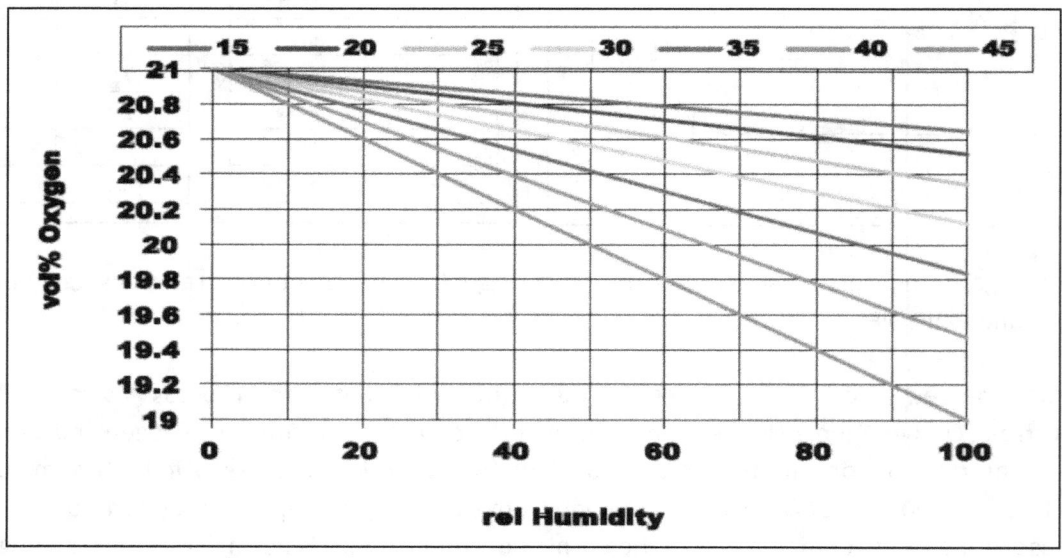

Fig.45 Effect of Humidity w.r.t oxygen percentage at different temperature

Advantage of Zirconia Gas Analyzer:
a) Response time is low.
b) Absence of moving parts.

Disadvantage of Zirconia Gas Analyzer:
a) Zirconia cannot be used in the process where process gas contains hydrocarbons. As hydrocarbons will burn the at the tip of cell due to the presence of oxygen. This causes low oxygen reading than the actual or in worst cases it may leads to explosion. The sensor acts as an ignition source, this makes Zirconia gas analyzer a hazard in any gas fired furnace during start up or shut down condition.

b) Zirconia gas analyzer cannot be used in application where process gas contains sulphur as it deforms the cell.

5.4 Troubleshooting and Maintenance of Gas Analyzers

Process Infrared gas analyzers are quite rugged, frequently exhibiting lifetimes in excess of 10 years. Detector failure will be exhibiting by a decay of sensitivity and selectivity.

Thermal Conductivity analyzers are normally simple to maintain. Preventive maintenance should be conducted according to the recommendations of the equipment manufacture.

In most of the cases, problem related with analyzer occurs less; mostly problem related with sample handling system are quite more frequent. A general guideline for sample handling system maintenance is shown in fig.46.

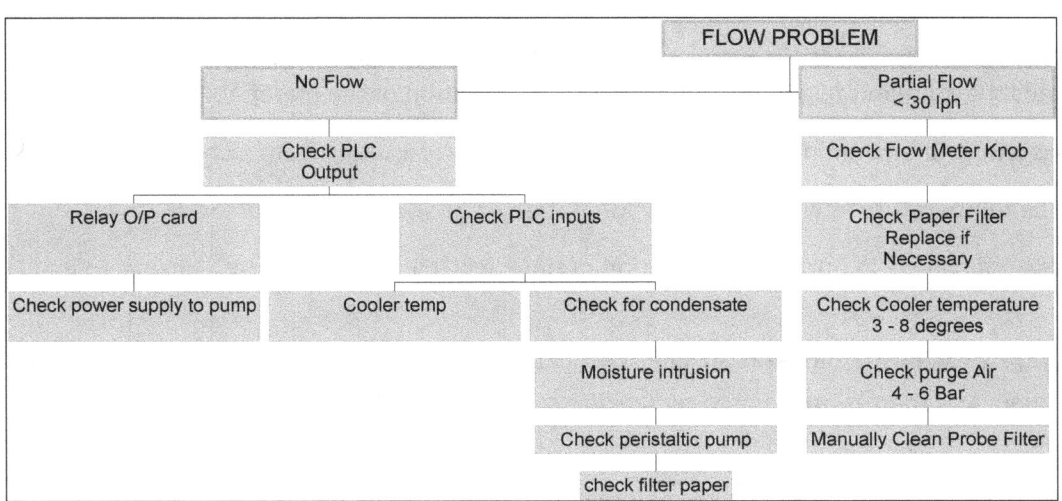

Fig.46 Troubleshooting flow chat for SHS

SL NO	Description of Measuring Points	Characteristic	Value Based		
			LCL	UCL	VALUE
1	Probe Heater Temperature	TEMPERATURE(degc)	100	150	
2	Heat Traced Tube Temperature	TEMPERATURE(degc)	70	90	
3	Sample gas cooler temperature	TEMPERATURE(degc)	2	6	
4	Flow of Gas in Rotameter 1 —Module 1	FLOW(lpm)	50	80	
5	Flow of Gas in Rotameter 2 – Module 2	FLOW(lpm)	50	80	
6	Flow of N2 in Rotameter 3 –Purge N2	FLOW(lpm)	100	150	
7	Reading of the Analyzer Module 1	PPM/%	0	XXX	
8	Reading of the Analyzer Module 2	PPM/%	0	XXX	
9	Reading of the Analyzer Module 3	PPM/%	0	XXX	
11	Condition of Probe Filter				
12	Current of the Heat Traced Tube	CURRENT	0	10	
13	Purging System of Probes and its Solenoid Valves				
14	Tripping circuit of Condensate Monitor				
15	N2/Air Pressure of the Purging system	PRESSURE(bar)	2	5	
16	Tripping ckt of Rotameter 1				
17	Tripping ckt of Rotameter 2				
18	Tripping ckt of Rotameter 3				
19	Condition of the Peristaltic Pump				
20	Condition of the Main Pump				
21	Condition of Rotameter 1				
22	Condition of Rotameter 2				
23	Condition of Rotameter 3				
24	Condition of By-pass Pump				
25	Analyzer Purging vent Lines				

Fig.47 General Sample Handling System Maintenance Points

5.5 General Terms to Know for Process Gas Analyzers

Measured Accuracy: The maximum positive and negative deviations observed in testing a device under specified condition and by a specified procedure.

Beer- Lambert Law: Relates to the transmission properties of the material through which light is travelling. In other words, it relates the amount of light absorbed to the concentration of the material of interest in a sample.

Chemiluminescence: Chemiluminescence is the emission of light resulting from a chemical reaction.

Dew Point: The temperature at which water condensation occurs when gas is cooled.

Drift: An undesired change in output over a period of time, which is not caused by a change in input, environment or load.

FFT (Fast Fourier Transforms): Algorithms convert signal from the time or space domain to a representation in the frequency domain and thereby manage to reduce the complexity of the calculations.

Luft Detector: Consist of two chambers, divided by a diaphragm, and the same intensity of pulsed infrared radiation is received by both chambers. When the gas of interest flows through the sample cell, a reduction in radiation energy is received by the detector chamber, which causes the temperature and pressure to drop in that chamber. The amount of pressure drop is in proportion to the gas concentration, and this pressure difference between the two chambers causes a movement of the diaphragm, which is detected by capacitance.

LEL (Lower Explosion Limit): The minimum concentration of combustible gas or vapour that supports combustion in air is called LEL. Below LEL combustion won't happen.

UEL (Upper Explosion Limit): The maximum concentration of gas or vapor that supports the combustion in air is defined as UEL. Above UEL, the mixture is too rich to burn.

5.6 Specification Form for Sample System

Sample System Specification Form					
Stream Name or Identification					
Stream Composition Data	Concentration in mol%, wt%, ppm		Range of Component to Be Measured		
Component 1					
Component 2					
Component n					
Operating Process Data					
Temperature _____			Pressure _____		
Phase: Liquid _____			Vapor _____		
Sample bubble point			Dew point		
Corrosive components/solids					
Stability (polymerize, decompose, etc.)					
SAMPLE CONDITIONS					
Maximum distance: Tap to analyzer _____			Analyzer to return _____		
Speed loop required: Yes			No		
Sample return pressure point					
Sample probe requirements: Connection size			Orientation		
Materials of construction:					
Stainless steel	Teflon	Viton	Glass	Other	
Electrical areas classification (shelter)			Sample point(s)		
Power supply					
Output signal					
Utilities available: Steam		Air		Cooling water	

1. Specification Form for Sample System by ISA

5.7 Data Sheet for Process Gas Analyzers

	RESPONSIBLE ORGANIZATION	ANALYSIS DEVICE		SPECIFICATION IDENTIFICATIONS	
1			6		
2	(ISA)	Operating Parameters	7	Document no	
3			8	Latest revision	Date
4			9	Issue status	
5			10		

	ADMINISTRATIVE IDENTIFICATIONS			SERVICE IDENTIFICATIONS Continued	
11			40		
12	Project number	Sub project no	41	Return conn matl type	
13	Project		42	Inline hazardous area cl	Div/Zon	Group
14	Enterprise		43	Inline area min ign temp	Temp ident number
15	Site		44	Remote hazardous area cl	Div/Zon	Group
16	Area	Cell	Unit	45	Remote area min ign temp	Temp ident number
17			46		
18	SERVICE IDENTIFICATIONS		47		
19	Tag no/Functional ident		48	COMPONENT DESIGN CRITERIA	
20	Related equipment		49	Component type	
21	Service		50	Component style	
22			51	Output signal type	
23	P&ID/Reference dwg		52	Characteristic curve	
24	Process line/nozzle no		53	Compensation style	
25	Process conn pipe spec		54	Type of protection	
26	Process conn nominal size	Rating	55	Criticality code	
27	Process conn termn type	Style	56	Max EMI susceptibility	Ref
28	Process conn schedule no	Wall thickness	57	Max temperature effect	Ref
29	Process connection length		58	Max sample time lag	
30	Process line matl type		59	Max response time	
31	Fast loop line number		60	Min required accuracy	Ref
32	Fast loop pipe spec		61	Avail nom power supply	Number wires
33	Fast loop conn nom size	Rating	62	Calibration method	
34	Fast loop conn termn type	Style	63	Testing/Listing agency	
35	Fast loop schedule no	Wall thickness	64	Test requirements	
36	Fast loop estimated lg		65	Supply loss failure mode	
37	Fast loop material type		66	Signal loss failure mode	
38	Return conn nominal size	Rating	67		
39	Return conn termn type	Style	68		

	PROCESS VARIABLES	MATERIAL FLOW CONDITIONS			PROCESS DESIGN CONDITIONS	
69			101			
70	Flow Case Identification	Units	102	Minimum	Maximum	Units
71	Process pressure		103			
72	Process temperature		104			
73	Process phase type		105			
74	Process liquid actl flow		106			
75	Process vapor actl flow		107			
76	Process vapor std flow		108			
77	Process liquid density		109			
78	Process vapor density		110			
79	Process liquid viscosity		111			
80	Sample return pressure		112			
81	Sample vent/drain press		113			
82	Sample temperature		114			
83	Sample phase type		115			
84	Fast loop liq actl flow		116			
85	Fast loop vapor actl flow		117			
86	Fast loop vapor std flow		118			
87	Fast loop vapor density		119			
88	Conductivity/Resistivity		120			
89	pH/ORP		121			
90	RH/Dewpoint		122			
91	Turbidity/Opacity		123			
92	Dissolved oxygen		124			
93	Corrosivity		125			
94	Particle size		126			
95			127			
96	CALCULATED VARIABLES		128			
97	Sample lag time		129			
98	Process fluid velocity		130			
99	Wake/natural freq ratio		131			
100			132			

	MATERIAL PROPERTIES			MATERIAL PROPERTIES Continued	
133			137		
134	Name		138	NFPA health hazard	Flammability	Reactivity
135	Density at ref temp	At	139		
136			140		

Rev	Date	Revision Description	By	Appv1	Appv2	Appv3	REMARKS

Form: 20A1001 Rev 0

© 2004 ISA

2. Data Sheet of Gas Analyzer by ISA

5.7 References

- Process Analyzer Sample-Conditioning System Technology – Robert E. Sherman
- Instrument and Automation Engineer Handbook Volume 2- Analysis and Analyzers – Bela G Liptak
- Adage Automation Private Limited
- Yokogawa India Limited
- ABB India Limited
- API RP -555 Process Gas Analyzer
- 60 years of CO Analysis by NDIR Gas Analyzers by Joseph W. Worthington
- Calibration and Validation Philosophy and Procedure for TDLS Analyzers in Process Applications by- Dr. Janardhan Madabushi, David Fahle and Dr. Christian Heinlein
- Analytical Instrumentation- Robert E. Sherman and L. Rhodes